AF522253

EVOLUTIONARY GENETICS

ENCYCLOPAEDIA OF GENE EVOLUTION

Vol. III

EVOLUTIONARY GENETICS

By

Dr. M. Prakash

DISCOVERY PUBLISHING HOUSE
NEW DELHI-110002

First Published-2006

ISBN 81-8356-169-1 (Set)

Published by

DISCOVERY PUBLISHING HOUSE
4831/24, Ansari Road, Prahlad Street,
Darya Ganj, New Delhi-110002 (India)
Phone: 23279245 • Fax: 91-11-23253475
E-mail:dphtemp@indiatimes.com

Printed at:
Amit Enterprises
Delhi-53

PREFACE

The present title **Encyclopaedia of Gene Evolution** encompasses diverse areas of study from the origin of life and is subsequent record to the molecular and population concepts that explain these events. All biological phenomena derive from evolutionary relationships and past interactions. As the great evolutionary geneticist Theodosius Dobzhansky remarked, "Nothing in biology makes sense except in the light of evolution" yet unifying all biology under an evolutionary theme is still difficult, although the explosive increase in molecular, organismic, and populations information makes the realization of this goal more possible now than ever before.

The aim in writing this text was to bring together the highly dispersed literature on gene structure, function and expression, to integrate this with our emerging knowledge of chromosome and genome structure, and to draw comparison both within and between paralogous and orthologous gene sequences in order to establish the nature of the mutational mechanisms responsible for their evolutionary divergence. Such an approach, encompassing a large number of human genes and their extended families, is essential for any attempt to discern underlying evolutionary principles.

The purpose of this book is to bring together some prevailing knowledge and ideas about evolution, to provide an informed framework of together for under graduates, post-graduate and those of related fields.

Since evolution is the broadest of biological fields, covering the greatest range of disciplines, even the brief survey of evolution offered here has errors and ambiguities. To the extent that this book has been spared many such failings the author owes thanks to many reviewers we commented on one or more sections of this text.

There can be no claim to originality except in the manner of treatment and much of the information has been obtained from the books and scientific journals available in the different libraries.

The author expresses his thanks to his friends and colleagues whose continue inspirations have initiated him to bring out this book.

The author expresses his gratitude to Mr. Wasan and staff of M/s Discovery Publishing House for their whole hearted co-operation in the publication of this book.

Author

CONTENTS

1

INTRODUCTION

A great deal of water has flowed under the bridges of biology since the early days of mendelian work, when mendelian factors were rigidly equated with mendelizing characters when only two states of a given gene were recognized, the dominant being supposed to represent its presence and the recessive its absence, and when all mutations and all mendelian genes were supposed to have considerable and obviously discontinuous effects. To-day the notion of mendelian characters has been entirely dropped. The term may occasionally serve as a useful piece of shorthand notation, but is in point of fact a false conception.

In the first place, a single gene may affect a number of characters, a phenomenon known as pleiotropism. Grüneberg (1938), in an illuminating analysis, points out that pleiotropic effects may be realized in three different ways. In the first place, a gene may exert a direct effect on two or more distinct processes. The example of the effect of the series of white eye-colour allelomorphs in *Drosophila*, which also exert an effect on the shape of the spermatheca, is probably an example of this category. Secondly, a gene may exert a direct effect on a single process, but in many different sites and conditions. This holds for the primary action of the gene studied by Gruneberg (1938) in the rat, which causes hyperplasia and abnormal growth of cartilage in the ribs, trachea, and elsewhere.

Another and even more striking case is the array of anomalies in such different organs as eyes and feet, found in a particular strain of mice, which Bonnevie (1934) has shown are due to alteration in a single developmental mechanism, namely the causing of embryonic blebs of fluid at a particular stage of embryonic development. The most interesting examples for our purpose, however, belong to the third category, of indirect effects. A gene exerts a primary direct effect, and this then causes numerous secondary effects. Grüneberg's gene in the rat has a hyperplastic anomaly of cartilage as its primary effect. But among the secondary effects are such varied "characters" as emphysema, hypertrophy of the right ventricle, blocked nostrils, and incompletely occluded incisor teeth.

An equally good example is that of the frizzled breed of fowl. Here the primary effect is entirely on the feathers; as the secondary effect of the resultant abnormal heat loss, we find (in temperate climates) marked thyroid and adrenal enlargement, subnormal body-temperature, and much increased food-intake. Such secondary effects are excellent examples of what Darwin called "correlated characters:, which may be of great evolutionary importance. Furthermore, any given character represents the end-result of a great number of genes interacting with the environment during development, and is not inherited as such. What is investigated in any genetic experiment is the inherited basis for a constant character-difference. Thus a *character-difference* may be said to be inherited in mendelian fashion, while the *character* cannot but even so the differential effect of a particular gene on the character need not by any means always be the same. It may alter according to differences in the environment, and also according to differences in the remainder of the gene-complex.

As an example of the first, we may take the well worked out case of "abnormal abdomen" in *Drosophila*. This effect depends on a single partially dominant sex-linked gene: but it is only manifested in moist conditions. In dry conditions flies pure for the gene appear perfectly normal, while intermediates are produced by varying degrees of moisture. An equally striking botanical case is that of a type of albinism in barley, dependent on a single gene. When grown below 6.5°C. the plants entirely lack chlorophyll, while above 18°C. they are quite normal. Between these limits the mutants produce a graded amount of chlorophyll. Baur's classical case (1923) in *Primula siensis* has now been shown to be due to faulty experimentation.

In Himalayan rabbits and Siamese cats (both simple recessives) black pigment is produced only below a certain threshold temperature. Normally only the extremities fall below this; but Iljin (1927, 1930) has experimentally produced pale extremities, and black on the body. Another *Drosophila* example is *short-wing*, a sex-linked recessive which at 27.5°C. markedly reduces wing-length and affects eye-development.

However, the effect falls away with temperature and is absent at 14°C. Thus environmental changes may either mask or bring out the results of genetic difference. We must therefore distinguish carefully between the *nature* of the gene and its *expression*. The gene itself can only alter by mutation; but its expression can be affected in a number of ways. The most revolutionary change has come in regard to the way in which the expression of a gene can be altered by other genes.

The discovery of this fact has given us the two fundamental concepts of genic balance and the gene-complex. Thus the internal or genetic environment, of a gene may produce effects upon its expression which are as striking as those induced by the external environment, and of course very much more important from the point of view of evolution. By genic balance we imply that individual genes act, not absolutely, in virtue solely of their inherent qualities, but relatively, in virtue of their interaction with other genes. The concept was first reached by studies on sex. It was at one time supposed that in *Drosophila* and other forms with male heterogamety, one X-chromosome automatically determined maleness, and two femaleness. It has since been shown, however, that it is the balance of the X-chromosomes to the autosomes (A) which is operative. A ratio of 1 X to 2 A produces maleness, of 1 X to 1 A produces femaleness; while one of 1 X to about 1.5 A produces intersexuality: Sterile "super-males" and "super-females" are produced by ratios of 1 X to over 2A and under 1A respectively. Here we can deduce that sex-determination is effected by the quantitative ratio between sets of male-determining and female-determining genes, though we do not know how many separate genes are involved in each set.

The principle, however, appears to be of universal application: the effect produced by any gene depends on other genes with which it happens to be co-operating. The effects of modifying genes are the most striking examples. A classical case of the kind is the alteration in the hooded pattern of rats by modifiers. The basic gene remains the same, but its effects may be reduced to a few specks of black on the head, or progressively extended over the whole back and most of the belly, by the agency of accessory genes. In cotton (*Gossypium*) differences in leaf-shape have been evolved in a precisely similar way. Extending this concept, we reach that of the gene-complex. The environment of a gene must include many, perhaps all other genes, in all the chromosomes. This gene-complex may be altered in numerous ways by mutation or recombination so as to modify the effects and mode of action of particular genes, whether well-established ones or new mutations. We can thus distinguish between the genetic and the somatic environment of genes. Further, it is now known that a gene can exist in a great variety of allelomorphic forms (alleles), up to a dozen or more being known for single loci.

The effects of these usually differ in a quantitative way (though occasionally in a qualitative way as well), and the steps between the various alleles may be very slight. Multiple alleles are, in general, taken to represent different states of a homologous material unit. They thus constitute one type of gene-differences with quite small effects. Many modifiers and cumulative factors such as those involved in quantitative characters also have small effects. In many cases the actual origin of such small differentials by mutation has been observed. Further, where a gene's effect is small, the variations of expression, due to environment and to other genes, may readily cause an overlap with the phenotypic expression of an allele or of another gene with similar type of expression. Thus though genetic variability must be discontinuous, its expression in measurable characters may become continuous.

Even mutations which in one gene-complex are pathological, in another may be perfectly harmless, and in yet another advantageous. A striking example of this is provided by the work of M. Gordon (1931) on generic hybrids between the viviparous freshwater fishes *Platypoecilus* and *Xiphophorus*. Some strains of the former possess a gene for the production of large pigment-cells with black pigment, responsible for a certain type of spotted pattern. When, however, this gene interacts with certain genes in the sword-tail (*Xiphophorus*), the pigment-cells produce cancer-like melanotic tumours.

Equally striking and curious results may occur as the result of the interaction of two gene-complexes in species-hybridization. As an example of this, we may cite an intergeneric pheasant cross recently described. The cross was between a Lady Amherst pheasant (*Chrysolophus ampherstiae*) and an Impeyan pheasant (*Lophophorus impeyanus*), and the hybrid was a male. The coloration of the males of both parent species is not only brilliant but varied. Thus the Lady Amherst cock has a black-and-white extensible "cape" on the head, a striking regional pattern on the body, and elaborately barred central tail feathers; while the Impeyan cock is distinguished by brilliant patches of burnished bronze, green, and blue-back on its upper parts, with white rump and buff tail. The hybrid, however, has most of its upper parts uniformly black, with mere traces of green and bronze iridescence, but neither regional patterning nor brilliant colouring. The lower parts and central tail feathers are mottled with brown, grey and white in various ways. This simple colour-scheme, by the way, cannot be considered to have any reversionary or "atavistic" significance whatever: it is simply that the delicately balanced gene-systems responsible for the two elaborate patterns have cancelled out, so to speak, to produce a wholly different and almost uniform coloration.

It may perhaps be mentioned that in other characters the hybrid is intermediate (e.g. the shape and size of the cape), and in still others shows obvious dominance (e.g. the blue face-skin of the Impeyan). What we may call partial gene-complexes may also arise in relation to the separate chromosomes into which the total gene-complex is divided. This has been demonstrated by Mather (1941) as regards what he terms polygenic characters—i.e. quantitative characters dependent on the co-operation and interaction of numerous genes. In *Drosophila*, the number of ventral abdominal hairs can be changed by selection so as to trangress the limits of normal variability in both plus and minus directions (cf. Castle's hooded rats). But the effect of selection is exerted in two main stages.

During the first two generations, a marked change is effected, which Mather interprets as being due to recombination of whole chromosomes. Then, after a period of relative stability for two or three further generations, a further and more marked change is produced, continuing for a number of generations; this appears to be due to recombination of originally linked genes forming polygenic combinations for hair-number. Different polygenic combinations for this character have arisen in homologous chromosomes in different straits, each combination being balanced in that it contains both plus and minus modifiers of the character. Furthermore, a number of such combinations will tend to co-exist in a species with considerable outcrossing, since the delicacy of the balance (see below) is improved when the genes for a polygenic character are heterozygous. When selection is practised, crossing-over provides new and extreme combinations.

No such balanced polygenic combinations can be detected in the modifiers of abnormal mutant characters, such as bar eye. Mather suggests that they will arise by natural selection in respect of wild-type characters, in order to prevent too great deviation from the normal, while at the same time affording the possibility of change under selection, through crossing-over providing a limited number of extreme recombinations. Such polygenic combinations, like other features of genetic systems are thus a compromise between immediate individual fitness and long-term evolutionary plasticity.

To sum up the evolutionary bearings of recent discoveries about gene-complexes, we may say that evolution not only need not occur by a series of sharp single steps, but is not likely to do so; each such step is immediately *buffered*, as it were, by ancillary changes in genes and gene-combinations which can act as modifiers for the major mutating gene and adjust it more or less completely to the needs of the organism, though final adjustment may have to wait upon further mutation. In any case, what evolves is the gene-complex; and it can do so in a series of small, if irregular steps, so finely graded as to constitute a continuous ramp.

When we reflect further that it is theoretically possible for a gene to alter its character radically by mutating step by small step from one member of a multiple allelomorph series to another, we shall see that the discontinuity inherent in Mendelian genetics is no obstacle to the visible continuity revealed in paleontological evolution. Discontinuous germinal changes are perfectly capable of producing continuous changes in somatic characters. Nor, as we shall set forth more fully later, is the pathological character of many mutations at their first appearance necessarily a bar to their final evolutionary utilization by the species.

The divergence of two stocks will always involve the accumulation of different genes in the two lines, each buffered by special modifiers and adjusted in its own way to other genes; and this will inevitably lead to a certain amount of disharmony on crossing, the F1 or later generations being less fertile or less viable, or both. The fact of internal adaptation within the gene-complex thus automatically helps to bring about the inter-sterility of species.

Alteration of Genic Expression

Let us take some examples of mutations, at first deleterious, being rendered innocuous. One of the most striking cases occurs in the meal-moth *Ephestia kuhniella*. Here a red-eyed mutant is known which shows considerably lowered viability. But when the recessive gene for red eyes is combined with another recessive gene for transparency of eyes, the double recessive is as viable as the normal wild type.

A somewhat similar example comes from *Drosophila*. The mutation *purple* (eye-colour) causes the duration of life to be considerably reduced.

Another mutation, *arc,* affecting wing-shape, produces nearly as great a reduction. But the two in combination cause much less reduction than either separately. The figures (for both sexes together) are as follows:

Genes	*Length of life (days)*
Purple	24.54 ± 0.18
Arc	26.81 ± 0.29
Purple/arc	33.71 ± 0.34
(Wild-type	39.47 ± 0.28)

Considering what severe effects are exerted by the to genes separately, the favourable result of their combination is very striking. Brierley (1938) has worked out a means of determining the "selective index" of any gene-combination, and has obtained some suggestive preliminary results, also in *Drosophila*, on the general viability interactions of numerous genes. In a number of cases, the restoration of viability occurs gradually in the mere course of maintaining the mutant stock. The classical analysis of this phenomenon is that of the eyeless mutant of *Drosophila*.

Eyeless is due to a 4th-chromosome recessive gene. Its characteristics on its first discovery were that it considerably reduced the size of the eyes, in some cases to complete absence, decreased fertility markedly, and had a depressing effect on viability. After, however, a stock for eyeless had been inbred without any artificial selection for a number of generations, it was found that practically all the flics had normal eyes and showed little reduction in either fertility or viability. On outcrossing to the normal wild type and re-extracting the recessives in F2, it was found that these once more manifested the original characters of eyeless, though in even more variable degree.

The explanation of these facts is that the manifestations of eyeless are readily influenced by other genes, and that in general those modifiers which make for normal viability and fertility also make for normality in eye-size. Thus natural selection acting upon the recombinations of modifiers present in the stock speedily saw to it that the combination making for the manifestation of reduced eyes was eliminated. In competition with its wild-type allelomorph, eyeless would be eliminated; but in stocks pure for eyeless, the genes to be eliminated will be the plus modifiers of the mutation. In broadest terms, there has been a selection of the most favourable gene-complex.

A similar genetic modification of recessive mutations towards wild-type expression was found by W. W. Marshall and Muller (1917) in *Drosophila melanogaster* for the wing-characters *balloon* and *curved*. Still another example from *Drosophila* is the sex-linked mutation *vesiculated*, affecting the wings: this can be brought back to normal expression by means of autosomal modifiers. A very similar botanical case is recorded by Harland (1932) for chlorophyll deficiency in cotton; in a way this is even more striking, since the original manifestation of the genes (in this case three pairs are involved) was markedly lethal. Here again inbreeding and selection led to the production of a reasonably viable form, while outcrossing of this to normals caused the reappearance of lethal segregants. The genetic mechanism is similar to that operative in the classical case of Castle's hooded rats and the alteration of their pattern in either plus or minus direction by an accumulation of modifiers, though the selective implications are of course different.

Selection of this type, it now appears, is a constant and indeed normal process. It has become almost a commonplace in animal used for genetic analysis to find that mutant types, which at first are extremely difficult to keep going, after a few generations become quite viable. This has repeatedly occurred in *Gammarus*, for instance, and Mr. E. B. Ford tells me that it has often occurred in his cultures of other mutant strains of the same species. A recently-described example from *Drosophila* is that of white eye in *D. obscura*. This recessive mutant was at first very delicate but its viability improved progressively on inbreeding. A precisely similar course of events was observed in the hairless mutant of mice, showing that the phenomenon occurs in mammals. In plants, we have referred to cotton: an analogous case has been found in the nasturtium. In all cases, the explanation is basically similar to that for eyeless. The experimenter, however, will also attempt to keep the mutant character sharp: he will therefore be selecting for combinations which keep the viability up without altering the visible expression of the gene, so that the process may take a little longer. R. L. Berg (1941) finds that, owing presumably to this form of selection, dominance become more intense in laboratory stocks than in the wild.

A converse effect is found when a gene, which in one species or variety is harmless, becomes deleterious on outcrossing. The explanation is that the expression of the gene in its normal situation has become so conditioned by favourable modifiers that it exerts no ill-effects; on outcrossing, however, it finds itself in a genic environment lacking some or all of these modifiers, and consequently expresses itself in ways unfavourable to viability. We have mentioned a case of this sort in Gordon's fish crosses. A striking example, particularly relevant to our present discussion, comes from Stockard's work (1931, 1941) on dogs. The St. Benard breed shows various symptoms of hyperpituitarism that simulate the pathological condition known as acromegaly. Matings between St. Bernards give normal litters; but when the St. Bernard is crossed with the Great Dane, a breed that may be regarded as a simple giant type with no hyperpituitary characters, a considerable proportion of the F1 (and later) offspring show serious disturbances during their growth, notably hydrocephalus and paralysis of the hind limbs, these effects being clearly of genetic origin.

Man, it seems has pushed the St. Bernard breed as far as it can go in the direction of large size, heavy jowl, and other effects of extreme or one-sided pituitary action; and in the process has amassed those combinations of modifiers which will protect the organism against the harmful effects of its exaggerated glandular development. When the breed is outcrossed, the protective genes are diluted to a greater or lesser extent, with corresponding ill-effects. The modern show type of bulldog has similarly been produced by selection for genes causing abnormal thyroid structure and function; here the breed has been pushed still further towards the glandular limit, since a considerable proportion of males are partially or wholly sterile.

An example in which genic expression is altered without noticeable effects on viability is that of the hybrids between *Drosophila melanogaster* and *D. simulans*. About 50 per cent of these lack bristles present in both parent species—i.e. certain combinations of modifiers from the two parent species suppress the expression of certain genes controlling bristle development.

Excellent examples involving artificial election have resulted from the work of R. A. Fisher (1935, 1938), who by repeated back-crossing introduced dominant or semi-dominant genes from domestic breeds of poultry, into the unselected gene-complex of the wild jungle fowl. Polydactyly varies in its single-dose expression in domestic breeds. Punnett and Pease (1929) found it to behave usually as a complete dominant, but sometimes as a recessive and sometimes as an irregular partial dominant. To explain these facts, they postulated one dominant inhibitor preventing the expression of the gene for polydactyly, and a second capable of suppressing the action of the first. Hutchinson (1931) pointed out that a simpler explanation is provided by Fisher's theory, according to which there is a single gene which remains constant, but whose dominance-relations differ in different gene-complexes. In the wild gene-complex, moreover, as Fisher showed, the homozygote can be definitely distinguished from the heterozygote by possessing larger extra toes, with more bones. What has happened is that an originally intermediate single-dose expression has, in most domestic breeds, become more complete. The gene for barred plumage behaves in a somewhat similar way. Most interesting are the results with the gene for crest on the head. In the wild gene-complex the crested gene in single dose produces crest alone; in double dose, however, it produces not only an unusually large crest but a cerebral hernia of deleterious character. In the Japanese silky fowl, no hernia is ever produced, and the effect of the crested gene is the same in single and in double dose. Thus firstly, the gene has become fully dominant in domestication in place of intermediate; secondly, its effect on hernia has been suppressed by modifying factors (cf.); and thirdly, in the wild gene complex, its two effects are of different type, the harmful hernia being fully recessive, the neutral crest being partially dominant.

Another aspect of the adaptation of genic expression to the needs of the organism concerns the stability of expression of genes. Plunkett (1932), for instance, has analysed the fact, well known in general terms, that wild-type characters are usually much less modifiable by changes in environment than are those determined by mutant genes. His analysis was for the most part restricted to temperature-effects in *Drosophila*, but the principle can be widely generalized. The explanation is based on the fact that genes are in most cases concerned with the rates of processes. The curves expression the rate of the processes are in general obliquely S-shaped, tending to a final horizontal equilibrium-position.

In wild-type genes, the flattening is normally completed before the imaginal state (or corresponding definitive stage) is reached, whereas in the majority of mutant genes, the curve is still in the ascending phase at this stage. Thus quite small disturbances of the curve will have marked effects in mutant genes, but very slight ones on wild-type characters. There can be no doubt that selection has been at work to adjust the rates of gene-controlled processes so as to produce this result in wild-type genes, thus conferring a high degree of stability on the characters concerned.

As Plunkett further points out, the evolution of complete dominance, with which our next section deals, can be regarded as a special case of this principle, viz. that in general natural selection favours the genotype which produces the most stable and therefore uniform phenotype. Where special circumstances demand the contrary effect, that different conditions shall be met by quite distinct phenotypes, selection has often operated to produce plasticity of genic expression. This plasticity, however, is usually of a special type, operating by some sort of switch mechanism, so that two or a few contrasted phenotypes, each of them relatively stable, are produced. The classical case is that of the environmental control of caste in social hymenoptera, whereby the same genotype can be made to produce either neuter or fertile females, and intermediates are rare aberrations. The same sort of mechanism seems to be at work in regard to the winged and wingless condition of aphids, and in environmental sex-determination.

Evolution of Dominance

R. A. Fisher (1928, 1931, 1934) has extended this concept of the alterability of gene-expression by modifiers to account for dominance in general, or at least for many features of dominance as found in nature. His argument runs as follows. Mutation is always throwing up new genes; the majority of these will inevitably be deleterious, since in a delicately-adjusted system like the gene-complex most changes are likely to be for the worse unless compensated. Further, we know as an empirical fact that the majority of mutations are repeatedly produced. Obviously the great majority of mutant genes will be carried in single dose, so that it will be an advantage to minimize any activity shown by them while in the heterozygous state. Thus, even when a harmful mutation at its first appearance shows considerable single-dose expression, i.e. manifests some or all of its effects when in the heterozygous state, then, if it be repeatedly produced (which is the case with most mutations), the way is open for it to be forced into recessivity by selection action on the rest of the gene-complex.

If it is relatively abundant (and recent studies of wild populations—e.g. C. Gordon (1936), Dubinin and others (1936), and Sexton and Clark (1936*a*)—have shown the surprisingly high incidence of various recessives which they carry in the fly *Drosophila* and the shrimp *Gammarus* respectively), selection may get to work on the homozygous condition, and render it also inactive. In such a way, as Fisher points out, mutations may be reduced to the rank of specific modifiers, normally inoperative, but exerting effects in abnormal gene-situations. In support of this view, we will cite some of the array of facts that show how readily the degree of dominance of a gene may be altered by the presence of other genes.

The classical case is that of horns in domestic sheep. The difference between horned and hornless breeds depends on a single gene-difference, but whereas a single dose of the gene for horns will produce horns in rams, a double dose is necessary in ewes. The same gene is thus dominant, in the internal environment of males, but recessive in that of females. In fowls, Landauer (1937) found that the gene for frizzled plumage, whose effects are normally incompletely dominant is converted into an almost complete recessive by the presence of a particular recessive modifier in double dose. Dunn and Landauer (1234, 1936), with the gene *rumpless*, which reduces the tail and posterior end of the body, were able to go further and to show that this could be converted either into a dominant or a recessive by crossing to different stocks, followed by selection for dominance or recessivity respectively. In mice, on the other hand, the gene for black, which is normally a complete recessive, can be converted into an incomplete dominant by modifiers. In *Drosopila virilis*, the dominant gene for rounded win converts the gene for ruffled bristles from a recessive into an incomplete dominant. Mather and North (1940) describe gene in mice whose only known effect is to modify the dominance-relations of the *agouti* gene.

A case of some historical interest is that described by Federley (1911) of the behaviour of white spotting in the larvae of the moth *Phgaera*. In *P. anachoreta* an unspotted variety is found, and this behaves as a simple recessive to the normal spotted condition. In *P. curtula*, however, only the unspotted condition exists. In the F1 of a cross between the two species, using the spotted form of *P. anachoreta*, spotting is expressed in an intermediate form, i.e. its dominance has been partially abolished. Twenty-five years ago, this fact seemed so remarkable that an authority such as Sturtevant (1912) was disinclined to accept it: to-day, however, such alteration of gene-expression by modifiers has become a commonplace.

We need not multiply examples. It is clear that the dominance of a gene can be radically modified according to the genic environment in which it happens to find itself. The next step is to show that this undoubted fact of the modifiability of the degree of dominance has been utilized in the course of evolution to make most commonly-recurring mutations recessive, so as to reduce the degree of their heterozygous expression, including that of the decreased viability which accompanies most mutations at their first appearance.

In cotton, Harland (1933) and Hutchinson and Ghose (1937) have studied the mutation *crinkled dwarf*. This occurs in the Sea Island variety of *Gossypium barbadense*, and is there a complete recessive. When crossed to unrelated strains of the same species, it is not completely recessive but shows a low degree of single-dose expression.

When, however, the mutation was introduced from *G. barbadense* into the related species *G. hirsutum* (upland cotton) there proved to be a complete absence of dominance of the normal type. The F1 is intermediate, so that at first sight we might imagine the single-dose expression to be about 50 per cent; but the fact that the F2 gives a large and unclassifiable range shows that the degree of dominance must be under the influence of a number of modifying genes. This is confirmed by the results of back-crossing the F1 species-hybrid bearing the crinkled dwarf gene to various strains of *G. hirsutum*. In certain lines, complete or almost complete recessivity of crinkled dwarf was re-established, while in others the single-dose expressivity was rendered accurately intermediate, the heterozygote class being clearly separable from either homozygote. As Hutchinson and Ghose have clearly shown, the results entirely support R. A. Fisher's views. Later work has shown that it also occurs as a very rare mutant in *G. hirsutum*. The *barbadense* crinkled show intermediate single-dose expression in the F1 with *hirsutum*, but is recessive-when transferred to a pure *hirsutum* gene-complex. When, however, *hirsutum* crinkled is transferred to a pure *barbadense* gene-complex, it behaves as an intermediate.

Harland (1941) develops the thesis that the dominance of normal alleles in cotton may be built up in a considerable variety of ways. Thus the character *petal spot* in both *barbadense* and *hirsutum* is based primarily on a main gene which, however, exists in different allelic forms in the two species. But whereas in *barbadense* the action of this main gene must be reinforced by a number of plus modifiers to produce its full effect, in *hirsutum* the main gene is stronger, and requires no (or few) modifiers. As corollaries of these facts we find (1) that in *barbadense*, the modifiers exert some positive action (a small spot) even in the absence of the main gene; (2) the gene for *barbadense* spot transferred to the *hirsutum* gene-complex has a very weak effect, and the petals are barely spotted; (3) the gene for *hirsutum* spot transferred to the *barbadense* gene-complex is reinforced by the modifiers there present, and produces a spot which is larger and more intense than any previously known; (4) crosses of the spotted forms of the one species with the unspotted of the other give a graded F2 with all intensities of spotting; (5) the F2 from the unspotted forms of the same two species produces some spots as large as normal *barbadense*—

i.e. due to recombination of modifiers only, presumably from both species.

In *G. arboreum* yet a third method of producing the spotted character is found: there is no main gene, but spotting depends on the genes that in *barbadense* act as modifiers, but here must be called a multiple factor series or polygenic combination. Haraland suggests that if a character is of advantage to the species, it can be more readily retained, in spite of recessive mutation, if its expression depends on several genes. However, this conclusion does no seem justified. The advantage of the method of using a single main gene together with modifiers would rather seem to lie in keeping the expression of the character relatively constant in the normal range of environmental conditions, but retaining a considerable reserve of potential variability to meet new or extreme conditions. The single-gene control will give greater stability, the purely multifactorial control (by "modifiers" only) will give greater plasticity. Single-gene control will permit the character to be more readily lost under the influence of selection or by drift. Presumably in correlation with this, in *hirsutum*, where spot depends on one gene only, the spot character has been lost in all but a few rare varieties.

To return to crinkled dwarf, the same principles apply. Normality (non-crinkled) in *barbadense* depends on a single strong allele, like spot in *hirsutum*. Normality in *hirsutum* on the other hand, depends on a weaker main non-crinkled allele, together with a number of modifiers. These not only encourage the dominance of normal over crinkles, but make the pure crinkled forms more normal, both in appearance and viability. By rigorous selection, new combinations of modifiers have been obtained which render the *hirsutum* crinkled practically indistinguishable from normal. When the two species are crossed, using the normal of one and the crinkled form of the other, F2 ranges from forms which are phenotypically normal through all grades of crinkling to "super-crinkled" types which are almost lethal. The petal spot experiments demonstrate two types of dominance. In *hirsutum* the Fisher effect is clearly operative, with modifiers aiding a weak "normal" gene, and also modifying the recessive towards normality. Recessive modification is much harder where a nearly dominant main gene exists, as in *barbadense*. This may then be due to an extension of the Haldane effect, by selection of "stronger" normal alleles. Harland and Atteck (1941) consider that this also operates for *crinkled dwarf* in some species, but the evidence is not decisive. They further point out that the Haldane effect is likely to be more important in self-fertilized forms, where the Fisher effect cannot so readily be produced. Where both types occur, as in *Gossypium*, doubtless "accidental" events such as the time of occurrence of suitable mutations, will determine which mechanism evolves.

An important empirical fact which was among those that led Fisher to promulgate his theory is that of the behaviour of multiple alleles. In almost every case so far investigated, the wild-type allele shows complete dominance over all the lower members of the series, whereas these when crossed with each other show intermediate expression. Thus the normal red colour of the eye in wild *Drosophila* is completely dominant over white, ivory, eosin, cherry, and all the other members of the white allelomorphic series: but white crossed with eosin, or ivory with cherry, gives an F1 with eyes intermediate in colour between the two parents.

The exception proves the rule: and the exception to this rule, to which attention was first drawn by Ford (1930), concerns the effect of this same series of genes upon an internal character of apparently no selective value. Dobzhansky (1927) had shown that the genes of the white-eye series affect the shape of the spermatheca to a small but constant degree. Whereas, as we have seen, the genes are all recessive to wild-type as regards their effects on the eye, they show intermediate expression as regards this secondary effect on the spermatheca. The body-colour genes of the ebony-sooty series show the same effect on the spermatheca, and the same differential expression as regards their effect on the external and the internal character (except that ebony body-colour is not wholly recessive). The most obvious explanation is that selection has been operative in modifying the expression of the disadvantageous external character, whereas on such effect was called for, or has been produced, with regard to the harmless secondary internal effect. We may also compare the different dominance-relations of the deleterious and harmless effects of the *crest* factor in fowls.

It is worth, nothing that this differential expression of the gene as regards two characters affected by it cannot be reconciled with any rigid form of the "inactivation" theory of recessiveness. This extension of the old Presence and Absence theory, which is obviously unternable in its original form, claims that the degree of recessivity corresponds to the degree of partial loss or inactivation suffered by the gene in mutating to one or other of its recessive allelomorphs. It is clear from what we know of actual deficiency-mutations, in which a portion of the chromosome is missing, that loss may produce effects of the same nature as gene-mutations (e.g., Mohr, 1920, who showed that that loss of the *white* locus produces an *ultra-white* effect more intense than *white* itself), so that the inactivation theory may sometimes apply. But the demonstration that some genes can become dominant or recessive according to the gene-complex shows that it can at best have a partial application. It is also noteworthy that the true absence of a gene in *ultra-white* produces some effects (spermatheca shape) which are not recessive!

As R.A. Fisher (1931) points out, the majority of the characters of most domestic breeds, or at least of the most obvious characters, especially of pattern and colour, depend on mutations recessive to the wild type. It would seem clear that man has here taken advantage of two facts, first that more or less recessive mutations are commoner in nature and secondly that they can be readily fixed by mating two similar individuals, in order to utilize striking and more or less recessive characters during his selection. Furthermore the natural tendency to concentrate for breeding purposes on individuals showing a character in more extreme form will, in the case of genes originally largely recessive, then automatically encourage the production of complete recessivity. If, however, the gene had more than intermediate single-dose expression, selection would tend to make it more dominant. Thus in general man's artificial selection will tend to encourage either complete dominance or complete recessivity, though for reasons quite other than those operative in nature. But, as mentioned, the distinguishing characters of domestic breeds in most species are usually recessive.

The chief exception occurs in poultry, where the majority of "domestic" characters are partially or wholly dominant. Fisher (1931) suggests that this difference is due to the fact that in the countries of their origin, the domestic forms, even to-day and more so in earlier times, would frequently be mated by wild cocks. In such a case, recessive, could not readily be fixed, whereas partial dominants would at least reappear in every generation; thus dominants would tend to be bred into the race by a natural selection of man's selective processes. A further effect would be that the degree of their dominance would be increased, through the new varieties being almost wholly heterozygous, and through man selecting the most striking individuals from which to breed. We have seen that his experimental tests have confirmed this hypothesis.

Marchlewski (1941) has recently confirmed Fisher's theory in dogs. Here, black was originally dominant over yellow; however, in the dingo and in various domestic breeds, yellow has become dominant through the selection of modifiers.

The recessivity of characters in other domestic species is not so nearly universal as Fisher was at first inclined to think. It seems clear, however, that man, by his breeding methods, has modified the single-dose expression of wild-type genes in his domestic animals, accentuating the recessivity of those with low, and the dominance of those with high single-dose expression; and that while this will result in most species in a preponderance of recessive breed-characters, in poultry it will tend to a preponderance of dominant ones.

It should be stated that Wright (1934*a*) does not agree with Fisher's views on the evolution of dominance, but wishes to ascribe recessivity to partial inactivation of the gene. An alternative hypothesis for the origin of the recessive character of most mutations has been given by Haldane (1930). This is based on an interesting view as to the mode of action of genes, namely, that different multiple allelomorphs produce different amounts of some substance, but only up to a certain saturation value: beyond this they can produce no phenotypic effect. Thus any mutation in a minus direction below this level can be detected, but those in a plus direction cannot. In consequence a number of different multiple alleles of different strength, but all above that needed to give the saturation value, may readily accumulate without being phenotypically detected. If now minus mutations occur, Haldane suggests that, in order (I speak teleological shorthand) to prevent their visible and viability effects from being manifested in the heterozygote, those higher alleles will be selected which in combination with the mutation will not fall below saturation level.

In other words, higher members of the series will be selected, and visible dominance will be the result. Ford, however, has pointed out that the saturation level itself will be determined in relation to the residual gene-complex, so that even here the action postulated by R. A. Fisher may be operative, though in some cases in addition to the Haldane mechanism. In any case both suggestions involve selection acting upon other genes than the mutant. Since the above was written, Ford (1940*b*) has shown that the Fisher effect can be artificially produced. The currant-moth, *Abraxas grossulariata*, has a single-gene wild variety (*lutea*) with yellow instead of white ground-colour, which normally gives an intermediate F1 with wild type. By four generations of plus and minus selection, Ford has conferred both complete dominance and complete recessivity upon the gene.

Sewall Wright (1929, 1934*a*) attacks Fisher's general conception on the ground that the selection-pressure available will be inadequate to achieve the results envisaged. However, there seems little doubt that dominance of the "normal" wild-type allele has been evolved; and Plunkett and Muller independently have shown how the need for stability of gene-expression in development will secondarily result in the evolution of dominance.

Whatever the precise method employed, it seems clear that dominance and recessiveness must be regarded as modifiable characters, not as unalterable inherent properties of genes. Dominant genes, or many of them, are not born dominant: they have dominance thrust upon them. Mutations may *become* dominant or recessive, through the action of other genes in the gene-complex. The evolution of dominance is thus seen to be in large measure an adaptation to the deleterious nature of most mutations.

R. L. Berg (1941) points out that the intensity of dominance will be selectively balanced against the accumulation of deleterious recessives which it makes possible, and that it will tend to be decreased in species consisting of incompletely isolated groups. The extra plasticity thus conferred upon such species will be in addition to that deduced by Wright.

As another example of an adaptation of the genetic mechanism itself, Fisher (1930*a*) cites the plasticity conferred by sexuality. For one thing, it will permit evolutionary advance by the combination of new mutations. If several favourable mutations occur in a population in a given time, then in a sexual cross-fertilizing species they can be combined. But if the species is asexual, they will almost certainly remain isolated, each confined to one line; for them to be combined, one mutant must be selected until it has become the main type, and only then will a second favourable mutation have a chance of becoming combined with the first.

In the second place, it will permit recombination to throw up new gene-combinations and so to use the existing genetic variance of the species to alter the type quickly in relation to changed conditions. Thus it promotes both progressive specialization and plasticity in response to the changes and chances of the environment. In addition, as Fisher (1932) stresses, it has a function to perform in relation to the deleterious nature of most mutations. For, by allowing recombination, it permits mutations to appear in homozygous form, and thus facilitates the elimination of the more deleterious. Elimination will be greater when the frequency of homozygosis is increased by inbreeding or self-fertilization. Thus variations in the type of sexual reproduction will alter the emphasis of its evolutionary function: evolutionary plasticity will be more encouraged by crossbreeding, evolutionary stability by inbreeding. Inbreeding will also promote both the rejection of unfavourable and the spread of favourable mutations.

Recombinational plasticity will be especially valuable when conditions vary and become less favourable. This is doubtless the reason why so many organisms adopt some method of asexual reproduction (which is more efficient *qua* reproduction) so long as environmental conditions are favourable, but resort to a sexual process as soon as they become unfavourable. This is so, for instance, with many protozoa, rotifers, lower crustacea, and aphids.

The biological meaning of this has been clearly brought out by careful genetic studies on *Paramecium* and other ciliates. It has been found in general that conjugation causes an increase of genetic variability in the resultant population, and that while (as in higher organisms) the majority of the new type may be regarded as unfavourable, some are actually or potentially better-adapted than those prevailing before conjugation. Thus conjugation will in many cases provide an increased chance of throwing up a recombination better able to cope with unusual and unfavourable conditions. Precisely similar results have been obtained with lower crustacea by Banta and his colleagues.

Attention is elsewhere drawn to the action of the impulse to migrate in unfavourable conditions. This also confers plasticity on a species, but in this case by increasing the range of environmental opportunities available to a given hereditary constitution, instead of increasing the range of hereditary constitutions available to cope with given environmental conditions. This, however, can hardly be called an adaptation of the genetic mechanism. On the other hand, the peculiar reactions of the crossing-over mechanism to temperatures may well, as Mr. E. B. Ford has suggested to me, fall into this category.

In *Drosophila*, the best-investigated case, crossing-over is least at temperatures close to the optimum for the species, and increases rapidly both with increase and with decrease of temperature. Increased crossing-over will, of course, have the effect of increasing the recombination of the genes located in a single kind of chromosome, and this will have a considerable effect in a form like *Drosophila* where the chromosomes are few in number. Unfavourable temperatures will thus increase the genetic variance available to a population. The discovery of the position-effect allows us to deduce certain ways, previously quite unsuspected, in which the evolutionary mechanism must itself have evolved. If, as now seems established, it is the case with some or all genes that interaction with near neighbours in the same chromosome affects their expression in an important way, then it is clear that all the genes within a given chromosome must be delicately adjusted to each other so as to produce a harmoniously functioning whole.

Any given gene must be adjusted to its neighbours within a certain chromosome-distance either way; the genes at the limit of this range will be adjusted both to the gene at our hypothetical starting-point and to others further away, thus conferring a certain organization on the chromosome as a whole. This involves a new conception of chromosomes. Up till quite recently, it was possible and usual to regard them as mere vehicles for the carriage and distribution of the hereditary constitution, without any functional organization of the genes they contained. These were assumed to be arranged at random, like coloured beads picked up haphazard by a blind man and threaded on a string; and their positions in the chromosomes were not supposed to have any relation with their effects on visible or other characters. With the discovery of the position-effect, however, this assumption, as a hard-and-fast principle, has gone by the board.

Although many genes affecting one character are scattered irregularly through the chromosomes, and genes affecting different characters are often contiguous, yet some degree of non-random arrangement does occur. Basically, and in origin, their arrangement doubtless is random, and what we know of the frequency of sectional rearrangements shows that genes must often change their neighbours in an essentially accidental way. But this randomness must then be given a functional polish: neighbouring genes must be adjusted to each other by new mutation and by recombination.

To continue our metaphor, the blind man's necklace is looked at, and colour disharmonies are got rid of by choosing new beads of the same general colour but slightly different shade. The same general type of adaptation to position-effects has been necessary as with dominance in relation to gene-mutation. Indeed, the "Weakening" of genes in abnormal positions indicates that a disproportionate fraction of the single-dose expression of dominant genes is determined by their relations with their immediate neighbours. In addition, functionally-balanced groups of genes affecting polygenic characters will be evolved within the separate chromosomes. We must to-day consider chromosomes not as being purely mechanical gene-vehicles, but to a certain degree as organic gene-arrangements.

SPECIAL CASES: MELANISM; POLYMORPHISM FLUCTUATING POPULATIONS

Before proceeding further in our main argument, however, we must turn aside to discuss certain special cases which illustrate various points concerning neo-mendelism and selection.

Melanism in Moths

The first of these concerns the phenomenon of melanism in moths, which has played a prominent art in recent evolutionary discussions. The facts may be summarized as follows: In the first place, there is no doubt that melanism among many species has become much more frequent during the last century, and that this change has been associated with industrialization: the predominance or abundance of melanic occurs in large cities and in industrial areas. In some cases the entire population of an area has become melanic. Descriptions of the historical course of events have been given for Britain by Harrison (1920*b*, 1932), and for the continent by Walther (1927), and by Hasebroek (1934). A summary of their genetic basis is given by Ford (1937).

As regards their origin, Harrison (1928, 1935) claimed that he had been able to cause melanic forms to appear, in two species belonging to different genera, by means of incorporating lead and manganese salts in the food, and that the induced melanism behaved as mendelian recessive, as does the naturally-occurring but very rare melanism of these species; (in a third species in which natural melanism is dominant, and has shown industrial spread, he abandoned the work after only preliminary results).

However, repetition of the work on a large scale by Hughes (1932) and Thomsen and Lemcke (1933) has failed to confirm these results. R. A. Fisher (1933*a*) has also criticized Harrison's views as involving a mutation-rate much higher than any obtainable by X-ray treatment. Further, all industrial melanism is due to dominant genes. It would seem best to assume that some error has been at work, and that no true induction occurred. If so, then melanic mutations must, like other recurrent mutations, have been thrown up sporadically for a long period, but have spread owing to the altered selective conditions of an industrial environment. Numerous cases of sporadic melanism which have not become more frequent recently are known in moths as in other groups. An interesting case is *Boarmia extersaria*. This shows industrial melanism in Germany; but in Britain it does not enter industrial areas, and the melanic type has remained sporadic. A survey of industrial melanism reveals that the intensity of darkening varies from species to species. As regards its genetic basis, no case of recessive melanism is known to have shown industrial spread. Industrial melanism always depends either on a single dominant gene, or on multiple factors each exhibiting complete or partial dominance. As regards its physiological effects, numerous authors have shown that dominant or partially dominant melanism confers extra hardiness and viability.

Ford (1940*b*) has shown in *Boarmia repandata* that in highly unfavourable conditions (feeding on alternate days) the ratio of melanics emerging (on an expectation of 50 per cent) goes up from about 54 to about 70 per cent. He has also (1937) pointed out that, in spite of this physiological advantage, dominant melanic forms in non-industrial areas have not shown any spread or increase in frequency. He accordingly concludes that recessive melanism is due to genes which have been forced to become recessive by selection of modifiers, on account of their deleterious effects on viability. Dominant melanism, on the other hand, has favourable effects on viability, but in normal conditions is kept in check by counter-selection operating through natural enemies, the type forms being definitely cryptic in coloration, while the melanics stand out sharply against the normal background. A balance is thus reached, with a low percentage of melanics.

In industrial areas, however, the counter-selection in favour of the type is not so strong, since the background is darker. It has been suggested that in some cases ecological selection would here be reversed, and the dark forms become better protected by background resemblance. Detailed counts by Harrison, however, have shown that in some species at least this is not the case. In industrial areas it is further to be expected that many natural enemies of the adult would be reduced in numbers or absent, so that selection for concealment would be less stringent. As a still further possibility, it appears probable that in the chemically unfavourable environment of industrialism, the greater hardiness of melanics would have an increased advantage.

In any case, according to Ford's general hypothesis, the balance between opposed selective forces is in industrial areas tilted in favour of the melanic variety, with the result that this has speedily increased and in some cases has completely ousted the original type. The strength of the selection acting against melanics on account of their coloration is shown by the fact that in one species, although the melanic is more cold-resistant, and has spread in industrial areas, it is not found so far north as the normal. It would appear that on no other hypothesis can the lack of spread of dominant melanics in rural areas and their spread in industrial areas be reconciled. If so, we have one of the most striking demonstrations of the efficacy of selection.

Genetic Polymorphism

Genetic polymorphism, or the co-existence of two or more genetically-determined and well-defined forms ("phases") of a species in the same area, presents certain peculiar problems. We speak of polymorphism when the difference between the various forms is sharp, or at least expressed as a bi- or multi-modality in a frequency curve of variability; when the equilibrium between them is relatively stable; and when the frequency of the least abundant is high enough to make it certain that it is not due merely to mutation-pressure.

Polymorphism must be clearly distinguished from normal variability, however extensive, which will be grouped in a single unimodal frequency curve. The existence of separate forms or distinct modes is an essential characteristic of polymorphism. Since we are here concerned only with genetic polymorphism, we can neglect all such cases as those of the social hymenoptera, the seasonal forms of certain butterflies, etc., which are determined environmentally. We can also neglect the particular type of genetic polymorphism involved in the genetic determination of two sexes, since this is primarily maintained not by a selective balance but by the inherent nature of the genetic-reproductive mechanism: the same applies to heterostyly in plants.

Paulian (1936) applies the term *genetic* to certain cases of polymorphism in male insects where the forms are discontinuous and not to be explained by simple allometry (heterogony). Until, however, the developmental basis of this phenomenon has been ascertained, it is better not to assume that it must be genetically determined. The interest to the evolutionist of genetic polymorphism within a freely interbreeding population is that, as R. A. Fisher (1936*a*) was the first to point out in general terms, it must always involve a balance of selective advantages between the different types. For, ex *hypothesis*, mutation-pressure alone will not account for the facts, and it can readily be shown that in the absence of selective balance, one type would rapidly oust the other from any considerable representation in the population. There are two distinct methods by which this balance is actually effected—genetical and ecological.

Genetical balance

In the case of genetical balance, the heterozygote is more viable or enjoys some other selective advantage over either of the homozygotes. (For simplicity's sake we will consider only cases involving dimorphism: trimorphism will occur, as in certain species of foxes, when the heterozygote differs in appearance from either homozygote, or in other cases, as in *Papilio polytes* and *P. memnon*, when two interacting gene-pairs are involved, and polymorphism when two or more non-interacting gene-pairs are involved.) This may occur in two ways. Either the gene itself is less viable or even lethal in homozygous condition; or it is closely linked with a recessive lethal, which exert no effect in single dose but is lethal when homozygous.

Owing to the difficulty of proving a negative, no certain case of the former conditions is known, though the mutant *curly* in *Drosophila melanogaster* is a possible example. In the conditions of ordinary laboratory cultures, this maintains itself indefinitely, giving a dimorphism with wild-type. It does this because it is almost fully lethal when homozygous, but usually rather more viable than wild-type when heterozygous. As, however, it is located within an inversion, the possibility of linkage, either with a lethal or with the genes for the increased viability, is not excluded. Of linkage with a lethal, however, two examples have been thoroughly worked out. In the case of the butterfly *Argynnis paphia*, over most of its range an occasional second form of female occurs, known as *valesina*, in which the ground colour is dull green instead of rich brown. This form is usually rare, but in some localities constitutes 5 to 15 per cent of the female population. Goldschmidt and Fischer (1922) showed that *valesina* depends on a dominant gene closely linked with a recessive lethal; they were able to break the linkage, thus obtaining broods with all females of the *valesina* type. In China and neighbouring areas almost all the wild females are *valesina*. This well shows the relativity of the term *normal* as applied to organisms in nature.

A similar situation exists in the American Clouded Yellow Butterfly, *Colias philodice*. Here a white female variant exists, and normally constitutes 4 to 20 per cent of wild females. As with *Argynnis*, the rarer form is due to a dominant gene linked with a recessive lethal, and Gerould was able to separate the two genes. In one area the white type is the more abundant. The situation as regards selective advantage is complex. The lethal must have some advantage (*a*) when heterozygous, as it is (in *Colias*) more widely spread in the population than the dominant colour-gene; (*b*) in association with the dominant colour-gene, since the linkage, though not very close, survives in nature in most areas. The dominant colour-gene must also have some advantage in association with the lethal to balance the wastage arising from homozygosity. Where it is homozygous in nature, the advantage must come from association with some other gene.

Numerous cases, as yet unanalysed genetically, are known in nature where polymorphis, as with *Argynnis paphia*, exists in one area of the range of the species but not in another. A familiar example is the common red squirrel, *Sciurus vulgaris*, which is always red in some areas, e.g. Britain, but both red and black in mountainous parts of Europe. Other cases are mentioned. In some it would appear that a new area has been colonized exclusively by individuals of one type, which presumably then is not linked with a full lethal, but either is linked with a gene somewhat reducing viability, or is the type which is less advantageous than the lethal-linked form. In other cases the lethal linkage may have been broken, as with *Argynnis* and *Colias*.

Highly polymorphic species exist in nature among land-snails, such as the common *Cepaea nemoralis* and *C. hortensis*, with their vast range of ground-colour and degree of banding; grasshoppers and certain fresh-water fish such as *Lebistes*. In these cases the polymorphism appears to depend on the phenomenon of close linkage within each chromosome or of the obligatory association of many whole chromosomes to produce a similar effect to close linkage. In these circumstances, a recessive lethal will prevent the free recombination of any favourable mutations in the same chromosome. Thus, since recessive lethals are common types of mutation, whole chromosomes will have to compete with each other, instead of selection being able to act so as to produce an approximation to a single "best" combination of genes. Further, since homozygotes cannot live, there must be at least two different forms of each of the chromosomes which contain lethals; the different combinations of these will of necessity give a variety of forms; and this variety will itself be subjected to selection so as to give the best balance, and the least waste through excess mortality of one or some of the forms. In the land-snails, the interesting fact has been discovered that the frequency of the different types of bandling has remained about the same since the neolithic period, showing that the balance is an enduring one. The problem, of course, remains as to why the close linkage has become a characteristic of the species in the first instance, since no obvious advantage inheres in such a condition. Haldane has suggested that it is due to translocation of segments from one chromosome to another. The unusual phenomenon of the dominance of the mutant types over a "universal recessive" would then also be accounted for, since the mutants would possess the translocated segment in duplicate, both in its original and its new position. Further, if, as is to be expected and as appears actually to be the case, the translocation is less viable when homozygous, we could dispense with lethal genes as an explanation of the selective balance reached.

The common sea-anemone *Metridium senile* exists in a number of strinkingly different colour-varieties. D. L. Fox and Pantin (1941) enumerate eight, including white, red, brown, grey, and various combinations of these; in addition, there is much variation in the intensity of the colours. The different forms occur in different proportions in different localities. The various colour-types depend on the interrelation of (1) lipochrome, giving colours from red to yellow; (2) brown melanin, restricted to the ectoderm; (3) black melanin, restricted to the endodern. There seems no doubt that the main types are genetically determined, and differ in their metabolic properties, and also that the colour is here adaptively non-significant, but correlated with some basic physiological difference. Fox and Pantin conclude that selection is weak as between the different colour-varieties, and that this will account for the existence of the numerous phases. We have seen, however, that a selective balance is required for this, and experiments on the physiological and ecological differences between the varieties should yield interesting results. There are the additional complications that asexual reproductions occurs, and that single individuals can persist for great lengths of time, perhaps even indefinitely.

In the Mexican fresh-water fish *Platypoecilus maculatus*, M. Gordon (1939) has found well over 120 patterns in a state of nature, mostly dependent on the recombination of 15 gene-pairs. This is a remarkable degree of polymorphism for a wild species, especially as collections dating from 1867 indicate that it is a balanced one. Furthermore, as in the other examples we have cited, related species and genera (*Xiphophorus*) present much parallel variation. Some of the excessive variation, however, is apparently due to the fixation of "accidental" characters by drift. A quite different type of balance occurs in variable plant species which consist of numerous ecotypes.

Ecological balance

Ecological balance, on the other hand, depends either on a diminution in the degree of selective advantage due to increase of frequency of one or all of the polymorphic forms above a certain level, or on an alternation in the amount or type of selection due to alteration in the environmental conditions. The best examples of the former concern mimetic butterflies. Either all the forms are mimetic, or one is non-mimetic and the other or others mimetic. If a mimetic form happens to become too abundant relative to its model, the protection afforded by the resemblance will diminish. Where certain mimetic forms are wholly absent, the corresponding models are also fond to be missing. In any area, a balance will thus be struck, depending on relative abundance of models, intensity of predation, and viability factors.

Among the best-analysed examples are those of *Papilio polytes* and *Hypolimnas dubius*. In the former case, only the females are polymorphic, existing in three forms, two mimetic and one non-mimetic. The species is a successful one, able to live outside the range of its models: it is then, of course, monomorphic, all females being non-mimetic.

It seems clear that where models are available, mimicry confers a definite advantage, but one which diminishes rapidly with increased frequency of the mimetic forms. In *Hypolimnas dubius*, both sexes are alike, and are dimorphic, with two mimetic forms. For the details of the genetic basis of the condition, readers are referred to Ford (1937). In general the equilibrium due to ecological selective balance may be broadly compared with the effects of mass action in chemistry. As a matter of fact, *Papilio polytes* appears to illustrate a combination of genetic with ecological control, since the two mimetic forms, both of which are dominant, are more viable in the heterozygous than in the homozygous condition.

Various authors have regarded as a theoretical difficulty the fact that such enormous differences in pattern, and even shape and habit, obviously involving many independent characters, can be controlled by a single gene. It should, however, be clear that once a mutant type is established, conferring even a small mimetic advantage, the residual gene-complex can undergo evolution of the same nature as what we have discussed in previous sections. But here the extent of such modification may be pushed much further, with a result that is best described, not as a modification of the visible effects of the original gene, but as an addition of various new effects, all of which, however, are dependent for their expression on the presence of the original gene. Presumably this change in the gene-complex will depend more on new mutation and relatively less on recombination of previously existing genes than in e.g. alteration of dominance.

We may put it in another way by saying that the original gene-difference comes to act as a switch controlling the action of numerous mutant modifying and modifiable genes, precisely as occurs in the case of the primary sex-difference in regard to genetically sex-limited characters. There is no reason for, and every reason against, postulating the sudden origin of the whole pattern by one mutation. Further, the frequent superficiality of the characters by which mimetic resemblance is achieved shows that the resemblance cannot have arisen though parallel mutations occurring and being preserved owing to similarity of conditions. The apparently cryptic colour-polymorphism of certain stick-insects and mantids may have a similar ecological basis. The egg-colour, polymorphism of the cuckoo, *Cuculus canorus*, is largely related to the risk of ejection by the host. The different egg-colour strains are presumably balanced in relation to host-abundance.

In the ruff (*Machetes pugnax*) polymorphism is confined to males in the breeding season, and the number of distinctive types is extraordinary. No other birds rivals it Ford (1940*a*) suggests that the cumulative effect on females of display stimulation by numerous males at a common courting-ground, promotes maximum variability. Mayr and Rand (1937) cite a peculiar and striking dimorphism in tail-coloration in the bird *Rhipidura brachrhyncha*.

Polymorphism is much commoner than usually realized, e.g. in birds and mammals only, it exists among squirrels, foxes, bears, cats, owls, herons, hawks, skuas, etc. Next we come to cases where changes in external environment determine the ecological balance, not those in internal conditions (relative abundance of the separate types) We have mentioned the common squirrel (For plan ecotypes). Elton has found that in the red fox (*Vulpes fulva*), which shows trimorphism the "cross" type being apparently a heterozygote between red and silver, the rare silver type changes in its relative frequency in a regular way during each of the 10-year cycles of abundance to which the species is exposed. In this species, Cross (1941) finds a rough polymorph-ratio cline, red being commoner to the S., silver and cross to the N.

The arctic fox (*Alopex lagopus*) is dimorphic in winter (blue or white). Only the white type occurs in Kamchatka, and only the blue in certain Alaskan islands; while on the Alaskan mainland a N-S gradient is found, the white type decreasing in frequency with latitude.

Here any selective advantage afforded by the white coat in winter must presumably be offset by some disadvantage, probably connected with viability, for the blue type to be able to exist in numbers at all. The primary basis of the dimorphism (as in the red fox) is thus a genetic balance. But environment may somewhat alter this equilibrium, so that the balance is in part also ecological.

A remarkable case where the relation with environmental conditions is more direct is that of the hamster (*Cricetus cricetus*), as described by Kirikov (1934) and Timoféeff-Ressovsky (1940). About 150 years ago the naturalist-geographer Lepekhin noted that in a certain region of Russia black hamsters were unusually prevalent. Since then, the statistics of the Russian fur-markets have enabled biologists to trace the steady spread of the black type, until to-day, throughout a broad zone along the northern border of the range of the species, black forms are in the majority and in some areas are present to the total or virtual exclusion of the typical greys. This area coincides with the sub-steppe (wood-steppe) climatic zone, and is cooler and moister than the steppe region proper, which forms the main home of the species. In the steppes the black type occurs only as an occasional aberration. It seems clear that the black form enjoys some selective advantage in the sub-steppe area, while it is at a disadvantage in the drier steppes.

In the bird *Coereba*, the recent replacement of the typical by the melanic form on certain West Indian islands appears to be similar. In other cases, although the actual process of spread has not been followed, we can be certain that it is taking place. For instance, in the brush opossum (*Trichosurus vulpecula*), melanic variants are very rare on the Australian mainland. In the Tasmanian subspecies, however, they are common. Pearson (1938), from an examination of many thousand skins, was able to plot a contour ("phenocontour") map of the relative frequency of the black type, with contour lines ("isophenes") marking regions of a given frequency. In the first place, he was able to show that neither an isolated small island off the north-west, nor another on the north-east on the curse of the submerged land-bridge from Australia, contained black animals except at aberrations; this demonstrates that the abundance of the black type must have arisen after the isolation of the subspecies in Tasmania. The north-west corner of Tasmania contains only black opossums, while on parts of the east coast the proportion is under 25 per cent. There is no correlation of the frequency of blacks with climatic gradients. The suggestion that the black type appeared (whether by new mutation or by the crossing of two carriers of the black gene in single dose) in the north-west, and is gradually extending eastwards, is confirmed by conditions on the narrow-necked Tasman peninsula. This shows a markedly lower frequency of blacks than the adjacent zone of the main island: it seems clear that the extreme narrowness of its neck has hindered the spread of the black gene.

It is interesting to note that Tasmania is cooler and moister than Australia, so that the similarity to the case of the black hamster is very close. Humidity also favours melanism in *Coereba*.

Another case where a phenocontour map of a mutant type has been plotted, and examination of the map shows that spread of the mutant is occurring and is being impeded by geographical barriers, is that of the *simplex* condition of the teeth in the field-vole *Microtus arvalis* in north-central Germany. The aberration here consists in the absence of the last ridge of enamel on the molar. Here the *simplex* condition occurs in over 90 per cent of individuals in Schleswig-Holstei, with zones of decreasing frequency to west, south, and east. The mountains of central Germany have proved a complete obstacle, while certain large rivers have obviously hindered the spread of the character. This case is genetically slightly more complex, since various gradations in the expression of the character occur, and not only is the character more frequent in the presumed centre of origin, but also more extreme in type. It is further worth noting that the gradient in *simplex*-frequency bears no relation to another character-gradient or cline within the species, namely, the gradual east-west darkening and reddening of the coat-colour across the north German plain.

In the last two cases, the polymorphism may be regarded as transient, since the species or subspecies appears to be moving from a condition in which a given character is rare, maintained only by mutation-pressure, to one where it is universal, again apart from the rare and sporadic occurrence of its allele. This is theoretically to be distinguished from true polymorphism, in which a state of balance between the contrasting types is indefinitely maintained, but it will not always be possible to distinguish the one condition from the other. Thus Southern (1939) has shown that the bridled variety of the guillemot (*Uria aalge*) increases northwards from 0.5 per cent or lower to well over 50 per cent of the total population, in a fairly regular gradient. There is, however, no ways as yet of telling whether the two types are indefinitely balanced against a climatic gradient, as with the arctic fox, or whether the bridled type is spreading at the expense of the "normal". The owl *Megascops* shows a centrifugal cline between red and normal (grey) forms in U.S.A. while that between dark and light fulmar petrels is rather more complex.

Mather (1941) has recently propounded a possible genetic explanation of clines in such apparently non-adaptive characters. He suggests that the gene responsible is linked with one of two balanced polygenic combinations within homologous chromosomes, affecting some other character, and that this is related to some environmental condition. The mean development of the character will be adjusted by the relative frequency of the two combinations, and thus the frequency of the linked gene will also vary in relation to the environmental gradient concerned. One may add that the same result would be obtained if one of the genes in the polygenic combination were pleiotropic and also produced the visible effect (e.g. bridling).

A frequent condition found in nature is that of regional or geographical polymorphis, when two or more contrasted forms are confined to different regions. Thus in the moth *Spilosoma mendica*, the normal condition is for the male to be dark, the female white. In Ireland, however, the males as well as the females are white, and this condition (variety *rustica*) is known sporadically on the European continent. Such a state of affairs may represent the final stage of a transitory polymorphism, in an area which favours the spread of an alternative type (cf. the brush opossum), or be due to colonization by one only of the two forms. Elsewhere, the condition may represent the end of a dimorph-ratio cline. Thus, in *Accipiter novae-hollandiae* such a cline culminates at either end in an area exclusively inhabited by one of the two forms. Again, the white form of the palearctic moth *Leucodonta bicoloria*, abundant in the east, decreases westwards and is absent in the extreme west.

We may conclude by referring to the floral dimorphism of higher plants, even though that is solely or mainly environmental or developmental in its determination, not genetic. Floral dimorphism is concerned with the relation between cross- and self-pollination.

Strictly speaking, the term dimorphism should perhaps only be employed when whole plants of distinct type are found. This occur in most cases of gynodioecism, e.g. in the common plantain *Plantago lanceolata*, the viper's bugloss *Echium vulgare*, etc., where some plants produce normal hermaphrodite flowers, while others have only female (pistillate) flowers, which are sometimes smaller than the hermaphrodite ones; in a few cases of cleistogamy, such as the balsam *Impatiens noli-me-tangere,* where in addition to plants exclusively producing the normal showy flowers, others may occur bearing only the inconspicuous and permanently closed cleistogamous flowers adapted solely to self-pollination (as well as still others with both types); and in heterostyly. Mather (1941) has an interesting discussion on the evolution of heterostyly. He points out that an illegitimacy reaction appears to be compulsorily associated with heterostyly, and is the chief bar to inbreeding. Homozygous thrum plants, which normally do not occur in nature, are less viable than the heterozygotes or the recessives (pms). This reversal of the usual relations of dominance to viability must be due to the accumulation of deleterious mutations, in the region adjacent to the thrum gene, which will only exist in nature in a heterozygous condition; this is similar to the accumulation of loss mutations in the Y-chromosome, which has led to its almost total inertness.

If conditions demand greater stability and therefore increased inbreeding, in some cases selection may reverse the intensity of the illegitimacy reaction. Since commercial seed-raisers prefer selfing, this has occurred with cultivated species, e.g. *Primula sinensis*. Here the fertility of illegitimate relative to that of legitimate pollinations has almost doubled since the experiments of Hildebrand and Darwin in 1864 and 1877. Another possibility is the selection of mutants giving homostyle plants which are then capable of self-pollination. This has occurred in nature.

In a few cases of heterostyly,. such as certain species of *Primula*, the two types of flowers also have different-sized corollas. In some cases the "pin", in others the "thrum" type is thus distinguished. The meaning of these last conditions is obscure. Kerner and Oliver (1902) suggest that the smaller size of corollais associated with the type where autogamy is more prevalent, and cross-pollination accordingly less essential; but this Dr. Mather informs me is not in accord with experimental facts.

It is also quite logical to use the term dimorphism for types in which two distinct types of flower are produced on the same individual plant, as occurs in most cleistogamous forms (species of *Viola*, *Glechoma*, *Lamium*, *Oxalis*, *Helianthemum*, *Juncus*, many grasses, etc.). Further, it is perhaps even legitimate to extend the concept to cover the frequent combination, in one and the same flower at different times, of definite devices to secure cross-pollination and self-pollination.

This would then constitute what we may perhaps call a dimorphism in time, since in the great majority of cases the plant produces flowers which are adapted to ensure cross-pollination, but if this does not occur, it transforms these same organs into what is virtually a few type of flower adapted for self-pollination.

Furthermore, in all these cases, a selective balance is at work. However, the balance is a complex one. In plants whose flowers change from exogamy to autogamy, as well as those which produce both showy and cleistogamous flowers, it used to be supposed that all that was involved was what the experimental embryologists style "double assurance" (*doppelte Sicherung*) to secure pollination. Cross-pollination was assumed to be in some way better, but if, through lack of suitable insects or other reason, it chanced not to be effected, then the plant fell back on its second line of defence, in the shape of self-fertilization. This in itself would constitute a selective balance of an ecological nature. The advantages accruing from cross-fertilization are offset against the disadvantage of its not being always possible; the disadvantage of having to produce a second type of flower (or to develop new adaptations within the original flower) is offset against the advantage of assured seed-production.

Ecological factors may further complicate the picture. Thus certain species produce solely or almost solely cleistogamous flowers when growing in deep shade, where few insects are likely to visit them, but go to the reverse extreme when grown in sunny open localities. In *Viola*, cleistogamous flowers are much more abundant in high latitudes; this is due to a photo-periodic response. Along these lines, the course of evolution of the cleistogamous condition is readily envisaged. Cleistogamy in the strict sense implies a special type of flower which not only does not open, but shows other adaptations; usually the petals and stamen are reduced in size, and the pollen-grains are not liberated from the anthers but germinate *in situ*. This condition has doubtless followed on one of "pseudocleistogamy", where, in certain ecological conditions, the normal flowers simply fail to open and self-pollination occurs.

This, however, is not the whole story. Autogamy, it is now realized, is not merely a pis aller. It has certain advantages, in perpetuating unchanged a vigorous and well-balanced genetic constitution once this has been evolved. This stability, however, will only be advantageous so long as conditions also remain unchanged: in addition, an entirely stable type loses the possibility of invading new environments. Thus the provision of devices for both cross- and self-fertilization constitutes a balance between the advantages of plasticity (with its disadvantage of wasteful production of less well-adapted recombinations) and of stability (with its long-term disadvantage of absence of adjustment to new conditions). Some forms are exclusively of one or the other type; but in a large number, probably the majority, of flowering plants, the two have been brought into balance, with consequent floral dimorphism of one sort or another.

A third type of selective advantage concerns the degree of waste of gametes associated with a particular mode of reproduction. This wastage is especially marked in monoecious plants, and its implications are discussed in reference to the selective balance involved in gynodioecism.

It is difficult to evaluate the precise shares of these various advantages and disadvantages in floral dimorphism. We must be content to observe that a selective balance is involved, and that the dimorphism is thereby maintained.

In other cases the dimorphism is between sexual and vegetative reproduction. Dr. W. B. Turrill informs me in a letter that in the lesser celandine, *Ranunculus ficaria*, the following types of reproduction occur: normal amphimixis, apomixis, abundant vegetative multiplication (of two types; one variety is wholly or almost wholly vegetative in its reproduction), and plants with only male, only female, and only hermaphrodite flowers. Different populations show different proportions of these various types. Here, in the field of reproduction, we may perhaps have something akin to the balance of ecotypes in many plant species.

We may in fact conclude that polymorphism always involves a selective balance, whether it is determined genetically, or environmentally, or internally by the processes of normal ontogeny, as when two or more kinds of persons or organs, adapted to different functions, are formed by the same colony or individual (for social hymenoptera).

Finally in view of its peculiar evolutionary interest as inevitably involving a selective balance and as in some cases leading by way of ecological regional differences, to sharp geographical differentiation, genetic polymorphism deserves the most intensive study, especially in cases where the ratios of the types are geographically graded, since here we may hope not only to measure the intensity of the selective forces at work, but also to discover something as to their nature.

Selection in Fluctuating Populations

Elton (1930) has pointed out that the customary assumption of a population approximately constant from year to year is very far from the truth for many, if not most, species. A stable "balance of nature" does not exist. Fluctuation in numbers, rather than constancy, is the rule. This fluctuation may be broadly progressive towards increase or decrease, it may be irregular, or it may be cyclic and regular.

Animal species subject to such regular or periodic fluctuations include lemmings (*Lemmus*), snowshoe rabbits (*Lepus americanus*), mice, voles, jerbils and other Muridae, foxes (*Vulpes*, *Alopex*), lynxes (*Lynx*), and other fur-bearing carnivores, certain birds, such as ptarmigan (*Lagopus*) in Labrador and nutcrackers (*Nucifraga*) in Siberia, some invertebrates, land and marine, and probably certain antelopes and other larger mammals. The period of the fluctuation from crest to crest of abundance varies from 3 to 4 years in the smaller rodents, to 10 to 11 years in the snowshoe rabbit, and probably a good deal longer in certain ungulates. The difference in abundance between crest and trough may be very great. In the snowshoe rabbit the ratio of high to low population numbers in extreme cases must reach at least 100: 1. An interesting point made by Rowan (1931) for Canadian birds is that migrant species are not affected by these cycles. In years when the grouse population has been reduced to a minimum, the migrants are present in normal numbers. This fact must, in combination with others peculiar to migrants, have important evolutionary consequences restricted to migrant forms. However, it seems not to be of universal occurrence.

Elton has pointed our certain important evolutionary consequences of these facts. In the first place, both the intensity and the type of selection will vary continuously during the cycle. During the period of rapid increase, when numbers are low and conditions favourable, the intensity of selection will be very low. During the peak period, intraspecific selection due to pressure of competition will be high. Since the catastrophic fall in numbers is normally due to infectious disease, selection during this period will mainly concern disease-resistance. And in the subsequent period of unfavourable environmental conditions (for all these cycles seem to have an external determination) selection will be concerned with resistance to cold and hunger and similar aspects of the struggle for existence. To use Elton's metaphor, the species is put through a series of examinations, with easy times between-whiles, and the different examinations test different capacities. Apart from special resistances, such varying selection will promote a general elasticity of response.

Elton also made the suggestion that periodic fluctuations would allow greater scope for chance in evolution, since if a rare mutation or gene-combination happens to be present in the much-reduced minimum population, it will be automatically reproduced in the same proportion during the period of rapid increase when the struggle for existence is light and the intensity of selection low.

His views have been criticized on mathematical grounds by Haldane. However, the proof of the pudding is in the eating, and the studies of Ford and Ford (1930) make it clear, first that selection-intensity may actually be relaxed during a period of rapid increase, and secondly that when it is once more tightened up, the resultant type may differ from that obtaining in the proceeding period of abundance. They observed a sharply isolated colony of the small Greasy Fritillary butterfly, *Melitaea aurinia*, for thirteen years, and obtained records and specimens for a total period of forty-nine years. The population was increasingly abundant from 1881 to 1897; it then decreased, and became scarce by 1906 and extremely rare from 1912 to 1920. A rapid increase to abundance then took place to 1924, from when until 1930 it showed a progressive slight further increase.

Variability was slight during the first relatively stable period of abundance. During the period of rapid increase after scarcity, however, "an extraordinary outbrust of variation occurred. Hardly two specimens were alike and marked departures from the normal form of the species, both in size, shape and colour, were very common. A high proportion of these were deformed in various ways, the amount of deformity being closely correlated with the degree of variation."

With the colony entered on its second period of abundance, the abnormal types and extreme variants practically disappeared, and the population settled down again to a uniform type. This, however, was not the same as before, but recognizably distinct.

It seems clear that the relaxation of selection during the recovery period allowed an excess of variability; and that when it again became rigorous, the new stable type was slightly different, owing to the accidental incorporation of different genes. R. L. Berg (1941) has demonstrated a similar effect in micropopulations of *Drosophila melanogaster*, aberrations increasing with decreased intensity of selection.

Sewall Wright (1932), in discussing such problems in more general terms, concludes that there must be available to most species a number of gene-combinations all of about the same survival-value; he compares them to peaks, separated by "valleys" of intermediate combinations which are less favourable. Normally it is difficult or impossible for selection to shift the type from an established peak to another, although this might be equally satisfactory if reached: but when the intensity of selection is reduced (or when low size of population promotes the accidental survival of genes and gene-combinations), many "valley" combinations are realized, the species can cross dryshod to other peaks, and it will be a matter of chance on which Ararat the type eventually remains perched when the rising tide of selection again floods out the valleys.

Elton draws a further interesting conclusion from the facts of periodic fluctuation. He concludes that it will have promoted the migratory impulse which is so strong in so many types of animals when in unfavourable conditions. At first sight, the existence of this impulse seems hard to explain on any selective hypothesis, since, in the more spectacular mass emigrations, such as those of locusts, lemmings, or certain butterflies, all, or all but a negligible fraction of the migrants perish, while the population is renewed from among those which do not manifest the impulse and stay in their original habitat. One would thus suppose that migratory tendencies would be strongly antagonized by selection.

However, although such migrations are exceedingly striking and have thus received a disproportionate share of attention from biologists, they are, in fact, but extreme and in a sense abnormal manifestations of a much more widespread phenomenon, namely, an impulse to react to unfavourable conditions by changed behaviour, notably by movement away from an environment which has become unfavourable. This does not normally result in mass migration on a vast scale, but in an irregular movement that disperses the population over a wide area. When lemmings are scarce in the Arctic, snowy owls (*Nyctea nyctea*) descend into north temperate latitudes in search of food. When the cedar-cone crop fails in Siberian forests, the Siberian nutcrackers (*Nucifraga caryocatactes macrorhynchus*) leave their usual haunts and may reach western Europe. And many quite small and inconspicuous movements of animal population are going on all the time.

When migration is of this less extreme type, many individuals which would otherwise die will survive temporarily in regions beyond the normal range of the species and be able to return later to their original habitat, while others may survive by reaching and remaining in other parts of the normal range. In addition, some individuals may be able to survive and to remain in areas outside the normal habitat of the species, either by adopting slightly different habits and so colonizing different habitats within the original geographical range, or by colonizing areas outside this range. This extension of habitat may in the first instance be dependent on a non-inherited modification of behaviour, mutation and selection later stepping in to fix the change genetically (the "organic selection" of Baldwin and Lloyd Morgan); or genetic variants may find themselves in surroundings to which their constitution is better adapted than was the normal environment of the species (pre-adaptation). In either case, migration will have been advantageous to the species as well as to the individual. Elton (1930) draws an illuminating comparison between the sexual process and the migratory urge. Both are extremely widespread, and both confer additional evolutionary plasticity on a species.

The sexual process enables the species to exploit to the fullest extent the mutations, old and new, which are carried by the species or which crop up during its evolutionary career, by making possible every kind of recombination of them. The migratory impulse, in relation to unfavourable conditions, has a precisely analogous effect, in increasing the range of environmental conditions with which any genetic variation that exist can be brought into contact. The two are complementary and often mutually reinforcing processes, and both have their most important function in times of stress. Fluctuations in numbers, both irregular and periodic, thus may have important evolutionary consequences.

2

Chromosomal and Morphological Evolution

Rates of molecular evolution were assessed by two approaches, allozyme electrophoresis and immunology (microcomplement fixation). The former gives a measure of the proportion of genes that differ between taxa and the latter gives a measure of the proportion of DNA triplets that differ between taxa for a single gene. We have both electrophoretic and immunological data for dasyurid marsupials, Australian rodents and Australian *Rattus* snakes. The estimates of evolutionary rates for each group, based on the two data sets, are comparable.

Rates of chromosome evolution were assessed from G-banding analyses. Comparisons between rates of chromosome evolution and rates of molecular evolution were available for the family Dasyuridae (marsupial mice), Australian *Rattus* (rodents), and Australian Hydromyinae (rodents). The data show virtually no correlation between chromosomal and molecular evolution. Moreover, cladistic analysis of the data shows that rates of chromosome evolution vary enormously between different lineages not only between groups but also within groups.

Although rates of electrophoretic, immunological and chromosomal evolution can be measured objectively, the same cannot be said of morphological evolution. We have attempted to overcome this problem by choosing extremes of morphological divergence, from sibling species (which by definition are morphologically very similar) to species that are clearly very different morphologically even to the casual observer. Examples of sibling species that nevertheless differ greatly at the molecular level include bats of the *Eptesicus pumilis* complex in Australia and marsupial mice of the *Sminthopsis murina* complex. The other extreme is represented by Australian cockatoos of the genus *Cacatua* and Australian elapid snakes, where species that are very similar at the molecular level are nevertheless very different at the morphological level.

The data show that molecules, chromosomes and morphology evolve at very different rates. A large body of evidence suggests that, of the three, molecular evolution is the closest to being constant with time. If this is true, then rates of morphological and chromosomal evolution vary enormously with time.

The purpose of the symposium was to bring together workers studying rates of evolution from various points of view, including morphology, genes, chromosomes and developmental biology. In this chapter we compare and contrast rates of evolution at the genetic, chromosomal and morphological levels using various Australian vertebrates as examples. We show that, although different measures of genetic divergence correlate closely with each other, chromosomal and morphological divergence bear little relationship to genetic divergence. These results have important implications for palaeontologists, many of whom seem to accept that morphological stasis in the fossil record implies genomic stasis, and that rapid morphological shifts imply rapid genomic evolution. The present results show that this is not necessarily so.

Genetic Level

Genes can be studied directly in various ways such as DNA sequencing and DNA-DNA hybridization, or indirectly via their protein products. Although the technology exists for direct DNA sequencing, it is nevertheless an expensive and time-consuming process and is limited at this stage to very few genes. Most of the information available on rates of evolution at the gene level has therefore been derived from studies of proteins. Protein evolution can be best assessed by direct amino acid sequencing. This also is an expensive and time-consuming process. Two techniques that are relatively inexpensive and rapid are microcomplement fixation (MC'F) and allozyme electrophoresis.

Microcomplement Fixation

In this technique, a protein is purified from one species and injected into rabbits which raise antibodies against the protein. The antisera are then cross-reacted with the same protein in related taxa, and the 'strength of cross-reaction' is measured to yield an immunological distance (ID). The ID is directly proportional to the number of differences in amino acid content between the protein used to raise the antisera and the protein of the test species. Hence it is a measure of the divergence between the taxa for the gene encoding that protein. Provided that convergences do not exceed divergences at the amino acid level, the ID between two taxa will be a monotonically increasing function of time since divergence.

Allozyme Electrophoresis

The electrophoretic technique detects different mobility rates of proteins when individual samples are loaded on a support medium (usually a gel) and subjected to an electric current. These different rates are a consequence of charge differences resulting from amino acid substitutions. A typical study will use allozyme electrophoresis to compare about 30 gene loci.

In the vast majority of cases, two taxa that possess electrophoretically distinguishable proteins for a particular gene will have different structural alleles for that gene. However, two taxa that possess the same electrophoretic form of a particular protein do not necessarily possess the same structural allele for that protein since not all amino acid substitutions are electrophoretically detectable; not all nucleotide substitutions in the coding regions yield amino acid substitutions (i.e. the code is redundant); and some base substitutions occur in non-coding regions.

Nevertheless, it seems reasonable to assume that the genetic distances assessed by electrophoresis are proportional to the total genetic divergence for the suite of proteins studied.

Correlation between MC' F and Electrophoresis

Here we examine whether there is a correlation between these two methods of measuring genetic divergence between taxa. That is, do taxa that differ greatly at one gene (the gene encoding the protein being studied by MC' F) differ at a high proportion of genes as studied electrophoretically? Before addressing this question, however, it will be necessary to consider the limits of the two techniques.

There is an upper limit to the extent of genetic divergence that can be estimated by the two techniques. For MC' F, the limit is a technical one; at IDs much above 100, high antibody concentrations inhibit the reactions being measured. For allozyme electrophoresis the upper limit will theoretically occur when the taxa share no alleles at any loci, i.e. they are 100% different. In practice, saturation occurs at a lower level because of convergences and parallelisms. Therefore a positive correlation between genetic distances as estimated by MC' F and electrophoresis will be expected only for closely related taxa.

There is a second statistical problem that also needs to be considered when searching for such correlations. While for MC' F distances, reciprocal data are independent (i.e. the distance from taxon A to taxon B is measured independently of the distance from B to A), the same is not true for electrophoretic data. Moreover, the data are not totally independent in the sense that if the distance from A to B is small, and A is distant from C, then B will also be distant from C. To obviate these potential statistical artifacts, we have chosen to plot a subset of the data, i.e. where A is close to B and distant from C, the A-B and A-C distances are plotted but not the B-C distance.

Both electrophoretic and MC' F data are available for three groups of Australian vertebrates, the rodent genus *Rattus*, the marsupial family Dasyuridae and the rodent subfamily Hydromyinae. The electrophoretic data are given by Baverstock et al. (1985) for *Rattus*, by Baverstock et al. (1982) for Dasyuridae and by Baverstock et al. (1981) for the Hydromyinae. In the case of the Dasyuridae and Hydromyinae, only one member of each genus was used for the intergeneric comparisons to avoid the statistical problem outlined above. In the case of the

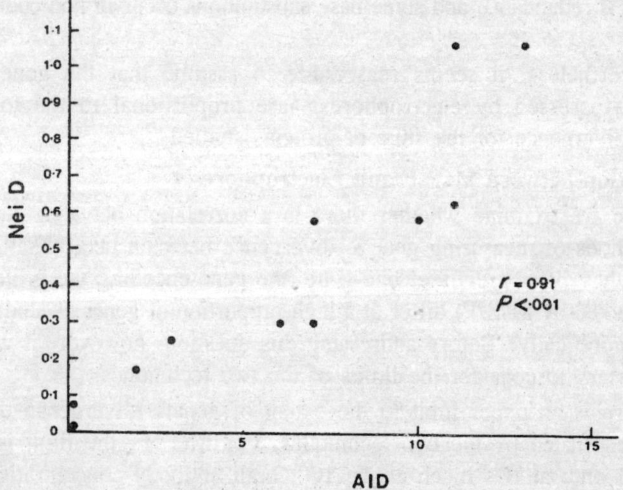

Fig. 2.1. Relationship between electrophoretic distance and albumen immunological distance for Australian Rattus.

Rattus data, *R. rattus*, *R. norvegicus* and the Australian *Rattus* were treated as three lineages and only one member of each lineage was used for interlineage comparisons.

The plot of genetic distance based on 55 electrophoretic loci against albumen immunological distance (AID) for the Australian *Rattus*. There is a strong correlation between the two of 0.91 (P < 0.001). The slope of the line through the origin is 0.074. Note that in this case the maximum AID is 13. Again there is a strong correlation between the two of 0.69 (P < 0.001). The points are scattered more widely in this case than in the *Rattus*, especially at higher AIDs. However, it is clear that we are here seeing the effects of saturation in the electrophoretic data, since the maximum AID is 34. The electrophoretic data are based on 20 loci. The scatter of points is considerable in this case, but this is clearly due to the saturation effect for the electrophoretic data since we are here dealing with AIDs of up to 60.

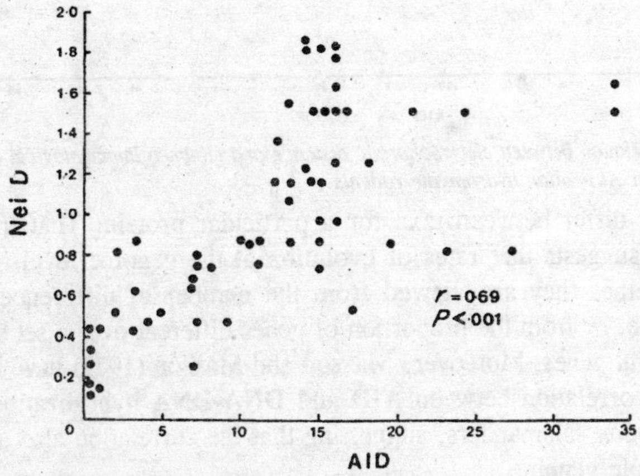

Fig. 2.2. Relation between electrophoretic distance and albumen immunological distance for Dasyurid marsupials.

For these three groups of Australian vertebrates, there is a strong correlation between genetic distance estimated from a single gene and genetic distance estimated from a suite of genes. A similar strong correlation between AID and Nei D has been reported for frogs, salamanders, lizards and mammals.

We emphasize that electrophoresis and MC' F measure genetic divergence between taxa in what would appear to be totally independent ways: electrophoresis measures the proportion of electrophoretic loci that differ between taxa, and MC' F measures the proportion of amino

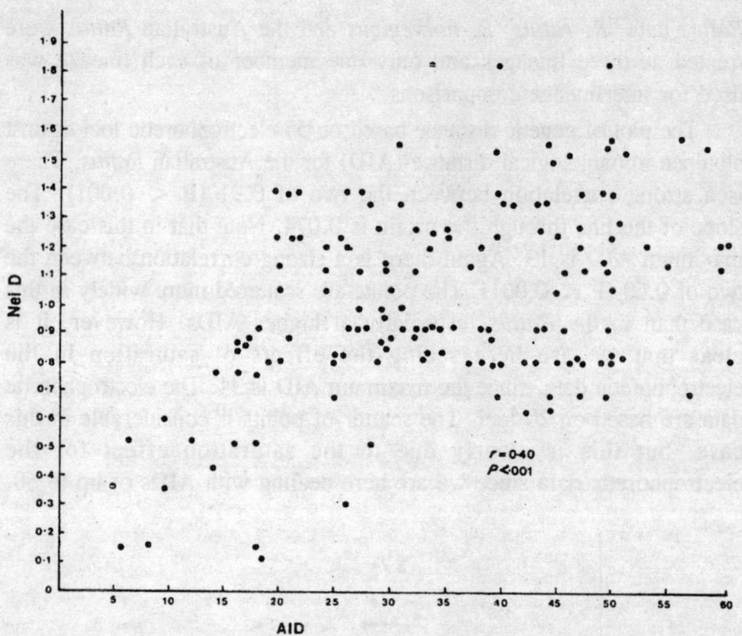

Fig. 2.3. Relation between electrophoretic distance and albumen immunological distance for Australian hydromyine rodents.

acids that differ between taxa for a particular protein. That they do correlate suggests that rates of evolution at the genome level are the same whether they are viewed from the number of differences in a single gene, or from the proportion of genes different over a set sample of structural genes. Moreover, Maxson and Maxson (1979) have shown a strong correlation between AID and DNA-DNA hybridization data for *Plethodon* salamanders, suggesting that the correlation also applies to the whole genome.

Thus if the immunological distance between two taxa is known, the electrophoretic distance can be predicted, at least approximately. As will become clear, this is not true for either chromosomal differences or morphological differences.

Chromosome Level

A number of studies have attempted to assess rates of chromosomal evolution from the differences seen in gross karyotypes. There are two general criticisms that may be made of such studies. First, some chromosomal rearrangements do not affect the gross karyotype, e.g. paracentric inversions, equal reciprocal translocations, independent fusions between chromosomes of the same size, etc. The gross karyotype can also be affected by modifications that do not involve rearrangement of the structural genome, e.g. addition or deletion of heterochromatin. Both these limitations can be overcome by using the techniques of G-banding and C-banding. We will therefore restrict ourselves to groups whose chromosome evolution has been assessed by these techniques. The second problem is to relate the chromosome differences to specific branches of the tree leading to extant taxa. Approaches that summarize karyotypic diversity of a group are fraught with statistical problems. Thus a speciose group may appear to have high karyotypic diversity simply because of the high number of species. Simply dividing by the number of species, however, overcompensates—if a high proportion of the chromosomal rearrangements have occurred in a branch leading to a highly speciose group, the amount of chromosome evolution will be underestimated. One way out of this dilemma is to work with groups whose cladistic relationships are fairly well established independently of the chromosome data, and plot the proposed chromosomal changes on the branches in a parsimonious fashion.

Thus an appropriate analysis requires groups whose chromosomes have been analyzed by G-banding and C-banding and whose cladistic relationships are established. Three such groups of Australian vertebrates are again the Australian Rattus, the dasyurids and the hydromyine rodents.

Cladistic relationships among the Australian *Rattus* are now firmly established. The group has been the subject of an intensive morphometric analysis, as well as an electrophoretic analysis based on 55 loci, and the two studies are in the main concordant. The group has also been the subject of intensive chromosomal analyses using G-banding and C-banding.

The cladistic relationships among the Australian *Rattus*, along with the minimum number of chromosomal rearrangements that have occurred along each branch. The rates of chromosome evolution along various branches are highly disparate. For example, from the common ancestor of the Australian *Rattus*, no chromosomal rearrangements have occurred in the line leading to *R. lutreolus*. However, the lineages leading to members of the *R. sordidus* group (*R. sordidus*, *R. villosissimus* and *R. colletti*), have undergone considerable chromosome evolution. Yet at the genetic level, these three taxa are very similar—*R. colletti* and *R. villosissimus* share alleles at all 55 loci screened and *R. sordidus* differs at only two of these loci. Clearly, in this case, extent of

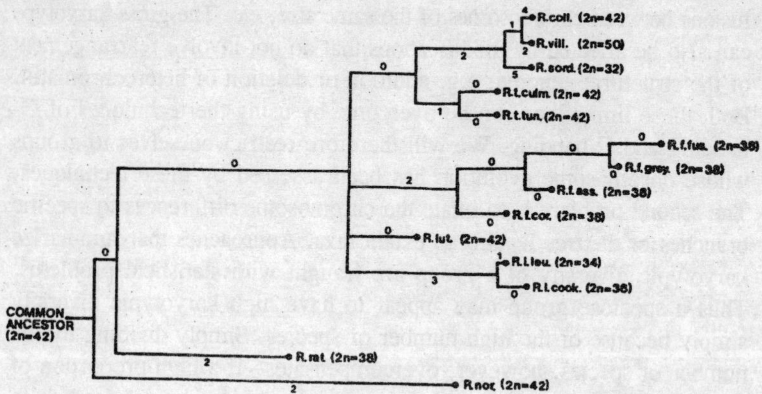

Fig. 2.4. Cladistic relationships among Australian Rattus showing minimum amount of chromosome evolution along each branch and 2n is the diploid number.

genetic divergence is a very poor predictor of the extent of chromosomal divergence.

Of the 35 or so species of Australian dasyurids, about half have been chromosomally analyzed by G-banding and C-banding. The cladistic relationships among the species for which such chromosome data are available. These relationships are based primarily on electrophoretic analysis of 32 loci and MC'F data, but are supported on morphological criteria.

The same karyotype is retained in II of the 13 species representing all major lineages. This karyotype is therefore the presumed ancestral state. *Antechinomys spenceri* possesses a unique paracentric inversion, and *Ningaui yvonneae* a unique pericentric inversion. The rates of chromosome evolution within the Dasyuridae are therefore seen to be fairly uniform, with no lineage showing a spurt of chromosome evolution. Moreover, compared with the Australian *Rattus*, the chromosomes of the dasyurids are highly conservative. This is all the more remarkable when it is remembered that the dasyurids are about three times as diverse at the molecular level.

The third group for which extensive G-banding data are available are the Australian hydromyine rodents. Of the 35 species in 12 genera, 25 species in 10 genera have had their chromosomes G-banded. Although the cladistic relationships among these 25 species are not known precisely, sufficient is known for the rates of chromosome evolution to be interpreted from the available data.

It is clear, for example, that there are three major lineages referable to the tribal level. A single G-banded karyotype occurs in

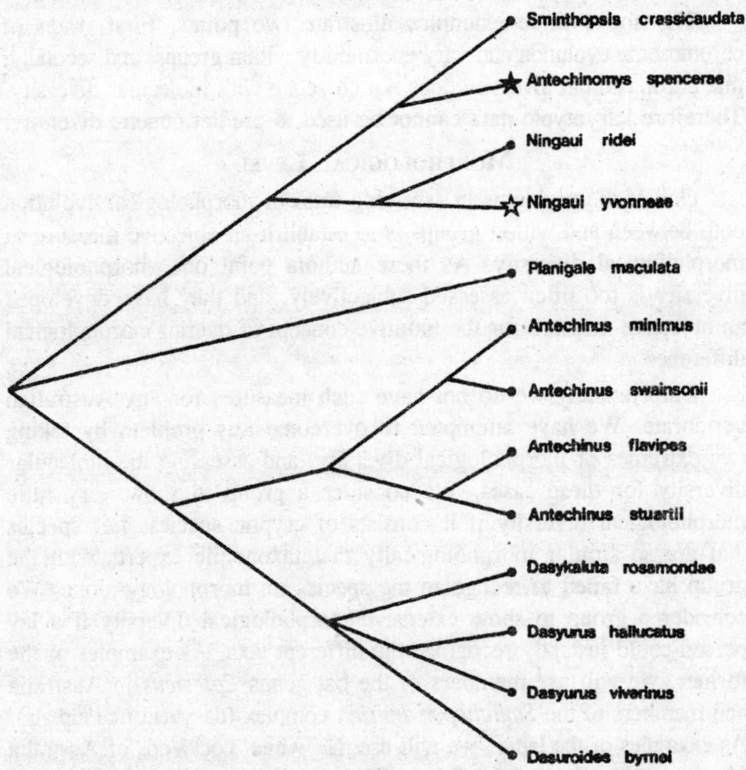

Fig. 2.5. Cladistic relationships among Australian dasyurid marsupials whose chromosomes have been G-banded.

all three tribes, suggesting that this is the ancestral karyotype. This ancestral karyotype is retained unmodified or only slightly modified in the vast majority of taxa studied. Thus the hydromyines are generally characterized by karyotypic conservatism. Yet at the molecular level, the hydromyines are about five times as diverse as the Australian *Rattus*.

One genus (*Zyzomys*) in one tribe has, however, undergone a high degree of karyotypic modification, to the extent that at least seven chromosomes show partial or no homology with the ancestral complement. Moreover, the rearrangements that characterize the *Zyzomys* complement are not explicable by simple inversions of translocations. Clearly the genus *Zyzomys* has undergone rapid chromosome evolution, whereas all other lineages in the subfamily have remained chromosomally conservative.

The above three examples illustrate two points. First, rates of chromosome evolution can vary enormously within groups, and secondly, that chromosomal diversity does not correlate with molecular diversity. Therefore karyotypic data cannot be used to predict genetic diversity.

Morphological Level

One of the problems in assessing rates of morphological evolution both between and within groups is to establish an objective measure of morphological diversity. As these authors point out, morphological diversity is too often assessed subjectively, and they have developed an objective measure for the intuitive concept of overall morphological difference.

Unfortunately we do not have such measures for any Australian vertebrate. We have attempted to overcome this problem by taking two extremes of morphological diversity, and assessing the molecular diversity for these cases. We consider a group to show very little morphological diversity if it consists of cryptic species, i.e. species that are so similar morphologically that taxonomic experts with the group have failed to recognize the species on morphology alone. We consider a group to show extensive morphological diversity if a lay person could instantly recognize the different taxa. As examples of the former, we will use members of the bat genus *Eptesicus* in Australia and members of the *Sminthopsis murina* complex (dasyurid marsupials). As examples of the latter, we will use the 'white' cockatoos of Australia (Aves), and snakes of the family Elapidae in Australia.

Bats of the Genus *Eptesicus*

Species of this genus in Australia are so similar morphologically that Ride (1970) recognized only one species. McKean et al. (1978) subsequently recognized four species separable on the shape of the penis bone in males. Discriminant function analysis was needed to separate males on other skeletal characters, but these same characters failed to discriminate females. Kitchener (1976) subsequently recognized a fifth species. Electrophoretic analysis has revealed, however, that there are at least nine species of *Eptesicus* in Australia, some of which are only differentiable electrophoretically. Clearly, by any criterion, the nine species of this genus in Australia are very similar at the morphological level.

The genetic relationships among these nine species based on 36 loci are shown phenetically. Nei D values range up to 0.9, this being higher than the genetic distances typifying species within a genus.

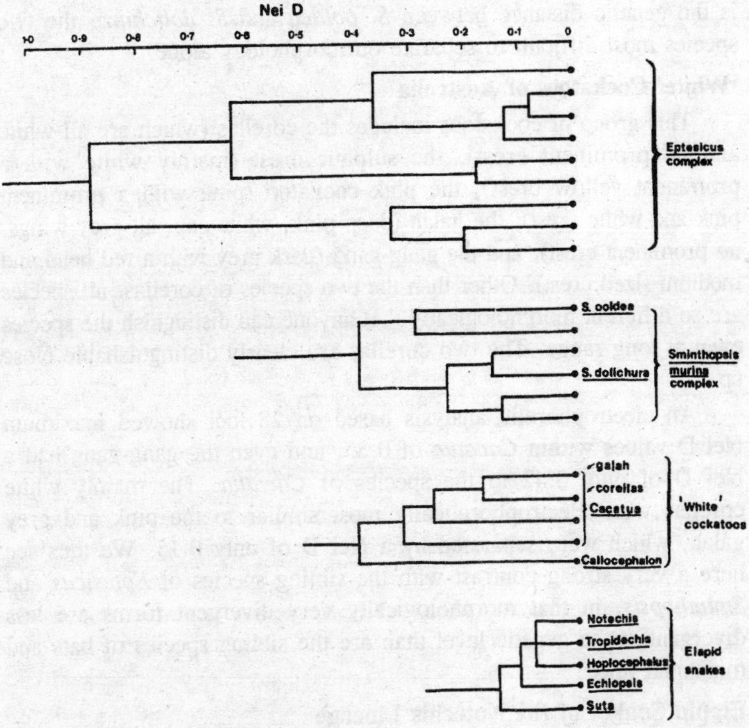

Fig. 2.6. Phenograms expressing genetic relationships among members of four groups of Australian vertebrate.

Sminthopsis murina Complex

The members of the complex are so similar morphologically that Troughton (1965) recognized only the species *S. murina*. In an extensive review of the genus, Archer (1981) recognized *S. ooldea* as distinct from *S. murina*, and provided evidence that *S. leucopus* also represented a distinct species. Subsequent electrophoretic analysis revealed that there were at least six species in the complex, which proved to be distinguishable by discriminant function analysis. However, it has recently become apparent that the characters used in this discriminant analysis fail to distinguish *S. ooldea* from *S. dolichura* where they are sympatric on the Nullabor Plain. Clearly the members of this group satisfy the criterion of cryptic species and are morphologically very similar.

The genetic distances based on 30 loci among six of the species of the complex are shown. Nei D values range up to 0.4. Indeed, this is the genetic distance between *S. ooldea* and *S. dolichura*, the two species most difficult to separate on morphology alone.

'White' Cockatoos of Australia

This group of cockatoos includes the corellas (which are all white and no prominent crest), the sulphur crest (mainly white with a prominent yellow crest), the pink cockatoo (pink with a prominent pink and white crest), the galah (deep pink, white cap, all grey wings, no prominent crest), and the gang-gang (dark grey with a red head and medium sized crest). Other than the two species of corellas, all species are so different morphologically that anyone can distinguish the species even at long range. The two corellas are clearly distinguishable close up.

An electrophoretic analysis based on 28 loci showed maximum Nei D values within *Cacatua* of 0.35, and even the gang-gang had a Nei D of only 0.42 to the species of *Cacatua*. The mainly white corellas were electrophoretically most similar to the pink and grey galah, which were separated by a Nei D of only 0.15. We thus see here a very strong contrast with the sibling species of *Eptesicus* and *Sminthopsis*, in that morphologically very divergent forms are less divergent at the genetic level than are the sibling species of bats and marsupial mice.

Elapid Snakes of the Notechis Lineage

The *Notechis* lineage of Australian snakes include the genera *Tropidechis*, *Hoplocephalus*, *Echiopsis* and *Suta*. Colour photographs of all but *Notechis* can be found in Cogger (1975). They vary considerably in size, head shape and colour to the extent that none could be confused by a lay person. Indeed these taxa are so distinct morphologically that they have traditionally been placed in separate genera.

Unfortunately, we do not have electrophoretic data for these snakes. However, we have made an analysis based on MC' F using the protein transferrin. It is known that the molecular diversity of transferrin is roughly twice that of albumen. Using the slope of the line relating Nei D to AID established earlier, it is possible to relate these transferrin IDs (TID) to expected Nei D. Although it must be admitted that the proposed relationship between TID and Nei D may be tenuous, these different genera are obviously very similar at the molecular level. Thus the snakes, like the cockatoos, are highly divergent at the morphological level, but at the genetic level are less divergent than the sibling species of *Eptesicus* and *Sminthopsis*.

The foregoing analysis demonstrates quite clearly that the extent of morphological diversity bears little relationship to the extent of diversity of the whole genome. This is not to deny the role of genes in determining morphological diversity. Clearly the morphological diversity exhibited among the cockatoos and among the snakes has a genetic basis. Rather, the proportion of the genome that is reflected in gross morphology may be a very small proportion of the total. Therefore morphological similarity cannot be taken to indicate genomic similarity.

3

Chromosomal Aberrations in Evolution

It has already been stated that chromosomes occasionally break and that the rate of breakage can be increased by the use of various mutagens such as ionizing radiation(s), ultraviolet light, and certain chemicals. When a chromosome breaks, the two broken ends usually undergo *restitution* or healing, and everything is as before. But this does not always happens, and the result can be the formation of (1) deficiencies, (2) duplications, (3) inversions, or (4) translocations. The first three of these ordinarily involve a single chromosome, while the fourth involves two or more chromosomes.

There is very strong evidence the DNA in chromosomes that extends as a double helical thread throughout the entire length of the structure. Furthermore, we know that the genes are linearly arranged in the DNA and that at least in some instances groups of genes under the control of common regulatory units are grouped together. It is of great interest then to ask about the genetic consequences of breaks in specific regions of chromosomes. One interesting observation made by George Lefevre on Drosophila chromosomes is that perhaps as many as 70 percent of chromosome break points that have successfully reunited with new neighbouring genes have no detectable genetic effect at all. In other words, only about 30 percent of healed breaks points have mutant effects associated with them. This could mean that only a portion of the DNA of chromosomes is informational, and breaks within the regions without information are inconsequential. It could also means that these regions may be more susceptible to breakage and rejoining than others. We do not yet know, but C. S. Lee has suggested that those break points that manage to reunite are likely to be in regions of the DNA that are represented repeatedly in the genome. The reasoning behind this suggestion has to do with the healing process itself rather than with a differential fragility of chromosome regions. The union of broken ends may involve some degree of base pairing at the joint in order for the DNA strands to be properly positioned and stabilized for ligase to be effective in forming covalent bonds between the two chains. It is of course possible that unions can be formed without a perfect match between the bases of the two strands. Experiments, with a λ phage by Wu and Taylor have shown that the ends of the λ chromosome in its linear form have complementary sequences of nucleotides at the protruding cohesive ends. The λ chromosome goes from the linear to the circular form by the formation of hydrogen bonds between the complementary bases as shown in the figure. The same sort of protruding single strand ends with complementary or mostly complementary sequences may cause the joining of broken ends of eukaryotic chromosomes. Broken chromosome ends have been described for many years by cytogeneticists as being sticky. This sickness may be nothing more than the complementarily that exists between repeated DNA sequences.

It is known that in some tissues of many organisms, such as the endosperm of maize kernels, where the broken end soft chromosomes remain unhealed at the time of chromosome replication, that there is often a fusion of the broken ends of the sister chromatids. The results in a dicentric chromosome that will produce a chromosome bridge at the anaphase of the next mitotic division. In plant cells, such a bridge will often break as the cell wall is laid down between newly formed daughter cells. This produces another broken end in each of the nuclei, and the cycle may begin again. This breakage-fusion-bridge cycle may be carried on for several cell generations, and it can be used to great advantage for manipulating gene dosage in such cells. As the bridge breaks, it may not do so at the original fusion point and the result will be a duplication of some genes in one cell and their deletion in the other. McClintock used such a scheme to study chromosome organization and gene regulation in the endosperm of maize.

For most cells, however, the rule is that aberrations that persist are those that result from rejoined ends. This is because a broken chromosome, unrejoined, will generally not persist through the zygote stage. The segment without the centromere, called the acentric fragment, will be lost. The segment that has the centromere may act as a dominant lethal, depending on how much and what is lost in the acentric fragment.

Deletions

Deletions are also referred to as *deficiencies*. Some geneticists, following the historical origins of the terms, use deletion to describe an intercalary (involving two breaks in the chromosome) loss of chromatin and deficiency to describe a terminal loss. But here we use the terms deficiency and deletion as synonyms, because in either case an acentric fragment is lost. A spindle fiber cannot attach to an acentri fragment during mitosis or meiosis and guide it to one of the poles.

Two pairs of chromosomes	1 2 3 4 5 6	7 8 9 10 11 12
	1 2 3 4 5 6	7 8 9 10 11 12
Translocation heterozygote	7 8 3 4 5 6	1 2 9 10 11 12
	1 2 3 4 5 6	7 8 9 10 11 12
Translocation homozygoyte	7 8 3 4 5 6	1 2 9 10 11 12
	7 8 3 4 5 6	1 2 9 10 11 12

Fig. 3.1. Chromosome constitution of a translocation heterozygote and a translocation homozygote.

If the DNA in the chromatin that is lost is crucial for viability, sterile gametes or nonfunctional somatic cells may result. Some deficiencies are known that are viable even when homozygous. For example, in *D. melanogaster*, a deficiency that removes the *white* locus produces, when homozygous, fully developed flies that appear normal except for their white eye colour. Most deficiencies, however, are lethal when homozygous, and these may involve no more loss of chromatin than in the example of white locus. A loss of a large segment of a chromosome may result in a dominant lethal effect even if a normal homologue is present with it in the heterozygote. If the dominant effect of a deletion is not lethal, it may result in an abnormal phenotype.

Genetic Significance of Deficiencies

The chromosomal deficiencies have following genetic effects :

(i) Lethal effect

Organisms with homozygous deficiency usually do not survive to an adult stage because a complete set of genes is lacking.

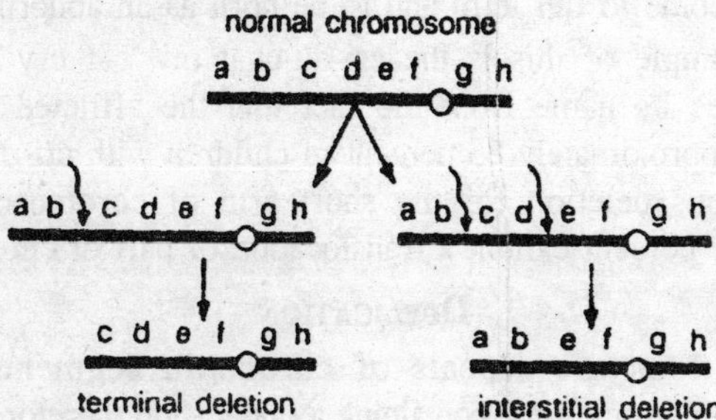

Fig. 3.2. Production of terminal and interstitial deletion. Chromosomes can be broken when struck by ionizing radiation (wavy arrows).

Examples

A mutant sex-linked condition called "notch" is lethal in *Drosophila* when hemizygous in males or when homozygous in females. This "notch phenotype is a sex-linked deletion (deficiency) which acts like a dominant mutation; a deletion at another sex-lined locus behaves as recessive mutation, producing yellow body colour when homozygous.

(ii) Pseudodominance

When an organism heterozygous for a pair of alleles, A and a, loses a small portion of the chromosome bearing the dominant allele (A), the recessive allele (a) on the other chromosome being hemizygous will become expressed phylogenetically. The phenotypic expression of recessive allele (a) due to deficiency of dominant allele (A) called *pseudodominance.* With a deficiency exchange (crossing over) becomes impossible. Disturbance of pairing causes a reduction of exchange in a segmen around the deficiency.

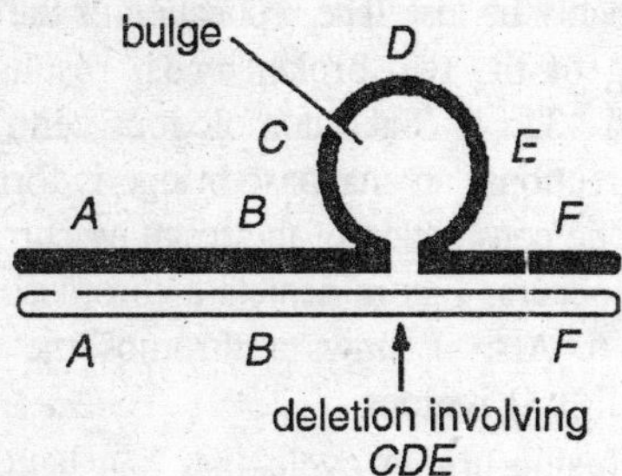

Fig. 3.3. Formation of a deletion loop during synapsis in a deletion heterozygote.

Occurrence of Deletions in Human Beings

Deletions are expected to occur in all organisms including humans. Some do not cause early spontaneous abortions, but unfortunately allow the fetus to come to full term and to be born as an abnormal child.

One example of this is the cri-du-chat or "cat cry" syndrome, which derives its name from the fact that the afflicted infant cries like a cat. Approximately 85 percent of children with cri-du-chat show a heterozygous deletion for the short arm of chromosome 5. The remaining 15 percent exhibit a translocation of part of chromosome 5.

DUPLICATION

Duplications are repeats of chromatin segments within a chromosome. Chromosome containing a duplication therefore possess more than the normal amount of DNA. There are several possible mechanisms for generating duplicate DNA segments in chromosomes.

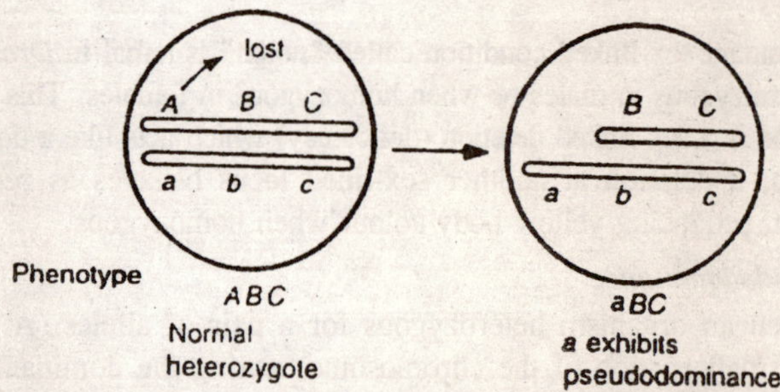

Fig. 3.4. Pseudodominance. A deficiency in the segment of chromosome bearing the dominant gene A allows the recessive allele a to become phenotypically expressed.

Duplications and deletions may also be generated by the breakage-fusion-bridge cycle. A chromosome with the normal sequence of 1.23456 has a break between 4 and 5. Since the deleted 5,6 is an acentric fragment it will probably be lost. The replication of the 1.2 3 4 fragment, followed by joining of the two broken ends, results in the dicentric chromosome 1. 234-432. 1. Since homologous centromeres generally move in opposite directions, an anaphase bridge is formed and breakage will occur. The genetic constitution of the resulting chromosomes depends on where the break occurs. Figure depicts a chromosome with deletions of regions 4,5 and 6. Also it shows a chromosome with both deleted (5,6) and duplicated (3,4) regions.

The breakage-fusion-bridge cycle that has been found in maize results in a wide variety of duplications and deletions. Note that

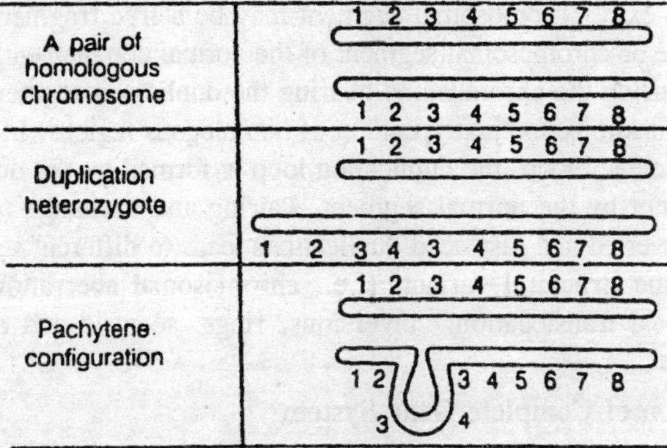

Fig. 3.5. Loop formation during chromosome pairing in a duplication heterozygote.

duplications and deletions. Note that duplications created in this manner often are displaced, that is, duplicated segments are not positioned side by side on the chromosome.

Duplications play an important role in the regulation of gene activity because the multiple copies of a gene allow large quantities of RNA to be generated quickly. Examples of this type of reiteration are the cistrons coding for the ribosomal RNA's and those coding for histones. Table 3.1. gives some examples of proteins and RNA's thought to be coded by repetitive DNA sequences.

Table 3.1. Examples of repetitive DNA sequences

Group	*No repetitive DNA sequences*	*Degree of DNA or protein homology, %*
Satellite DNA	103–106	80–100
18S-28S Ribosomal RNA	100–600	97–100
5S Ribosomal RNA	100–200	97–100
tRNA	6–400	
Histones	10–1200	87–99
Hemoglobins	=10	<75–100

Satellite DNA's contain large numbers of repeating sequences which apparently are never transcribed. The nature of the mechanism(s) that generate large numbers of DNA segments in tandem linkage is unknown but it may well be by unequal crossing over. Also unknown are the factors that protect the repeated sequences from mutational changes.

Duplication occurs when a segment of the chromosome is represented two or more times in a chromosome of a homologous pair. This extra-chromosomal segment may be a free fragment with a centromere or chromosomal segment of the normal complement. During meiotic pairing the chromosome bearing the duplicated segment forms a loop to maximize the juxtraposition of homologous regions. In contrast to the deficiency loop, the duplication loop is formed by the duplicated segment, not by the normal segment. Pairing and exchange (crossing over) is inverted and displaced duplications leads to different secondary chromosome structural variants (i.e., chromosomal aberrations) such as reciprocal translocation, inversions, rings, acentric and dicentric chromatids.

Duplication of Complete Gene System

In considering gene duplications, it is important to remember that a structural gene cannot be considered an entity in itself. The expression of any structural gene is controlled or regulated by DNA not directly involved in coding for proteins. Therefore, the duplication of a structural gene without the corresponding duplication of its regulator may lead to a large array of new phenotypes including lethality.

Gene Duplication as a source of New Genetic Material

In addition to permitting a cell or organism to vary its genetic capabilities through gene dosage effects, duplications may also provide a cell with a reservoir of extra DNA. The duplicated genes could be considered as spare parts in the sense that one gene is sufficient to enable normal cell function. With one gene functioning in the normal manner, the duplicate would be free to undergo mutation, since its gene product could no longer be essential to maintain normal function.

By taking this reasoning a step further, it may be postulated that the duplicate gene may mutate and eventually evolve into a different gene coding for a protein with a new function. If so, gene duplications offer a mechanism for the evolution of new genes. There are many examples in nature that appear to fit the hypothesis that gene duplication is followed by gene changes through point mutation. Base nucleotide additions or deletions and frameshift mutations may result in genes coding for new phenotypes. The following two examples illustrate this concept.

Some of the most extensively studied human proteins that are thought to have arisen through gene duplications are the multiple hemoglobins of vertebrates. Hemoglobins are tetramers composed of two different polypeptide chains. Adult humans possess hemoglobins A and A_2. These two hemoglobins, represented by the formulas $\alpha_2\beta_2$ and $\alpha_2\delta_2$, respectively, share a common α-chain, but have distinct β and δ-chains. Thus, the hemoglobins found in the red blood cells of adults are coded by three distinct genes. Hemoglobin studies in certain families with α-chain mutants indicate that there is probably more than one α-chain gene per haploid genome. In addition, during development, embryonic hemoglobins and fetal hemoglobins occur. The Gγ, Aγ, δ and β genes have been shown to be closely linked. As we have shoyn the amino acid sequences of the beta and delta chains are very similar, suggesting that the genes coding for these chains arose by means of gene duplication of a common ancestral gene. Furthermore, the gene that codes for the α polypeptide is structurally related to the other hemoglobin genes, but is not linked to them. The genes coding for the δ, γ, and β chains probably arose through gene duplications from the original a gene and subsequently moved to a different linkage group by means of translocation.

Since the myoglobin chain is also structurally related to the hemoglobin polypeptides, it is thought that its genes and all the hemoglobin genes are derived from a common ancestral gene by a series of duplications over a long period of time.

Haptoglobin, one of the iron-binding proteins in vertebrate serum, is another example of a protein that probably arose by gene duplication. Haptoglobin, like hemoglobin, is a tetramer. Apparently, two genes α and β are involved. The α locus has two alleles, α_1 and α_2. Three phenotypes occur in the population Hp1, Hp2, and Hp1-2. The sera of humans exhibit either the Hp1 type (α, β) or the Hp2 type (α_2,β_n), or the Hp–2 type (α, α_2, β_n). The Hp2 haptoglobin is unique to humans, suggesting that a recent gene duplication gave rise to the Hp2. Studies of amino acid sequences clearly show that a partial gene duplication (intracistronic duplication) us have arisen resulting in a large polypeptide chain, different from the original α gene product in size and ability to combine with several beta chains to yield the polymeric haptoglobin Hp1–2.

These two examples—hemoglobin and haptoglobin–clearly indicate that gene duplication has probably played an important role in evolution.

Inversions

Inversions result when two breaks occur within a chromosome and the broken fragment is rejoined to the original chromosome in a reversed order. In this example, a chromosome with a normal sequence of 12345678910 undergoes breakage where the arms are in the contact while folded over on itself, say between 3 and 4 and 8 and 9. An inversion takes place when the 3 end fuses with the 8 end and the 4

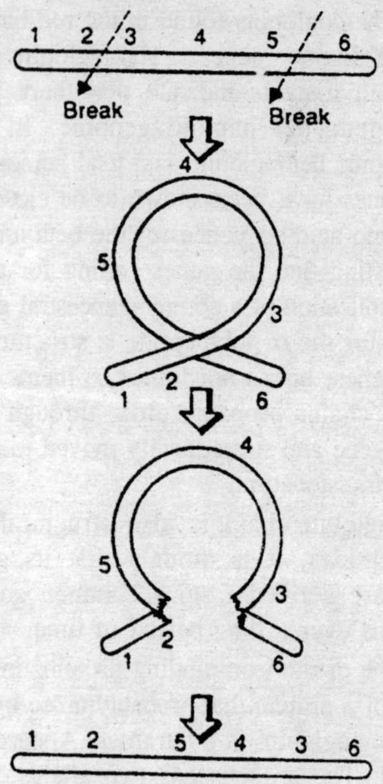

Fig. 3.6. Origin of an inversion in a chromosome.

end fuses with the 9 end to give the order 12387654910. Inversions are of two types, pericentric, with breaks flanking the centromere, and paracentric, with both breaks on the same side of the centromere.

Remember that the same polarity must be maintained along a strand of DNA. Proper integration of an inverted segment could never occur in a single strand of DNA. The proper integration of a double-stranded segment of DNA occurs if the inverted segment undergoes a 180° rotation.

Types of Inversions

The inversions are of following types :

(i) Pericentric Inversion

When the inverted segment of chromosome includes or contains centromere, then such inversions are called heterobranchial or pericentric inversions. If crossing over occurs within the loop of a pericentric inversion, the resulted chromatids include half of the chromatids with duplications and deficiencies forming nonfunction. The other half of the chromatids form functional gametes : 1/4 gametes have normal chromosome order, 1/4 gametes have the inverted arrangement.

(ii) Paracentric Inversions

When the inverted segment includes no centromere and the centromere remains located outside the segment, then such type of inversion is called homobranchial or paracentric inversion. Crossing over within the inverted segment of a paracentric inversion, produces a dicentric and an acentric chromosome. The dicentric chromosome contains two centromeres and forms a bridge from one pole to the other during first meiotic anaphase. When anaphase chromosomes separate towards poles, this bridge breaks somewhere along its length and the resulting fragments contain duplications and/or deficiencies. The acentric chromosome because lacks in centromere and fails to move to either pole and so, is not included in the meiotic products. Such, breakage-fusion bridge cycles of crossing over of paracentric inversions are most common in maize. The meiotic products includes half non-function, 1/4 functional normal and 1/4 functional inverted chromosomes.

Cytological and Genetical Methods for Detection and Identification of Inversions

Since the banding patterns of polytene chromosomes are visible, is an obvious way of detecting inversions. Figure shows a polytene chromosome in Drosophila with an inversion paired with a normal chromosome. The development of techniques that differentially stain metaphase chromosomes has also made the cytological characterization of inversions possible in species other than Drosophila.

The detection of inversions also involves the study of pairing relationships between a chromosome containing inverted DNA and its noninverted homologue during the prophase stage of meiosis. Pairing would homologous chromosomes on a point-by-point basis. However, homologous pairing between an inverted arrangement and its normal arrangement is difficult because they cannot line up properly. Therefore, in order for pairing to occur, an inversion loop must be formed.

Inversions have been dubbed crossover suppressors. The results of recombination within paracentric and pericentric inversions will clearly illustrate the reason for this name, which is, in fact, not entirely appropriate. Keeping in mind that crossing over occurs at the four-strand stage, it is probable that only two of the nonsister chromatids will be involved in the crossover while the other two remain unaffected. Let us consider two chromosomes that pair, one chromosome containing the normal gene sequence, ABCD and the other chromosome containing an inverted segment, ABCD. After replication and pairing occur, and while they are still in the four-strand stage, a single crossover occurs between nonhomologous chromatids within the inversion loop (e.g,. between genes C and B). Such crossing over results in the formation of a dicentric chromosome ABCD and an acentric fragment, DCBD, both of which contain duplications and deficiencies. Dicentric chromosomes undergo anaphase bridge formation which, in some types of cells, may be followed by breakage. The result will be chromosomes with deletions and duplications. The acentric treatment probably will remain at the metaphase plate and subsequently be lost. All of the recombination products that result from crossovers that occur within a paracentric inversion will form deficient gametes. Crossing over within the inversion loop region of a pericentric inversion also produces duplication and deficiency products.

Inversions are crossover suppressors, not because crossovers fail to occur within them in heterozygotes, but because the recombinant chromosome resulting from a crossover does not permit the offspring receiving it to survive if the crossover occurs within the inversion.

As crossover suppressors, inversion. Provide useful tools for geneticists. Inversions maintain particular seta of alleles of different loci within linkage groups by preventing recombinant offspring from surviving. Therefore, as for example in the isolation of sex-linked genes in Drosophila with the Basic stock. Inversions have also proved useful in the detection of linkage groups. A reduced number of recombinant types in a particular cross indicate that the genes are located within or near an inverted segment. And, conversely, a normal number of recombinant types indicate that the genes are located outside the inversion.

The number and type of inversions and the effects that they have in humans have not been extensively investigated until recently. Investigation has been limited by the difficulties in detecting inverted segments in human chromosomes. However, improved chromosome staining procedures now allow geneticists to detect paracentric and pericentric inversions in humans, but much work remains to be done in the field. One target area involves investigating the role of inversions in certain diseased states, such as heritable cancerous conditions.

TRANSLOCATIONS

Translocation is a broad term including all types of unilateral or bilateral transfer of chromosome segments from one chromosome to another. An important class of translocation having evolutionary significance is known as reciprocal translocations or segmental interchanges, which involve mutual exchange of chromosome segments between non-homologous chromosomes.

Cytology of a Translocation Heterozygote

If a translocation is present in one of the two sets of chromosomes, this will be a translocation heterozygote. In such a plant, the normal pairing into bivalents will not be possible among the chromosomes involved in translocation. Due to pairing between homologous segments of chromosomes, a cross-shaped (+) figure involving four chromosomes will be observed at pachytene. These four chromosomes at metaphase I can have one of the following three orientations:

1. *Alternate*. In alternate orientation, the alternate chromosomes will be oriented towards the same pole. In other words, the adjacent chromosome will orient towards opposite poles. This will be possible by the formation of a figure of eight.
2. *Adjacent I*. In adjacent I orientation, adjacent chromosomes having non-homologous centromeres will orient toward the same pole. In other words the chromosomes having homologous centromeres will orient towards opposite poles. A ring of four chromosomes will be observed.
3. *Adjacent II*. In adjacent II orientation, the adjacent chromosomes having homologous centromeres will orient towards the same pole. A ring of four chromosomes will be observed.

The alternate disjunction will give functional gametes. Adjacent I and adjacent II disjunctions will form gametes, which would carry duplications or deficiencies and as a result would be non-functional or sterile. Therefore, in a plant having translocation in heterozygous condition, there will be considerable pollen sterility.

A ring of four chromosomes as described above is found under conditions when a single interchange is found. If two interchanges are involving three non-homologous chromosomes, a ring of six chromosomes is found, and the size of the ring can increase with additional interchanges. More than one rings can also be found if two or more interchanges are independently found, each involving two different non-homologous chromosomes.

The first case of translocation was found in *Oenothera*, which was originally described as mutation by de Varies while working for his Mutation Theory. *Oenothera*, *Tradescantia* and *Rhoeo* are such cases, where translocations in the heterozygous condition are frequently found in nature. In many other crop plants they have been artificially induced.

Breeding Behaviour of a Translocation Heterozygote

The presence of translocation heterozygosity can be detected by the presence of semi-sterility and low seed set. This can then be confirmed at meiosis by quadrivalent formation. As shown above only two types of functional gametes are forced which result from alternate disjunction.

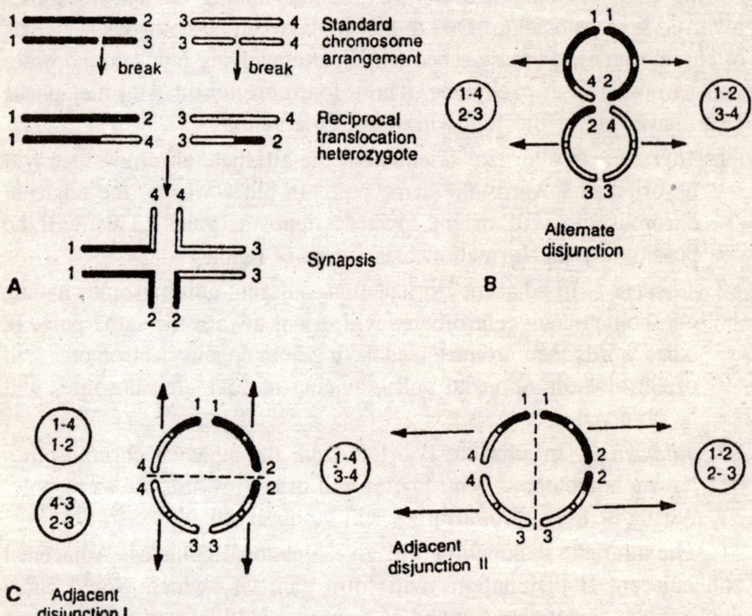

Fig. 3.7. A—Cross shaped figure formation during heterozygous reciprocal translocation; B—Alternate disjunction; C—Adjacent I and II disjunctions.

The functional gametes will give rise to three kinds of progeny namely (i) normal (ii) translocation heterozygote and (iii) translocation homozygote. These three types would be obtained in 1 : 2 : 1 ratio.

Interchange Heterozygosity in Oenothera

Subgenus *Euoenothera* of the genus *Oenothera* has been studied during 1920-1930 and cytogenetic structure leading to evolution in this group was examined. This group has 2 n = 14 and all 7 chromosomes of haploid complement have median chromosomes. There are three classes. (i) First is represented with species showing bivalents or small rings at meiosis (e.g. *O. hookeri*, *O. grandiflora*, *O. argillicola*). (ii) Second class is represented by species forming rings of various sizes at meiosis indicating the presence of interchanges. These rings are not permanent but are maintained due to their superiority in adaptive value (e.g., *O. irrigua*). (ii) The third class is represented by those having permanent translocation heterozygosity involving all chromosomes, so that a ring of 14 chromosomes is regularly formed (e.g., *O. biennis*, *O. strigosa*, *O. parviflora*).

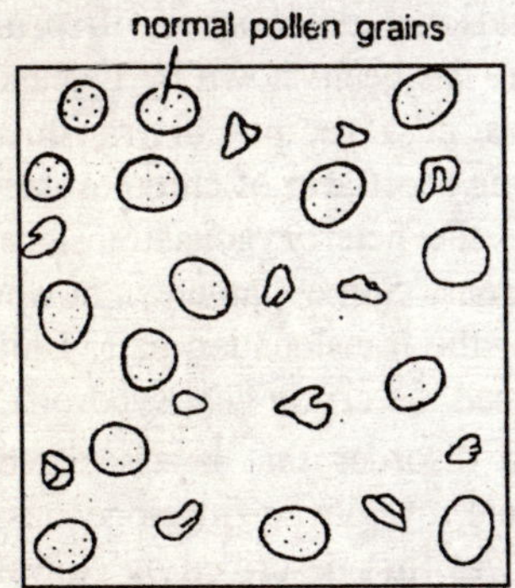

Fig. 3.8. Normal and aborted pollen in a semisterile corn plant. The small, shriveled pollen grains contain an euploid meiotic products of reciprocal translocation heterozygote.

The three classes described above also differ in phenotypes like flower size etc. and can be identified. The members of third category behave like pure lines and are actually permanent heterozygotes.

Balanced lethals and gametic complex-permanent hybridity in Oenothera. Permanent hybridity is maintained due to operation of a balanced lethal system, which may function due to gametic lethality or zygotic lethality. Since complete rings are formed and alternate disjunction is a rule, only two types of gametes are formed showing complete linkage between chromosomes. The gametic and zygotic lethality leads to survival of only heterozygotes.

Chromosome Aberrations in Humans

Chromosome deletions are usually lethal even as heterozygotes, resulting in zygotic loss, stillbirths, or infant deaths. Sometimes infants with small chromosome deficiencies, however, survive long enough to permit observation of some of the abnormal phenotypes they express.

Lejeune and his colleagues, for example, discovered a chromosome deficiency in humans that has been associated with the *cri-du-chat* (cat cry) *syndrome*. The name of this syndrome came from a plaintive catlike mewing cry from small weak infants with the disorder. Other characteristics microcephaly (small head), broad face and saddle nose, widely spaced eyes with epicanthic folds, unique facial features and physical and mental retardation IQs of cri-du-chat children studied are in the range of 20–40. The chromosome deficiency is in the short arm of chromosome 5 and is designated 5p.

Cri-du-chat patients die in infancy or early childhood and do not transfer the chromosome deletion to offspring. This chromosome deficiency, however, has been shown by Lejeune and others to become involved sometimes in a reciprocal translocation and thus to be transmitted. When the short arm of chromosome 5 became translocated to chromosome 15, the heterozygous translocation was carried in a normal healthy parent. Some gametes, however, carried only the deficient member of the translocation pair. Children inheriting the 5p chromosome expressed the cri-du-chat syndrome.

Another human disorder that is associated with a chromosome abnormality is chronic myelocytic (myelogenous) leukemia. A deletion of chromosome 22 was first described by Nowell and Hungerford and was called the Philadelphia (Ph) chromosome after the city in which the discovery was made. It was observed consistently in bone marrow preparations of patients with chronic myelocytic leukemia. Later, a translocation discovered by J. Rowley in a leukemia patient provided the correct chromosome rearrangement. A part of the long arm of chromosome 22 was translocated to another chromosome, usually chromosome 9 (46,XX.9q + 22q–), leaving a deficiency in the long arm of No.22.

More chromosomal anomalies are now being associated with malignancy. New staining methods permit more precise comparisons of (1) chromosome arms and (2) differentially stained sister chromatids. With balanced chromosomes, Burkitt lymphoma was related to a translocation of chromosome 14 [t(8q–; 14q+) (q24;q32). A deletion in chromosome 13q (band 14.1) is associated with the human embryonic tumor retinoblastoma. Williams tumor, an embryonic kidney tumor, is associated with a [illegible]tion in band 11p13. Sister chromatid exchanges, exchanges occurring between sister chromatids, occur with high frequency in somatic cells of patients with Bloom syndrome, indicating chromosome instability. Most people with this syndrome die with some form of cancer before age 30. Irregular chromosome numbers are also observed in malignant cells, particularly in later stages of cancer.

Like deficiencies, chromosome duplications are usually lethal even as heterozygotes, but sometimes they are sufficiently viable to permit observations of the abnormalities they produce. Like deficiencies, they may be associated with translocations and thus transmitted by normal healthy parents. Remember that a translocation is an exchange of parts between nonhomologous chromosomes or a transfer from one chromosome to a nonhomolog. Broken parts may be further divided and some may be lost or gained in a transfer. Thus, deficiencies and duplications may accompany a translocation carried in a parent. Such a parent may be a translocation heterozygote, with the long arm of one chromosome 21 attached at the centromere to the long arm of chromosome 14 along with normal chromosomes 14 and 21. Some gametes will contain a normal 21 and translocation 14q21q. When such a gamete is fertilized with a gamete from a normal individual, a translocated 21, along with two normal chromosomes 21, results in a viable infant with trisomy-21 and the Down syndrome.

With fluorescene microscope techniques, I.A. Uchida and C.C. Lin discovered a partial trisomy-12 that could not be identified by conventional methods. A body with some clinical features resembling the Down syndrome had been studied with conventional chromosome techniques. Because of continued lack of motor development at seven months age, further chromosomal investigations were carried out on lymphocyte cultures and slides stained with quinacrine dihydrochloride. With this technique, an additional band was identified in the short arm of chromosome 8. The mother's chromosomes were normal; but in the father, the short arm of No. 21 had become translocated to the short arm of No. 8 [46,XY, (8p+,12p-)]. The same chromosomal rearrangement was found in his daughter, an older sister of the patient. Since both father and daughter were clinically normal, the translocation was presumed to be reciprocal. The patients's No. 8 pair consisted of one normal and one translocation chromosome, similar to that of his father and sister, but both of his No. 12 chromosomes were normal. Thus, the boy had a duplication of part of the short arm of No. 12 and a deficiency of the tip of No. 8. Many translocations, addition to those cited above, have been detected in studies of human chromosomes, but most were apparently reciprocal and produced no phenotypic anomalies. At the Yale-New Haven Hospital, for example, cytological studies were conducted on 4500 infants born consecutively during one year. Lymphocytes from umbilical cord blood of each infant were grown in vitro and prepared for microscopic observation of chromosomes. Six translocations were detected. None of these was associated with a distinctive phenotypic anomaly. It is not the translocation process per se that produces abnormalities, but the imbalance of genetic material reflected in chromosome deficiencies and duplications that are produced by segregation of translocated chromosomes.

P. W. Allerdice et al. described a syndrome called the *chromosome 3 duplication-deletion syndrome*. Phenotypically, this syndrome includes a group of morbid symptoms; stillbirths, neonatal deaths, and spontaneous abortions. Most pregnancies are lost, but two children survived and became probands for the investigation. He cannot sit up, turn over, or eat solid food. Facial malformation of the living children includes a distorted head shape; thick, low eyebrows; low hair-line; long eyelashes; persistent lanugo; distended veins on scalp; hypertelorism; oblique palpebral fissures; a very short nose with a broad, depressed bridge and anteverted nares; protruding maxilla; thin upper lip; micrognathia; low-set ears; and short, webbed neck. Port-wine stains, congenital glaucoma, cloudy corneas, cleft palate, and harelip also occur frequently. Each infant had difficulty sucking and swallowing. Internal physical abnormalities were noted in infants who died neonatally.

Giemsa- and quinacrine-banded karyotypes from a parent of an affected child revealed an inversion [inv(3)(p25q21). Fetal cells cultured for prenatal diagnosis from a subsequent pregnancy of this couple carried a recombinant chromosome 3, with the long arm described as rec (3) del p, dup q, inv(3)(p25q21). With a banded chromosome 3, the inversion was analyzed. The inverted segment included the centromere (pericentric). Breaks had occurred in the short arm (p) at band 25 and in the long arm (q) at band 21. The central part of the chromosome including the centromere had rotated 180° and become reinserted into chromosome 3. Inv 3(p25q21) had been carried in the kindred for at least four generations and 35 kindered members were presumed to carry it. Karyotype of Case 1 is 46,XYrec(3), dup q, inv(3) (p25q21). He was the second child born to a 23-year-old carrier mother with karyotype 46,XX, inv(3)(p25q21) and a 26-year-old father, karyotype 46,XY.

Crossing over within the inversion loop is presumed to produce the imbalance of genetic material associated with duplications and deficiencies and to cause the symptoms of the syndrome. Odd numbers of crossovers within the inverted are of a chromosome 3 would be expected to result in genetic imbalance. Earlier in this chapter, crossing over in *Drosophila* inverted chromosomes was shown to create irregular combinations resulting in lethals. This, in turn, resulted in "suppressing" crossing over. In humans, unbalanced zygotes from inverted chromosome crossovers sometimes survive. Some infants carrying unbalanced chromosomes are kept alive, but with varying degrees of birth defect handicaps. The degree of imbalance of genetic material may be determining factor for life or death and degree of abnormality.

4

POLYPLOIDY

Somatic cells of higher plants and animals usually have chromosomes in pairs (2n); that is, two of each kind of chromosome are present in each cell. Mature germ cells, having undergone reduction division, normally have one number of each pair (n). Many individual plants and animals, however, have local areas of somatic tissue characterized by a multiple of the basic chromosome number. A doubling process in cell division is the usual explanation for these deviations.

With the exception of sex differences, somatic doubling, and minor variations that occur in natural and experimental populations, all members of a species of plants or animals have the same basic chromosome number. Chromosome number can be overemphasized as a criterion for species identification, but it represents a valid characteristic to be used along with others for distinguishing species. The range or reported chromosome numbers in animals extends from 2 pair in a rhabdocoel Gyratrix hermaphroditus and some mites, midges, and scale insects, to more than 100 in some butterflies and Crustacea. The Crustacean, *Paralithodes camtschatica*, for example, has 208 chromosomes or 104 pairs. The reported range in plants is from 2 pairs in the small composite plant Haplopappus gracilis to several hundred in some ferns. A species of fern-like plants of the genus Ophioglossum is reported to have 768 chromosomes.

Where all individuals within a species, with the exceptions noted above, have the same chromosome number, different species within a genus often have different numbers. Cytological investigations of chromosomes help to unravel problems of species formation. The evolutionary path of a certain species can be followed in some cases by comparing numerical and structural relations of its chromosomes with those of other species within the genus. The essential genetic material is a major factor in determining evolutionary patterns, but chromosome number itself represents merely the number of packages into which the genetic material is divided. Chromosome number is probably more constant, however, than any other single morphological characteristic that is available for species identification.

Table 4.1. Chromosome number of some common animals and plants

Species	*Number of chromosomes pair*
Plants	
Garden pea, *Pisum sativum*	7
Sorghum, *Sorghum vulgare*	10
Maize, *Zea mays*	10
Johnson Grass, *Sorghum halepense*	20
Alfalfa, *Medicago sativa*	16
Barley, *Hordeum vulgare*	7
Oats, *Avena sativa*	21
Tomato, *Lycopersicon esculentum*	12
Tobacco, *Nicotiana tabacum*	24
Trillium, *Trillium erectum*	5
Animals	
Gypsy moth, *Lymantria dispar*	31
Mouse, *Mus musculus*	20
Rabbit, *Oryctolagus cuniculus*	22
Cow, *Boss tarus*	30
Horse, *Equus caballus*	32
Donkey, (ass) *Equus asinus*	31
Dog, *Canis familaris*	39
Monkey, *Macaca rhesus*	21
Chimpanzee, *Pan troglodytes*	24

Changes in the number of chromosomes may be reflected in phenotypic variations, which constitute a useful tool for identifying the influence of individual chromosomes. If, for example, phenotypically distinguishable individuals with different chromosome numbers can be identified in natural populations or produced experimentally, it is sometimes possible to determine the effect of adding or removing certain chromosomes. Some plants with increased chromosome numbers have phenotypic changes in morphological or physiological characteristics which are of practical importance to man. Grape and tomato plants with chromosome numbers above 2n are larger and produce more desirable fruit than do corresponding varieties with the usual 2n number.

Classifications of chromosome changes are arbitrary and superficial because these changes are necessarily interpreted in terms of obvious additions or eliminations of parts of chromosomes, whole chromo-somes, or whole chromosome sets. The presently accepted classification system, therefore, is merely a working tool. Two main classes are euploidy and aneuploidy ("ploid," Greek for unit; "eu," true or even; and "aneu," uneven). Euploids have chromosome complements consisting of whole sets or genomes. The chromosome number of euploid organisms is basically represented by the monoploid (n). Euploids with chromosome numbers above the monoploid level may be diploid (2n), triploid (3n), tetraploid (4n), or have some other "polyploid" number.

Aneuploidy

The first critical study of aneuploid plants was made by Blakeslee and Bellingusing the common Jimson weed Datura stramonium, which normally has 12 pairs of chromosomes in the somatic cells. In 1924, these investigators announced the discovery of a "mutant type" having 25 rather than 24 chromosomes. At the meiotic metaphase, one of the 12 pairs was found to have an extra member; that is, one trisome was present along with 11 disomes. This originally discovered trisomic plant differed from wild-type plants in several specific ways. Conspicuous deviations were observed in shape and spine characteristics of seed capsules. The chromosome complement with one extra member in addition to the regular set was illustrated by the formula 2n + 1. Theoretically, because the some have, however, been divided successfully into 7 groups identified with letters A to G. Numbers 1 to 22 are associated with the autosomes in descending order by length. All autosomes can be placed satisfactorily within a group but the numbering within the groups is more or less tentative. The X chromosome is difficult to distinguish from members of the C group and the Y chromosome shows considerable variability in different preparations.

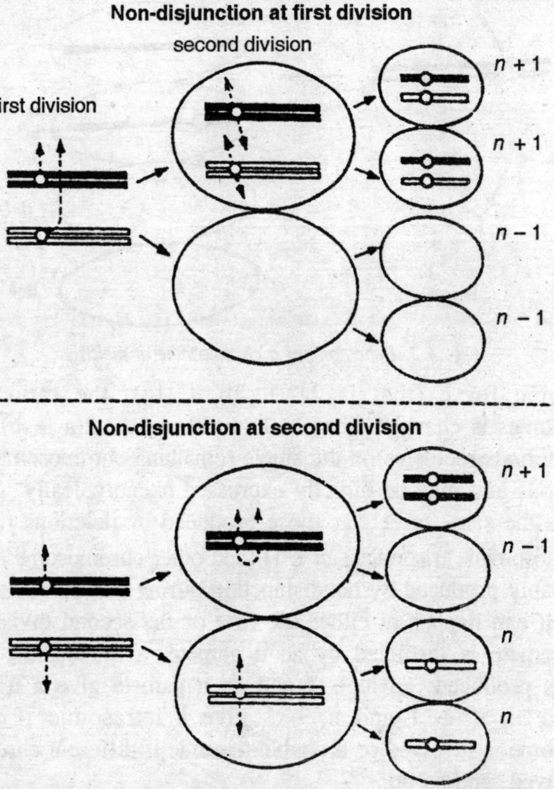

Non-disjunction at first division

second division

first division

n + 1

n + 1

n – 1

n – 1

Non-disjunction at second division

n + 1

n – 1

n

n

Fig. 4.1. The mode of origin of aneuploid gametes by non-disjunction at either the first or second meiotic division.

Nullisomics (2n – 2)

Although nullisomy is a lethal condition in regular diploids, an organism like wheat (which "pretends" to be diploid but is fundamentally hexaploid) can tolerate nullisomy. In fact, all of the possible 21 wheat nullisomics have been produced. Their appearances are different from normal wheat; furthermore, most of them show less vigorous growth.

Monosomics (2n – 1)

Monosomic chromosome complements are generally deleterious for two main reasons. First, the balance of chromosomes that is necessary for a finely tuned cellular homeostasis, carefully put together during evolution, is grossly disturbed. For example, if a genome consisting of two of each of chromosomes, a, b, and c becomes

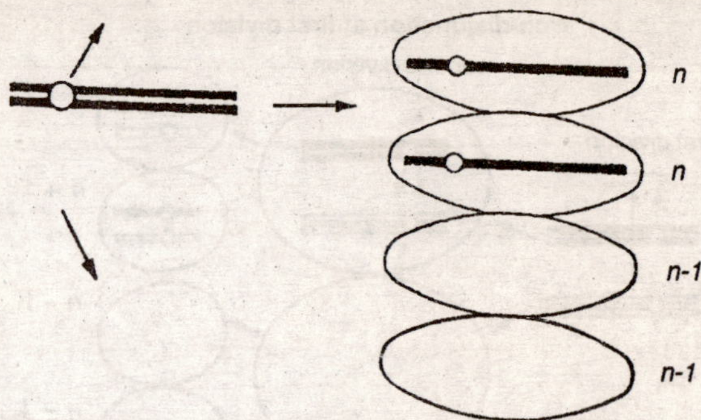

Fig. 4.2. Behaviour of a chromosome at meiosis.

monosomic for c (that is 2a + 2b + 1c), the ratio of these chromosomes is changed from 1c:1 (a+b) to 1c : 2(a + b). Second, any deleterious recessive on the single remaining chromosome becomes hemizygous and may be directly expressed phenotypically. (Note that these are the same effects as those produced by deletions.)

Monosomics, trisomics (2n +1), and other chromosome aneuploids are probably produced by nondisjunction during mitosis or meiosis. In meiosis it can happen at either the first or the second division. If an n – 1 gamete is fertilized by an n gamete, a monosomic (2n – 1) zygote is produced; an n + 1 and an n gamete give a trisomic 2n +1; and an n + 1 and n + 1 give a tetrasomic if the same chromosome is involved or a double trisomic if different chromosomes are involved, and so on.

In Neurospora (a haploid), the n – 1 meiotic products abort, and do not darken like the normal ascospore; so MI and MII nondisjunctions are detected as asci with 4:4 and 6:2 ratios of normal to aborted spores. For loci on the aneuploid chromosomes, what ascus genotypes are produced?

In humans, the sex-chromosome monosomic (44 autosomes + 1X) produces a phenotype known as Turner's syndrome. Affected people have a characteristic, easily recognizable, phenotype: they are sterile females, are short in stature, and often have a web of skin extending between the neck and shoulders. Their intelligence is near-normal, although some specific cognitive functions are defective. Their frequency is about 1 in 5000 female births. Monosomics for all autosomes die in utero.

If viable, nullisomics and monosomics are useful in locating newly found recessive genes on specific chromosomes in plants. In one such method, different monosomic lines lacking a different chromosome in each line are obtained. Homozygotes for the new gene are crossed with each monosomic line, and the progeny of each cross are inspected for expression of the recessive phenotype. The cross in which the phenotype appears identifies its chromosomal location. In nullisomics and monosomics, of course, n - 1 gametes are produced. In general, these gametes tend to be more viable in a female parent than in a male. It is the union of these n – 1 gametes within n gametes, bearing the new mutation, that provides the crucial progeny types for the linkage test.

A similar approach can be used in humans. For example, two people whose vision is normal may produce a daughter who has Turner's syndrome and is also red-green colour-blind. This shows that the allele for red-green colour blindness is recessive, that it was on the X chromosome of the mother, and the nondisjunction must have occurred in the father.

Trisomics (2n + 1)

In trisomics, trivalents are regularly seen. For genes that are tightly linked to the centromere of a trisomic chromosome set, the random segregations can be represented a trisomic Aaa. All types occur equally frequently, and a gametic ratio of 1A:2Aa:2a:1aa is produced. Trisomics are sometimes recongnised by these ratios, which are also useful in locating genes on chromosomes. We have already observed a complete set of trisomic lines in Datura. Once again, note that chromosome imbalance produces highly chromosome-specific deviations from the normal appearance.

In humans there are several examples of viable trisomics. The combination XXX (1/1000 male births) results in Klinefelter's syndrome,

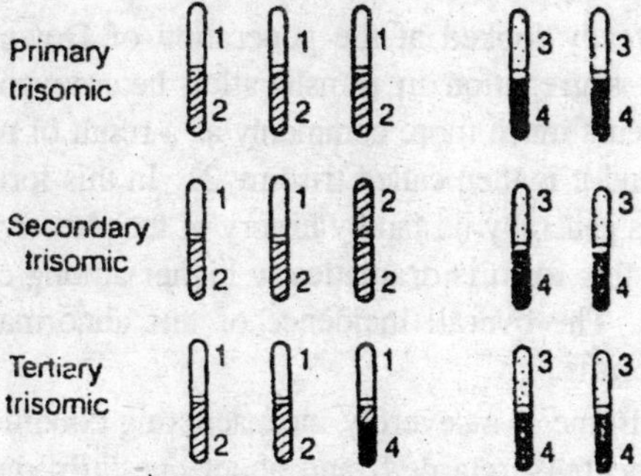

Fig. 4.3. Three kinds of trisomics.

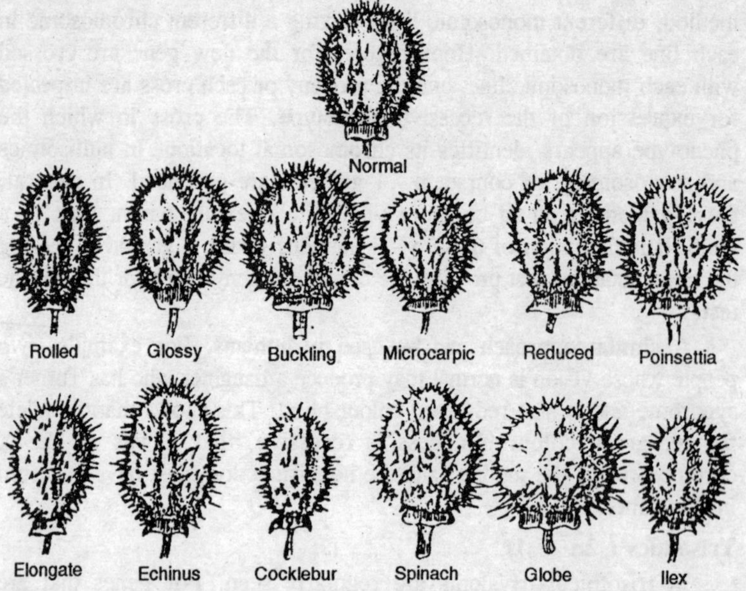

Fig. 4.4. Fruit capsules of the 12 primary trisomics of Datura stramonium, each with its particular phenotype.

producing males that have lanky builds, are mentally retarded, and sterile. Another combination, XYY, also occurs in about in 1000 male births. A lot of excitement was aroused when an attempt was made to link the XYY condition with a predisposition toward violence. This is still hotly debated, although it is now clear that an XYY condition in no way guarantees such behaviour. Nevertheless, several enterprising lawyers have attempted to use the XYY genotype as grounds for acquittal or compassion in crimes of violence. The XYY males are usually fertile.

We have already looked at the generation of Down's syndrome through adjacent segregation in translocation heterozygotes. Down's syndrome also occurs much more commonly as a result of nondisjunction during meiosis, and it is then called trisomy 21. In this form of Down's syndrome, there is generally no family history of the phenotype; however, the frequency of this form is dramatically higher among children born to older mothers. The overall incidence of this abnormality is about 0.15 percent of births.

Down's syndrome is a severely incapacitating condition. Affected individuals are mentally retarded, and about one-third die by the age of 10 years. Recent advances in mapping the human genome will allow the identification of precisely those genes on the long arm of chromosome 21 that must be trisomic to produce this syndrome; these advances offer some hope of a more precise understanding of its chemical nature and possible therapy. In humans, the only other two autosomal trisomics known to survive past birth are individuals with trisomy 13 and trisomy 18. Affected children are even more severely handicapped, both mentally and physically, and rarely survive to 1 year of age.

Chromosome mutation in general plays a prominent role in determining genetic ill health in humans. Figure summarizes the surprisingly high levels of various chromosomal abnormalities at different development stages of the human organism. In fact, the incidence of chromosome mutations ranks close to that of gene mutations in human livebirths. This is particularly surprising when we realize that virtually all chromosome mutation arise a new with each generation. This is in contrast to gene mutations, which owe their level of incidence to a complex interplay of mutation rates and environmental selection, acting over many generations of the history of the human species.

Somatic Aneuploids

Aneuploids can arise spontaneously in somatic tissue or tissue culture. In such cases, the initial result is a genetic mosaic of cell types. Good examples are provided by certain conditions in humans.

Table 4.2. Relative incidence of human ill health due to gene mutation and to chromosome mutation

Type of mutation	*Percentage of live birth*
Gene mutation	
Autosomal dominant	0.90
Autosomal recessive	0.25
X-linked	0.05
Total gene mutation	1.20
Chromosome mutation	
Autosomal trisomies (mainly Down's syndrome)	0.14
Other unbalanced autosomal aberrations	0.06
Balanced autosomal aberrations	0.19
Sex chromosomes	
XYY, XXY, and other ♂♂	0.17
XO, XXX, and other ♀♀	0.05
Total chromosome mutation	0.61

Sexual mosaics provide the first example. These are people whose bodies are a mixture of male and female tissue. One type of sexual mosaic is XO/XYY. This mosaic can be explained by postulating an XY zygote in which an early mitotic division involved a nondisjunction of the Y chromosomes, so that both went to one pole:

The phenotypic sex of such individuals depends upon where in the body the male and female sectors end up. In this case, if nondisjunction occurred at a later mitotic division, there would be a three-way XY/XO/XYY.

Aneuploidy in Man

More recent studies on individuals have related chromosome numbers below and above 46 with intersex conditions and other irregularities manifesting physical, reproductive, and mental abnormalities. Participants in the Chicago Conference agreed on a system of nomenclature for identifying numerical and structural chromosome alterations. In a description of a karyotype, the first item to be recorded is the total number of chromosomes, including the sex chromosomes, followed by the sex chromosome constitution and any autosomal aberrations. A complement of 44 autosomes and one X, for example, is symbolized 45, X. This monosomic chromosome complement has been associated with an abnormal female condition known for many years as Turner's syndrome and which occurs in one of about 5000 people in the general population.

The Turner Syndrome (45,X)

This monosomic has a chromosome complement of 44 autosomes and one X chromosome. The chromosome anomaly is associated with an abnormal female phenotype described in 1938 by H. H. Turner and associates and known as the Turner syndrome. It occurs in about 1 per 2500 live female births. More than 90 percent abort spontaneously. A rough estimate for 45,X adults in the general population is in 1 in 5000. These adults have virtually no ovaries, have limited secondary sexual characteristics, and are sterile. Microscopic sections of the ovaries show fibrous streaks of tissue representing remnants of ovaries. Affected females have short stature, low-set ears, webbed neck, and shieldlike chest. Mental deficiency is not usually associated with this syndrome. Epithelial cells of 45,X patients are X chromatin negative, as expected when only X chromosome is present.

X monosomic probably originate from exceptional eggs or sperm with no sex chromosome or from the loss of a sex chromosome in mitosis during early cleavage stages, after an XX or XY zygote has been formed. This latter probability is supported by the high frequency of mosaics that result from postzygotic events in patients with the Turner syndrome. Mosaics with X/XX sex chromosomes show symptoms of the Turner syndrome but are usually taller than X and have fewer anomalies that nonmosaic 45,X females. They show more feminization, more normal menstruation, and may be fertile. Many cases of the somatic Turner phenotype without the typical 45,X chromosome constitution are now known. Most of these have one normal X chromosome and a fragment of a second X chromosome. Both arms of the second X chromosome are apparently necessary for normal ovarian differentiation. Individuals with only the long arm of the second X are short in stature and show other somatic symptoms of the Turner syndrome, whereas those with only the short arm of the second X have normal stature and do not show as many signs of the Turner syndrome. This indicates that the Turner phenotype is mostly controlled by genes on the short arm of the X chromosome.

Patients with partial deletion of one X chromosome are X chromatin positive and therefore may be misdiagnosed if a buccal smear is the only test used. The deficient X chromosome always forms the X chromatin body. A Y chromosome also occurs in some individuals with the Turner phenotype. These patients are usually mosaic for 45,X/46,XY, with a normal Y. People with one X and a Y fragment, not including the Y short arm, have only streak ovaries but are normal in phenotype. This suggests that male-determining genes are in the short arm of the Y chromosome, and those that prevent Turner phenotype are in the Y long arm as well as the X short arm. Major features of the Turner phenotype occur in some males as well as in females. The male Turner syndrome is characterized by defective development of the testes, sterility, and limited male secondary sexual characteristics, along with somatic features of the Turner phenotype. These people have normal male karyotypes.

The Klinefelter Syndrome (47,XXY)

An extra X chromosome in addition to the usual male (XY) chromosome complement (47,XXY) has been associated with the abnormal male syndrome described (in 1942) by H.F. Klinefelter and known as the *Klinefelter syndrome*. It is estimated to occur in 1 per 500 live male births. Individuals with this syndrome are phenotypically males but with some tendency toward femaleness, particularly in secondary sex characteristics. Such features as enlarged breasts, underdeveloped body hair, small testes, and small prostate glands are a part of the syndrome. Presumably, the XXY constitution originates either by fertilization of an exceptional orginates either by fertilization of an exceptional XX egg by a Y sperm or of an X egg by an exceptional XY sperm. Studies of Klinefelter syndrome and Turner syndrome indicate that the Y chromosome in human beings, unlike that in *Drosophila*, determines male sex.

The most common karyotype (about three-quarters of the cases) for the Klinefelter syndrome is 47,XXY, but the symptoms of the syndrome will usually occur whenever more than one X chromosome is present along with a Y chromosome. More complex karyotypes associated with the Klinefelter syndrome include: XXYY, XXXY, XXXYY, XXXXY, XXXXYY, and XXXXXY. All patients with the Klinefelter syndrome have one or more X chromatin bodies in the their cells. Mental retardation is usually found when there are more than two X chromosomes. The XY/XXY mosaicism in patients with Klinefelter syndrome is associated with less severe physical and reproductive anomalies.

Aneuploidy of X Chromosomes and Mental Deficiency

Other irregular combination of X chromosomes have also been recognized among females with X chromosome aberrations. About 1 percent of all mentally defective women in institutions have been shown to have one or more extra X chromosomes. This chromosome abnormality occurs in about 1 in 700 live births in the general population. Individuals with the "triple X syndrome" are comparable in some ways with Drosophila metafemales (XXX). In Drosophila, however, such individuals are usually lethal, and those that survive are strikingly abnormal and sterile. Human XXX individuals are sometimes visibly undistinguished from normal XX females, but there is considerable range in phenotypic expression. They may be mentally abnormal.

The best-known symptoms in this syndrome are abnormalities associated with functional processes such as menstruation. One patient cited by P. A. Jacobs was a 37-year-old female who reported that the first suggestion of an abnormality was highly irregular menstruation. When the abdominal wall was opened, the ovaries appeared as if they were postmensopausal. Microscopically, they showed deficient ovarian follicle formation. Of 63 cells observed, 51 had 47 chromosomes; the extra chromosome was an X. Nondisjunction in the production of the egg from which this woman developed was postulated as the mechanism for the occurrence of extra chromosome. Buccal smears showed two sex chromatin bodies in the epithelial cells as expected when three X chromosomes are present. Individuals with tetrasomic X chromosomes (48,XXXX) are all mentally defective. The degree of mental deficiency increases with the number of X chromosomes present.

47,XYY and Behaviour

P. A. Jacobs and her associates reported in 1965 that seven XYY males were detected in a population of 197 male, mentally subnormal inmates of a penal institution in Scotland. The XYY men were unusually tall, with an average height of 73.1 inches, compared with 64 inches for XY men in the same prison. Numerous other studies, mostly in institutionalized populations, have since confirmed that a high proportion of XYY individuals are tall, subnormal in intelligence (with IQs individuals are tall, subnormal in intelligence (with IQs ranging from 80 to 95), and antisocial. The aggressive behaviour that brought them into conflict with the law was usually against property rather than people.

XYY trisomy occurs about once in 1000 live male births in the general European population. Only a few of these can be accounted for in the criminal population. Furthermore, most XYY men have been described as perfectly normal in behaviour. Hook has shown that only 3.6 percent of all XYY men are institutionalized for any reason.

Environmental factors are presumed to be involved in the development of aggressiveness. Since some XYY men are subnormal in intelligence and excessively tall in stature, particular environmental situations in childhood or adulthoood may lead to withdrawal from society or aggressive behaviour. Unfavourable social conditions such as frustration in personal accomplishment and taunting from associates may encourage physical aggression as a means of adaptation. Males with this sex trisomy have not been found to transmit the extra Y chromosome to their sons. This extra chromosome seems to be weeded out in gametogenesis. A wide range of physical and mental abnormalities has been detected in institutionalized XYY men, but most of these are irregular in occurrence and do not form a syndrome. Tallness of stature and mental dullness, however, are fairly constant characteristics among institutionalized XYY men.

Chromosomal Mosaics

Individuals who have at least two cell lines, with different karyotypes derived from one zygote, originate from nondisjunction in a cleavage division after fertilization. One daughter cell would thus receive one too many and the other would be one deficient. Each cell would give rise to cell line with its irregular chromosome number. Proportions of cells representing the different cell lines would vary in different tissues, making the extent of the mosaicism and the effect on the organism difficult to predict.

Many sex chromosome mosaics have been detected in human beings. The main phenotypic characteristic is extreme variability. Some sex chromosome mosaics have been reported —X/XX, X/XY, XX/XY, XXY/XX, XX/XXX, XXX/X XXX/XXXXY—and several other combinations reflecting two or three cell lines. Mid to severe phenotypic symptoms have been associated with these sex chromosome mosaics.

ABNORMAL EUPLOIDY

Monoploids

In this section we shall consider monoploidy as an unusual condition. Monoploid individuals can arise spontaneously in natural populations as rare aberrations, but in several forms (such as bees, wasps, and ants) the males are normally monoploid, having derived from unfertilized eggs.

In the germ cells of a monoploid, meiosis cannot occur normally because the chromosomes having no pairing partners. Thus monoploids are characteristically sterile. (However, meiosis can be bypassed in some monoploid animals, such as male honeybees, which produce gametes essentially by mitotic division). If meiosis occurs and the single chromosomes segregate randomly, then the probability of their all going to one pole is $(1/2)^{x-1}$, where x is the number of chromosomes. This will determine the frequency of viable (whole-set) gametes, obviously a vanishingly small number if x is large.

Monoploids have a major role in modern approaches to plant breeding. Diploidy is an inherent nuisance in the induction and selection of new favourable plant mutations and of new selection of new favourable plant mutations and of new combinations of genes already present. Monoploids provide a way around some of these problems. In some plants, monoploids may be artificially derived from the products of meiosis in the plant's anthers. A cell destined to become a pollen grain may be induced by cold treatment to grow instead into an embryoid, a small dividing mass of cells. The embryoid may be grown on agar to form a monoploid planter, which can then be potted in soil to mature.

Monoploids may be exploited in several ways. In one method, they are first examined for favourable traits or gene combinations.

These may arise from heterozygosity already present in the parent or induced in the parent by mutagens. The monoploid can then be subjected to chromosome doubling to achieve a completely homozygous diploid with a normal meiosis, capable of providing seed. How is this achieved? Quite simply, by the application of a compound called colchicine to meristematic tissue. Colchicine, an alkaloid drug extracted from the autumn crocus, inhibits the mitotic spindle, so that cells with two chromosomes sets are produced. These may proliferate to form a sector of diploid tissue that can be identified cytologically.

Another way the monoploid may be used is to treat its cells, basically like a population of haploid organisms, in a mutagenesis-and-selection procedure. The cells are isolated, their walls are removed by enzyme treatment, and they are treated with mutagen. They are then plated on selective medium— perhaps a toxic compound normally produced by one of the plant's parasites, or an insecticide— to select resistant cells. Resistant plantlets grow eventually into haploid plants, which can then be doubled (using colchicine) into a pure-breeding resistant type. These are potentially powerful techniques that can circumvent the normally slow process of what is basically meiotic plant breeding. The techniques have been successfully applied in several important crop plants, such as soybeans and tobacco. This is, of course, another aspect of somatic-cell genetics in higher organisms.

The another technique of or producing monoploids does not work in all organism or in all genotypes of an organism. Another useful technique has been developed in barley, an important crop plant. When diploid barley, Hordeum vulgare, is pollinated using a diploid wild

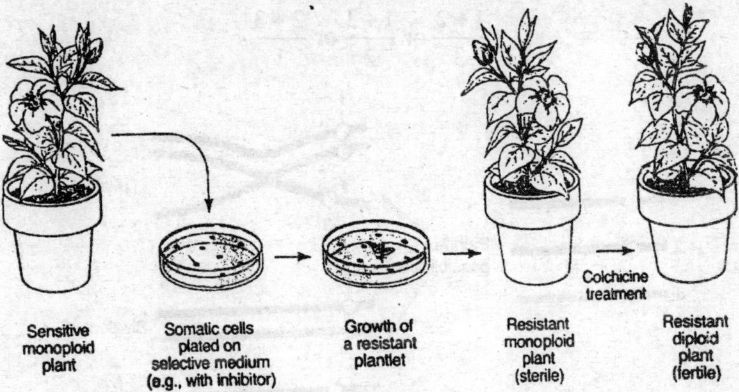

Fig. 4.5. Using microbial techniques in plant engineering. Haploid cells have their cell walls removed enzymatically.

relative called Hordeum bulbosum, fertilization occurs, but during the ensuing somatic cell divisions, the chromosomes of H. Bulbosum are preferentially eliminated from the zygote, resulting in a haploid embryo. (The haploidization process appears to be caused by a genetic incompatibility between the chromosomes of the different species.) The resulting haploids can be doubled with colchicine. This approach has led to the rapid production and widespread planting of several new barley varieties. It is being used successfully in other species too.

Polyploids

Once into the realm of polyploids, we must distinguish between autopolyploids and allopolyploids. *Autopolyploids* are composed of multiple sets from within one species, whereas *allopolyploids* are composed of sets from different species. Allopolyploids form only between closely related species; however, the different chromosome sets are homeologous (only partially homologous), not fully homologous as they are in autopolyploids.

Triploids

Triploids are usually autopolyploids. They are constructed from the cross of a 4x (tetraploid) and a 2x (diploid). the 2x and the x gametes unite to form a 3x triploid.

Triploids also are characteristically sterile. The problem again involves pairing at meiosis. Although pairing can take place in several ways, it usually occurs between only two chromosomes at a time. The net result is always the same, an unbalanced segregation of one of the following types:

$$\frac{1+2}{3} \text{ or } \frac{1+3}{2} \text{ or } \frac{2+3}{1}$$

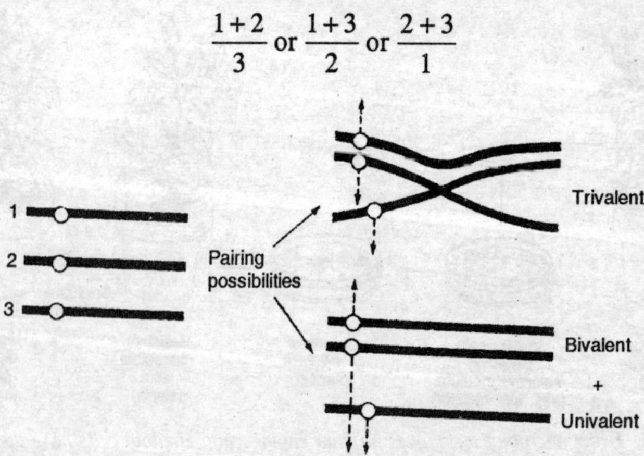

Fig. 4.6. Meiotic pairing possibilities in tetraploids.

This happens for every chromosome threesome, and the probability of obtaining either a 2 x or x gamete is $(\frac{1}{2})^{x-1}$, where x is the number of chromosomes in a set. The other will be unbalanced gametes, having two of one chromosome type, one of another, two of another, and so on, and most will be nonfunctional. Even if the gametes are functional, the resulting zygotes will be unbalanced. A practical application of the sterility associated with triploidy lies in the production of seedless varieties of watermelons and bananas.

Autotetraploids

Autotetraploids occur either naturally, by the spontaneous accidental doubling of a 2x genome to 4x, or artificially, through the use of colchicine. Autotetraploids are evident in many commercially important crop plants because, as with other polyploids, the larger number of chromosome sets is often associated with increased size of the plant. This is manifested in increased cell size, fruit size, stomata size, and so on. Because 4 is an even number, autotetraploids can have a regular meiosis, although this is by no means always the case. The crucial factor is how the four chromosomes of one type pair and segregate. The two bivalent and the quadrivalent pairing modes tend to be most regular in segregation, but even here there is no guarantee of a 2 → 2 segregation. If a regular 2 → 2 segregation is achieved at each chromosome type, as is the case in some species, than a formal genetic analysis can be developed for autotetraploids.

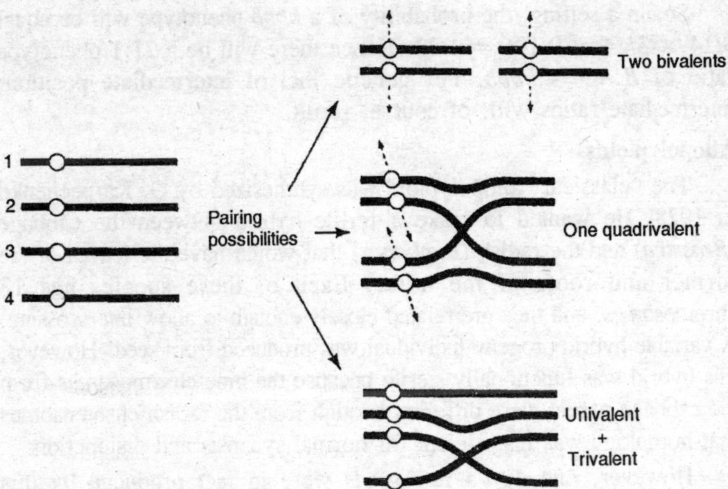

Fig. 4.7. Meiotic pairing possibilities in tetraploids.

Let's hypothesize an experiment in which colchicine is used to double the chromosomes of an *Aa* plant into an *AAaa* autotetraploid, which we will assume shows 2 → 2 segregation. We now have a further worry because autotetraploids give different genetic results in their progeny, depending on whether or not the locus concerned is tightly linked to the centromere. First we consider a centromeric gene. The three possible pairing and segregation patterns are presented in Figures, these occur by chance and with equal frequency. As the figure shows, the 2x gametes will be *Aa, AA, or aa,* and these will be produced in a ratio of 8:2:2, or 4:1:1. If such a plant is offspring is obviously 1/6 × 1/6 = 1/36. In other words, a 35:1 phenotypic ratio will be observed if *A* is fully dominant over three *a* alleles.

If, in the same kind of plant, a genetic locus *B/b* is very far removed from the centromere, crossing-over must be considered. This forces us to think in terms of chromatids instead of chromosomes, and we have for *B* chromatids and for *b* chromatids. Because the number of crossovers in a such a long region will be large, the genes will become effectively unlinked from their original centromeres, and the packaging of genes two at a time into gametes is very much like grabbing two balls at random from a bag of eight balls, four of one kind and four of another. The probability of picking two *b* genes is then

4/8 (the first one) × 3/7 (the second one) = 12/56
= 3/14

So, in a selfing, the probability of a *bbbb* phenotype will be equal 3/14 × 3/14 = 9/196 = 1/22. Hence there will be a 21:1 phenotype ratio of *B – – –:bbbb*. For genetic loci of intermediate position, intermediate ratios will, of course, result.

Allopolyploids

The "classical" allopolyploid was synthesized by G. Karpechenko in 1928. He wanted to make a fertile hybrid between the cabbage (*Brassica*) and the radish (*Raphanus*) that would have the leaves of the former and roots of the latter. Each of these species has 18 chromosomes, and they are related closely enough to allow intercrossing. A variable hybrid progeny individual was produced from seed. However, this hybrid was functionally sterile because the nine chromosomes from the cabbage parent were different enough from the radish chromosomes that homology was insufficient for normal synapsis and disjunction.

However, one day a few seeds were in fact produced by this (almost) sterile hybrid. On planting, these seeds produced fertile individuals with 36 chromosomes. These individuals were allopolyploids. They had apparently been derived from spontaneous accidental chromosome doubling in the sterile hybrid, presumably in tissue that eventually became germinal and underwent meiosis. Thus, in $2n_1 + 2n_2$ tissue, there is a pairing partner for each chromosome, and balanced gametes of the type $n_1 + n_2$ are produced. These fuse to given $2n_1 + 2n_2$ allopolyploid progeny, which are in turn fertile also. This kind of allopolyploid is sometimes called an amphidiploid. (Unfortunately for Karpechenko, his amphidiploid had the roots of a cabbage and the leaves of a radish.)

If the allopolyploid is crossed to either parent species, sterile offspring result. In the case of the cross to radish, these offspring would be $2n_1 + n_2$, constituted from an $n_1 + n_2$ gamete from the allopolyploid, and an n_1 gamete from the radish. Obviously, the n_2 chromosomes will have no pairing partners, so sterility will result. Consequently, Karpechenko had effectively created a new species, with no possibility of gene exchange with its parents. He called his new species *Raphanobrassica*.

Nowadays, allopolyploids are routinely synthesized as a major tool in plant breeding. The goal obviously is to combine some of the worthwhile features of both parental species into one type. This kind of endeavor is very uncertain, as Karpechenko found out. In fact, only one amphidiploid has ever been intentionally produced that is of potentially widespread use. This is *Triticale*, an amphidiploid between wheat (*Triticum, 2n = 6x = 42*) and rye (*Secale,* 2n = 2x = 14). *Tricale* combines the high yields of wheat with the ruggedness of rye. A massive international *Triticale* testing program is now under way, and many breeders have great hopes for the future of this artificial amphidiploid.

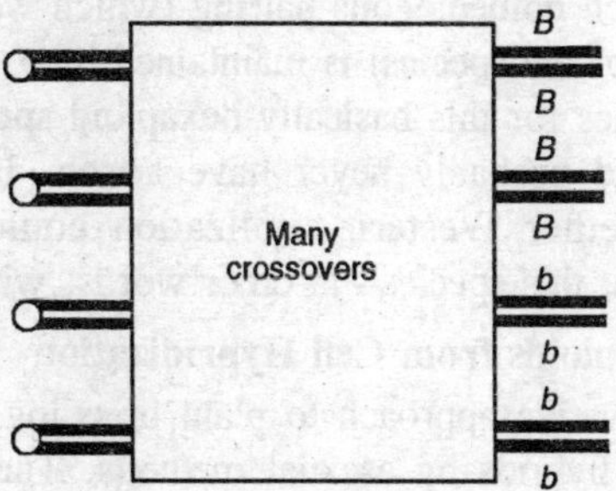

Fig. 4.8. Highly diagrammatic representation of a tetraploid meiosis involving a heterozygous locus distant from the centromere.

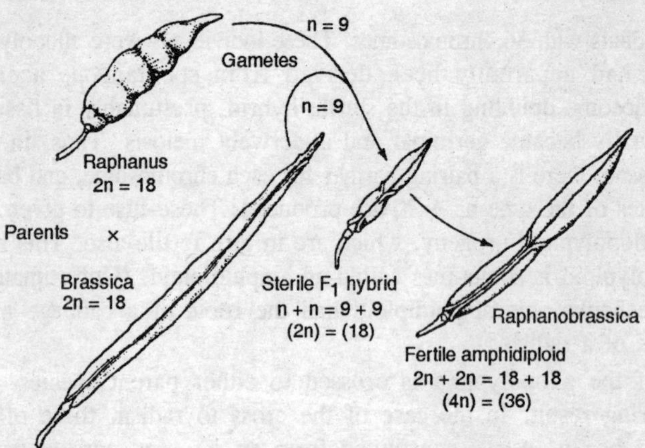

Fig. 4.9. The origin of the amphidiploid (Raphanobrassica) formed from cabbage (Brassica) and radish (Raphanus).

In mature, allopolyploidy seems to have been a major force in speciation of plants. There are many different examples. One particularly satisfying one is shown by the genus *Brassica*. Here three different parent species have been hybridized in all possible pair combinations to form new amphidiploid species. This has all taken place in nature, but *Brassica* amphidiploids also have been artificially synthesized.

A particularly interesting natural allopolyploid is bread wheat, *Triticum aestivum* (2n = 6x = 42). By a study of various wild relatives, it has been possible to reconstruct a probable evolutionary history of breat wheat. In a wheat meiosis, there are always 21 pairs of chromosomes. Furthermore, it has been possible to establish that any given chromosome has only one specific pairing partner (homologous pairing)—not five other potential ones (homologous pairing). The suppression of such homeologous pairing (which would lead to much reduced stability of the species) is maintained by a gene *Ph* ensures a diploid-like genetics for this basically hexaploid species. Without *Ph*, bread wheat could probably never have arisen. It is interesting to speculate on whether Western civilization could have arisen or progressed without this species—in other words, without *Ph*.

Somatic Allopolyploids from Cell Hybridization

Another innovative approach to plant breeding is to try to make allopolyploid-like hybrids by asexual methods. Theoretically, such a technique would permit combination of widely differing parental species. The technique does indeed work, but so far the only allopolyploids that have been produced are those thaι can also be made by the sexual methods we have considered already. The procedure is as follows. Cell suspensions of the two parental species are prepared and stripped of their cell walls by special enzyme treatments. The stripped cells are called *protoplasts*. The two suspensions (protoplast suspensions) are combined with polyethylene glycol, which enhances protoplast fusion. The parental cells and the fused cells will proliferate to form colonies (in much the same way as microbes) on agar medium. If these colonies, or calluses, are examined, a fair percentage of them are found to be allopolyploid-like hybrids with chromosome number equal to the sum of the parental types. Thus, not only do the protoplast cell membranes fuse to form a kind of heterokaryon, but the nuclei fuse too.

A good example of an allopolyploid-like hybrid is commercial tobacco, *Nicotiana tabacum*, which has 48 chromosomes. This species of tobacco was originally found in nature as a spontaneously occurring amphidiploid. The two probable parents are N. Sylves and N. Tomentosiformis, each of which has 24 chromosomes. A sexual cross between N. Tabacum and either of the other two gives a 36-chromosome hybrid n which there are 12 chromosome pairs plus 12 unpaired chromosomes. A cross between N. Sylvestris and N. Tomentosiformis gives a 24-chromosome hybrid n which there is no pairing at all. Hence, it appears that part of the N. Tabacum genome is from N. Sylvestris and part from N. Tomentosiformis. This amphidiploid can be re-created either sexually, by processing involving colchicine as described previously, or somatically by cell fusion. When cells or the prospective parental species are fused, a 48-chromosome hybrid cell line is produced from which may be grown plants whose behaviour is identical to that of N. Tabacum. (Note that in the latter method, colchicine is not required, since the fusion product is already amphidiploid.)

The recovery of somatic hybrids may be enhanced if a selective system is available. For example, two different monoploid lines of *N. Tabacum* had light-sensitive yellowish and light-resistant, as a result of complementation between the parental genotypes. The calluses can be grown into plantlets, which then either are grafted onto a mature plant to develop or are themselves potted.

Application of Polyploidy

Among the cultivated varieties of wheat, three different chromosome numbers are represented: 14, 28, and 42 (x = 7). For example, the primitive small-grained einkorn type of Europe and Asia, Triticum monococcum, has 14 chromosomes in its vegetative cells. Its yield is low and it is of comparatively little value. An emmer wheat (durum), T. dicoccum, grown chiefly in southern Europe but also in the United States, has 28 chromosomes. It has thick heads with large hard kernels and issued mainly for macaroni, spaghetti, and stock feed. The bread wheats, T. aestivum, with 42 chromosomes, were postulated by J. Percival in England to have come from a cross between emmer wheat and goat grass (Aegilops), both of which are native to the Babylonian region where bread wheat originated.

When techniques for artificial chromosome doubling became established, investigations of the origin of bread wheat confirmed Percival's theory. Experimental evidence obtained by E. S. McFadden and E. R. Sears and separately by H. Kihara traced the pathway for the origin of one type of bread wheat, T. Spelta.

Aegilops squarrosa (n = 7) was found to carry a group of major characteristics that distinguish the hexaploid (n=21) T. Spelta from the tetraploids (n = 1) T. dicoccum and T. dicoccoides. Hybrids between these tetraploid species of wheat and A. squarrosa proved to have all of the major taxonomic characters of T. spelta but the hybrids were completely or nearly sterile. When the F_1 hybrids of T. dicoccoides × A. squarrosa were treated with colchicine, highly fertile allopolyploids with 42 chromosomes were obtained. These synthetic hexaploids closely resembled the cultivated. T. spelta, and they produced highly fertile hybrids with that species and with T. vulgare, known to be in the ancestry of the bread wheats. This demonstrated that the genome of the hexaploid wheats corresponded to one chromosome set of A. squarrosa. It was postulated that T. spelta is the ancestral hexaploid wheat of Europe, having arisen, possibly in fairly recent times, in southeastern Europe or southwestern Asia following chromosome doubling of natural hybrids of T. dicoccoides (or its cultivated close relative, T. dicoccum) × A. squarrosa. T. spelta is believed to have been carried over the northerly route into central and western Europe. Experiments of McFadden, Sears, and Kihara reconstructed the pathway through which a moderately useful wheat and a goat grass hybridized in nature and produced forerunners of a most valuable crop, bread wheat.

New World Cotton

Crosses can be made between distinct species of cotton, members of the benes Gossypium. The hybrids show a wide range of vigor and fertility, making the material favourable for studies of origins. Three cytological groups have been found to correspond with the major world distributional areas. Old World cotton had 13 pairs of large chromosomes. American cotton, which originated in Central or South America, has 13 pairs of small chromosomes. New world cotton (the cultivated long-staple type) has 26 pairs, 13 large and 13 small. Evidently, hybridization and chromosome duplication occurred somewhere in the ancestry of the New World Cotton.

J. O. Beasley used the colchicine technique and succeeding in doubling the chromosomes of a hybrid between the Old World and American cotton. The resulting hybrids, with four set of chromosomes (amphidiploids), crossed readily among themselves and produced fertile plants resembling New World cotton. The process by which the valuable polyploid cotton may have originated in nature was thus duplicated in the laboratory.

Primrose Hybridization

The primrose, Primula kewensis, is an allotetraploid with 36 (2n) chromosomes. It was derived from a cross between two diploids, P. floribunda (x= 9) and P. verticillata (x= 9). Plants from these two species crossed readily, producing hybrids with 18 chromosomes in their vegetative cells. 9 from P. floribunda and 9 from P. verticillata, but the hybrids were sterile. Eventually, however, a branch on a hybrid plant developed from a cell in which the chromosome number was doubled (36), so that each chromosome had a homologous partner. This branch was propagated and gave rise to a fertile primrose plant with cells containing 36 chromosomes of the two diploid parents, the sterile diploid hybrid, and the fertile allotetraploid are shown.

Tobacco Resistance

Induced polyploidy has been exploited to a great extent. Practical applications may become more common as additional data are accumulated. By artificially induced polyploidy, disease resistance and other desirable qualities have been incorporated into some commercial crop plants. Tobacco, Nicotiana tabacum, for example, is susceptible to the tobacco mosaic virus (TMV), whereas N. glutinosa appeared at first observation to be resistant. Further investigation, however, showed that in N. glutinosa the virus killed the cells that were invaded and the virus particles became isolated in the dead cell. The apparent resistance thus was attributable to hypersensitivity. When the two tobacco species were crossed, the hybrid was found to be "resistant" to the virus, but totally sterile. When the chromosomes were doubled, it was possible to secure a fertile polyploid "resistant" to the virus.

Polyploid Fruits, Flowers, and Wheat

Some varieties of plants that serve human needs more effectively than others have now been identified as polyploids. Many polyploids were selected and cultivated because of their large size, vigor, and ornamental values, before their chromosome numbers were known. Giant "sports" from twings of McIntosh apple trees that were found to be tetraploid (4n) were propagated into whole trees, which produce extra-large fruit. The texture of the giant apples is as fine as that of diploids, but the yield is inferior. Mass selection of seedlings may overcome this difficulty. Bartlett pears, several varieties of grapes, and cranberries have also produced sports with giant fruits. Some of these show promise of practical usefulness. With colchicine treatment, a number of polyploids have been developed artificially. This technique has provided a way to explore the mechanism involved in polyploid formation and to make use of the good qualities of polyploids. Tetraploid (4n) maize is more vigorous than the ordinary diploid and produces some 20 percent more vitamin A. Its fertility is somewhat reduced, but this drawback responds to selection. Polyploid watermelons have been developed from colchicine treatment by Kihara and others. The tetraploid with 44 chromosomes is large and has practical value. Triploid watermelons with 33 chromosomes are especially desirable because they are sterile and have no seeds. Among the flower garden varieties, 4n marigolds and snapdragons are widely cultivated.

Polyploid plants respond to artificial selection and hybridization, as do diploid species. The recent history of plant breeding has been characterized by a marked improvement in many polyploid plant crops. The yield of wheat, for example, has increased appreciably. This has been accomplished by developing disease-resistant strains and breeding for increased hardiness and greater efficiency so that crops may survive under the various environmental conditions found in wheat-growing areas.

A constant threat to the wheat crop is rust—a fungus that attacks the stems and leaves of the growing plants and destroys the ripening grain. Spores are borne by wind and, when conditions are right, they spread like fire through wheat fields. The disease can be combated by developing rust-resistant strains and by eradicating barberry bushes, which are hosts to the spores during the spring months. But new varieties of rust that destroy previously resistant grain keep evolving, thus perpetuating the job of plant breeders. The larger kernels at the left are from a new strain of rust-resistant spring wheat. At the right are shown kernels of wheat, similar in other respects but not resistant, that are dwarfed from infection with stem rust. The number of kernels of grain per plant as well as the size of the kernels is decreased by rust infection. Investigators in agricultural experiment stations are constantly alert for new rusts. When a new one is found, the standard wheat varieties are tested against it. If they are not resistant, breeding programs are initiated immediately to develop new strains resistant to that particular rust.

Chromosome Anomalies in Spontaneous Abortions in Humans

A wide variety of chromosome numbers and chromosome structural aberrations is found in spontaneoulsy aborted human fetuses. The types of aberrations found vary according to differences in the age of the fetus at the time of abortion. For example, 40 percent of spontaneously aborted fetuses under 90 days of gestational age (i.e., length of time since the last menstrual period of the mother) exhibit chromosome anomalies. For 91-to 120-day-old fetuses, 25 percent show chromosomal anomalies, and only 5 percent of fetuses over 120 days old exhibit chromosome anomalies. Thus, chromosomal anomalies cause most fetuses to die and be aborted in early developmental stages.

Before the percentage of spontaneous abortions resulting from chromosomal anomalies can be calculated, one must first define what constitutes an aborted fetus. Apparently a large number of conceptions occur in humans, and the rechronic myelogenous leukemia possess two cell lines; one cell line has a normal chromosome complement, whereas the other appears to be missing the long arm of chromosome 2. This condition was originally thought to be a monosomy; however, using band staining techniques to identify the long arm, it was learned that the long arm was translocated to one of the larger chromosomes, usually chromosome 9 (referred to as the Philadelphia chromosome). The role of this chromosomal aberration in the induction, development, and progression of cancer is unknown.

Inherited autosomal recessive disorders, such as Bloom's syndrome, Fanconi's anemia, ataxia-telangiectasia, and xeroderma pigmentosum, have been associated with chromosomal instability and or deficiency in mutation repair mechanisms that result in chromosomal aberrations. These individuals have a high incidence of cancer.

5

GENE MUTATION

Mendel's work in 1900. In 1901 he published his accumulated data in a book entitle. The Mutation Theory. De Varies was careful to distinguish between hereditary and environmental variations, but his mutations are now known to have included several kinds of changes in the hereditary material. It took nearly half a century to clarify the situation in Oenothera and discover the causes of the variations that he observed. Chromosome changes, both structural and numerical, were eventually found to cause the phenotypic variation that de Vries had grouped under the heading of mutations. One complex alteration is described later on in this chapter in connection with the discussion of permanent heterozygotes. It was de Vries' concept of discontinuous variation rather than his precise observations and example that became significant. None of his examples would now be classified as gene mutations.

MUTATIONS AND PHENOTYPIC CHANGE

Because genes are chemical entities that cannot be observed and compared directly, some phenotypic alteration must be associated with the gene change (mutation) or it will go unrecognized. Minor changes presumably occur in genes as a matter of course without producing any phenotypic alterations. Some gene changes are known to be associated with only slight effects on the fertility or viability of the organism. These alterations ordinarily go undetected unless critical comparisons are made. They may, however, influence natural selection and thus represent a factor in evolution.

Mutations also represent in evolution, by which the presence of particular wild-type genes can be postulated. Normal development of any organism is influenced by numerous genes. An observable characteristic of the adult may be altered by an individual gene substitution, but because the entire organism develops as a unit, the action of individual wild-type genes is not always apparent. The existence of mutant and wild-type genes is substantiated when a gene is changed so that its influence on developmental reactions causes a visible difference from that exerted by the unmutated gene. Mutations thus provide the basis for postulating wild-type genes.

Mutated as well unmutated genes tend to give rise to other genes exactly like themselves. Once established, a mutated gene is as stable as the original gene from which the mutation occurred. This process of duplication is associated with mitotic cell division and occurs repeatedly as the number of cells increases. Duplication among genes has a crude parallel in the procedure of printing. After type is set and placed on the printing press; similar copies can be made repeatedly. If the type is changed slightly with an altered template replacing the original, the new pattern reproduces itself faithfully. Instead of a physical change, such as the substitution of one piece of metal for another, the gene in the process of mutation undergoes a chemical change. A part of the gene unit is altered by the chemical modification in such a way that it henceforth behaves differently from its unmutated ancestral gene.

Not all mutations are immediately detectable because many, perhaps the great majority, are recessive and must become homozygous before they can express themselves. Because the phenotypic effect is the only readily observable evidence of mutations, however, a sudden phenotypic change that subsequently proves to be heritable is an accepted indication that mutation has occurred in an organism. In everyday language, the phenotypic change is synomous with the gene change or actual mutation. This is quite natural because the term"mutation" was coined and used before the present gene concept was established. Historically, therefore, the word has been used to describe the perceptible change alteration. It is now known that most mutations that are not detectable by visible phenotypic change influence the viability of the organism in some indirect way.

Classification for Mutations

Dominance and recessiveness provided convenient, although perhaps superficial, criteria for early classifications of mutations. Mutations also may occur in either the autosomes or the X chromosomes, and thus may be classified as autosomal or sex-linked. Viability mutations have also been classified on the basis of their effect on the organism.

Various mutations of this nature in Drosophila have been expressed in the egg, larva, pupa, or adult. When disadvantageous, the mutations are called deleterious; when disastrous to the individual, they are called lethals. Relatively few mutations occur as dominants and even fewer are advantageous in the environments where they occur. The great majority of changes are recessive and detrimental under usual environmental conditions. In a changing environment, a mutation may happen to coincide with new environmental situations and be favourable.

Pleiotropy

A given mutation may alter the organism severely or have such slight effect that it can be detected only when associated with other genes, through cumulative action. The name and gene symbol associated with a mutation are usually taken from the most conspicuous phenotypic alteration that the mutation produces. Thus, white eye (w) is used in contrast to the wild type, red; and ebony (e) body designates the deviation of dark from wild type, gray body colour. This method of designation perpetuates an overly simplified concept of the effects of mutations. Because they exert their influence on basic chemical reactions in the developmental period, single mutant genes affect more than a single trait and may modify in some way every trait of the organism.

The term *pleiotropy* refers to the situation in which a gene is known to influence more than a single trait. For example, in the presence of the white-eye mutation in *Drosphila melanogaster*, the ocelli, malpighian tubules, and testicular envelopes are colourless, whereas these structures in wild flies are coloured. The balancers and wing muscles are altered and the general viability and fertility of the flies are changed. Sheldon Reed has demonstrated a discrimination against white-eyed flies in mating. Such discrimination indicates that the gene causes some effect on the cuticle that influences the mating stimulus. White eye is thus only one several effects of the single mutant gene (w). Many examples of genes with more than single effect have been discovered. All genes (mutants and nonmutants) may be pleiotropic, even though their various effects are not recognized at present. Even though a gene may have many end effects, probably it influences only one primary function in the chemistry of the developing individual.

Congenital hydrocephalus in the mouse, cited by Gruneberg, illustrates the manifold effects of a single gene, ch, on the developing mouse. The gene occurred as a spontaneous mutation and was found by preliminary studies to influence cartilage formation in early development of the mouse. Mice that carried the gene in homozygous condition were born alive but died immediately after birth because the lungs did not inflate properly. Other conspicuous abnormalities also occurred: the skull, forehead, and face were out of proportion; large protuberances filled with fluid extended out from the cerebral hemispheres because the skull bones were abnormal and only skin covered the forehead; eyelids were always open; sensory hairs of the face were abnormal; and the sternum was abnormal with little or no bone formation. Developmental studies showed that abnormalities occurred as early as the thirteenth day in the embryo. The manifold effects were traced back to a single cause, the abnormal cartilage formation. Various skeletal abnormalities were involved directly; physiological disturbances occurred secondarily. A single primary reaction concerned with cartilage formation was controlled by the mutant gene.

Somatic and Germinal Mutation

Mutations may occur in any cell and at any stage in the cell cycle. The immediate effect of the mutation and its ability to produce phenotypic change is determined by its dominance, the type of cell in which it occurs, and the stage in the cycle of the cell. If the mutation occurs in a somatic cell that can reproduce other cells like itself but not the whole organism, the mutant change would be perpetuated only in somatic cells that descended from the original cell in which the mutation occurred.

The "Delicious" apple and the naval orange have resulted from mutations that occurred in somatic tissues. Changes that gave these two fruits their desirable qualities apparently followed spontaneous mutation in single cells, which constituted only a very small part of the body of the individual apple and orange trees that were involved. In each case, the cell carrying the mutant gene gave rise to more of its own kind, eventually producing an entire branch on its respective tree which had the charactristics of the mutant type. Fortunately, it is possible to propagate desirable types of many plants by vegetative methods such as budding and grafting. Vegetative propagation was feasible for both the delicious apple and the navel orange, and today numerous progeny from grafts and buds have perpetuated the original mutation. Descendants of the mutant types are now widespread in apple orchards and orange groves.

Results of somatic mutations have been detected in animals as well as in plants. In Drosophila, white sectors are infrequently observed in an otherwise red eye. Sectors in some male flies have been demonstrated to be composed of descendants from a single cell in which a mutation had occurred (at the w locus in the X chromosome). If the same change had occurred in a germ cell, a white-eyed male might have been produced.

If dominant mutant genes occur in germ cells, their effects may be expressed immediately in the progeny. If they are recessive or hypostatic, their effects may be obscured by other genes. Germinal mutations, like somatic mutations, may occur at any stage in the cycle of the organism but they are more common in some stages, particularly during the gene replication process preceding gametogenesis. If the mutation arises in one of the gamètes produced by an individual, a single member of the progeny may receive the mutatant gene. If, on the other hand, a mutation occurs during an earlier stage of gametogenesis, several gametes may receive the mutant gene and therefore several individuals could perpetuate it. In any case, the dominance of the gene and the stage in the cycle when the mutation occurs are major factors in determining the extent of any expression that follows.

Wright noticed a peculiar male lamb with unusually short legs in this flock of sheep. It occurred to him that it would be an advantage to have a whole flock of these short-legged sheep, which could not get over the low stonefences in his New England neighbourhood. Wright used the new short-legged ram for breeding his 15 ewes in the next season. Two of the 15 lambs produced were short-legged. Short-legged sheep were then bred together and a line was developed in which the new trait was expressed in all individuals. The mutation that gave rise to the short-legged sheep was obviously of the germinal type because the cell carrying the mutation had the capacity to reproduce the entire organism. Other examples of germinal mutations have since been described in a wide variety of animals and plants.

Mutation by Deletion or Insertion of Nucleotide Bases

Consider the sequences of bases written in the RNA code shown in Figure 5.1. Now suppose that a base is deleted or an additional base is inserted. The net effect of either a deletion or an insertions that the reading frame is shifted, because different triplets are now read after the mutation site. These are called frameshift mutations. Their net effect is that the amino acids inserted after the deletion or insertion will be different from those in the unmutated cell, and hence the polypeptide will contain a grossly abnormal amino acid sequence

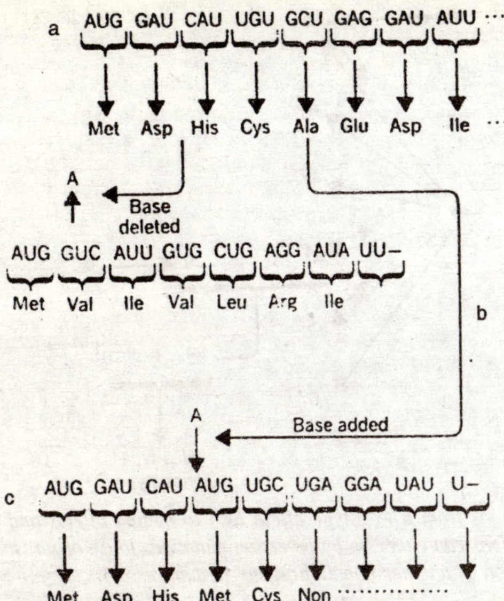

Fig. 5.1. Base deletion and base addition resulting in frameshift mutations. Translation is from left to right.

and be nonfunctional. In addition, somewhere along the line it is possible for a nonsense codon to be produced and the "nonsense" chain will be terminated.

Note that the addition or deletion of two bases will have essentially the same effect as a single base addition or deletion because the reading frame will shift. A three-base change will not shift the reading frame if three immediately adjacent nucleotides are added or deleted, but simply add or delete an amino acid. This is also true for duplication or insertion of adjacent bases in multiples of three. Two or more amino acids will be affected instead of one.

Like nonsense mutations, frameshift mutations can be expected to have drastic effects on the phenotype, unless they occur in multiples of three or cause a change at a nonessential end of a protein. In the latter case, the effect may be great or hardly noticeable, depending again on the position of the deleted or added amino acid(s) in the polypeptide chain. Frameshift mutations can be suppressed by other mutations.

The Reversal of the Mutant Phenotype

So far in this chapter we have been looking at mutation in a somewhat restricted way. It has been tacitly assumed that there is a wild-type allele (or group of wild-type alleles called *isoalleles*, which

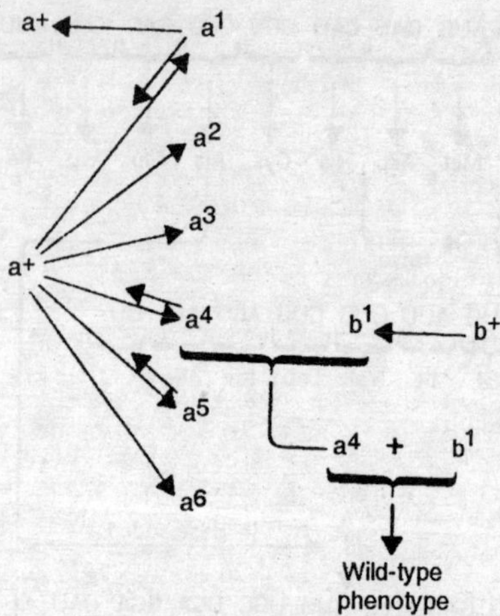

Fig. 5.2. Mutations from a wild-type allele (a^+) to mutant alleles and reversions. The reversions can either be by reversion mutations to a^+ or an isoallele (a^+) or by mutation of another nonallelic gene (a suppressor).

all produce a wild-type phenotype) for each gene that mutates to produce mutant alleles in an unidirectional way. However, mutation can occur either way. The mutant phenotype observed is a result of a forward mutation, and the change back to the original genotypic state is a reverse mutation, by definition. In addition, reversal of phenotype to the original situation, or almost so, may occur without a reversal of genotype but rather by a further mutation in the same gene or in another nonallelic gene. We consider these two situations in this section.

Reverse Mutation

If a missense mutation occurs by substitution of a nucleotide base, it should be expected that it can be reversed by a second substitution in the same codon so as to change it back to the condition in which it will code for the original amino acid, or one which will again give a wild-type phenotype. This has been demonstrated in a most convincing fashion by the use of the chemical mutagens about whose action we have some knowledge. Consider the action of 5-bromouracil, which, pointed out previously, causes transitions of the A:T–G:C type, if 5-bromouracil is able in its enol state to pair with guanine it should also be able to pair with adenine when not in the enol state, as the new chain is forming in the mutant. The net effect is another transition, G:C – A:T, the reverse of the first situation. Therefore, mutations caused by 5-bromouracil should be reversible by 5-bromouracil, and

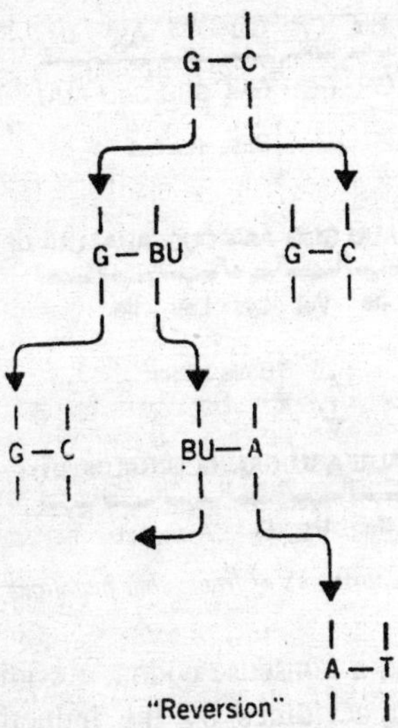

Fig. 5.3. The reversion of a mutation initially caused by 5-bromouracil.

this turns out to be the case. 2-Aminopurine and nitrous acid will also cause reversion of 5-bromouracil-induced mutations.

Hydroxylamine is somewhat different from 5-bromouracil, 2-aminopurine, an nitrous, acid; mutations caused by it are not truly reverted by it. All these results agree with the theory that 5-bromouracil, 2-aminopurine, and nitrous acid should cause transitions in both direction (A:T G:C). Hydroxylamine should cause transitions in only one direction (G:C A:T).

Ethyl ethanesulfonate (EES) apparently causes transitions in the G:C—A:T directiou and also transversions such as G:C T:A and G:C C:G. Some, but not all, EES-induced mutants are reverted by base analogues and nitrous acid. The reading frame mutations caused by acridines do not revert in the presence of base analogues or compounds that later bases. And this, of course, is precisely what is to be expected, because acridines cause the deletion and duplication of bases. On the other hand, mutations caused by acridines are for the most part revertible by acridines. In this fact there resides another interesting story; revertants of reading frame mutations may occur in several ways as related in the next several paragraphs.

Intragenic Supressors

An acridine-induced mutant will usually (if not always) result in a gene deficient or duplicated for bases. This shifts the reading frame

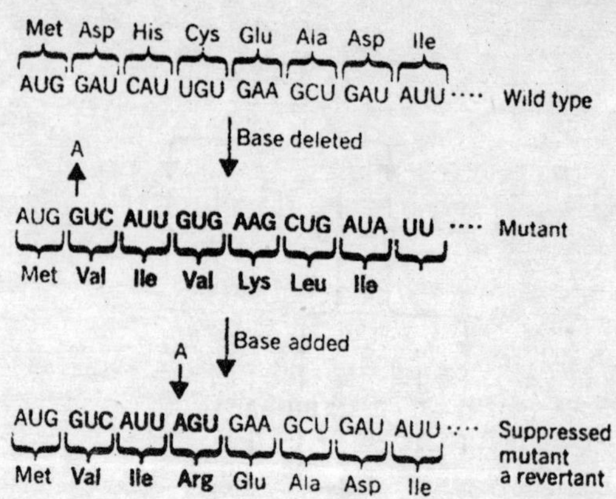

Fig. 5.4. How intragenic suppressors of frame-shift mutations act to produce a functional polypeptide.

and generally results in a nonsense codon, Acridine-induced mutations can be reverted with acridines by the induction of an intragenic suppressor. If the original mutation is a deletion of a base, the suppressor will be an insertion of a base occurring nearby in the DNA. As a result, the reading frame would be brought back to the state in which the proper amino acids are now coded, except for that region between the two mutations. In an analogous way an insertion initially caused by acridine can be suppressed by a deletion. If the amino acids coded for in this region do not form a segment that is important to the protein's ability to function, a true suppression of the mutant phenotype will occur.

A second type of intragenic suppressor involves missense mutations. As stated before, a single amino acid substitution caused by a missense mutation may so disrupt a polypeptide's tertiary conformation that is inactive as a protein. It has been shown that such a mutation may be rectified by a second missense mutation in the same gene. The second amino acid substitution in some way erases the "lethal" change in tertiary conformation caused by the first. This has been demonstrated in the tryptophan synthetase of E. coli and is probably a fairly common even in all organisms.

Extragenic Suppressors

When a nonsense codon is formed either by missense or frameshifts mutation, a drastic phenotypic change is expected because the polypeptide encoded by the affected gene will be incompletely synthesized during translation. Such mutants can be restored to approximately the wild-type phenotype by the occurrence of another mutation in a gene at another locus. Five such suppressors have been described in E. coli. They all suppress one or the other of the two nonsense codons: UAG and UAA or both. It has been shown that in the presence of a suppressor mutation amino acids are actually inserted during translation at points where nonsense codons exist. Hence the polypeptide chains are completed, although the completed chains may not function as well as the polypeptide chains coded in the wild type.

As might be expected, it has been found that the biochemical basis for the action of these suppressor genes is that they code for tRNA's. The sup D gene codes for serine tRNA. In its mutant condition, it produces an altered serine tRNA that introduces serine where a UAG occurs in the messenger. Similar explanations apply for the other suppressor genes, which presumably code for other tRNA's.

These kinds of suppressors have been *supersuppressors*. They are so named because they are capable of suppressing nonsense mutations in a wide range of different genes with quite different and unrelated functions. Theoretically, any gene can mutate by the formation of a nonsense codon. Hence, a single suppressor may effectively suppress specific alleles of a large number of different genes, provided the amino acid it introduces through its altered tRNA results in functional polypeptides. Suppressors have been reported in yeast, *Neurospora*, and perhaps barley. They probably also occur in *Drosophila*. The suppressor of Hairy Wing in *D. melanogaster* suppresses at least 10 other mutant genes, all of which are at different loci and have seemingly unrelated functions.

Suppressors in diploid organisms theoretically should always be dominant, because only one dose of altered tRNA gene should be necessary to produce the tRNA necessary to recognize a nonsense codon. However, suppressors do exist that are expressed only when homozygous. The suppressor of Hairy Wing is one example. Another example in Drosophila is the suppressor of vermilion su(s). It suppresses the mutant condition, sable(s). The action of this type of suppressor, other examples of which are known in *Drosophila*, is not clear.

Missense, Nonsense, and Frameshift Mutations

Various types of mutations can conveniently be discussed in relation to the b-globin gene. It has already been described that the human b-globin gene is split into three coding regions interrupted by two intervening sequences. The nucleotide sequence of the beginning of the first coding region, the sense strand is at the top, and transcription occurs from left to right (from the 3' end of the sense strand to the 5' end). The spaces that separate adjacent triplets of bases do not exist in the actual molecule. It shows the corresponding portion of the b-globin mRNA produced after transcription and RNA processing. Although it is not shown in the figure, the next upstream codon from GUG is the AUG initiation codon, which codes for methionine. Figure shows the first eight amino acids at the amino terminal end of normal b-globin; the initial methionine with which translation begins is not represented in the finished polypeptide because it is enzymatically removed from the chain. The underlined nucleotides represent the restriction site for the restriction enzyme *Mst*II, which cleaves the DNA at this position.

As noted in the previous section, a *base substitution* mutation is one in which a base pair in a DNA duplex is replaced with a different base pair. A base substitution that leads to an amino acid substitution in the corresponding polypeptide is called a missense mutation, and there are two principal types of missense mutations. Part (*a*) again shows a portion of the first coding region of the b-globin gene, and part (B) shows the result of a base substitution in which the normal TA pair indicated is replaced with an AT pair. In this substitution, a pyrimidine base in the sense strant (T) is replaced with a purine base (A), and a purine base in the antisense sense strant (A) is replaced with a pyrimidine base (T). Any base substitution in which a purine (A or G) is replaced with a pyrimidine (T or C), or in which a pyrimidine is replaced with a purine, is called a *transversion*. Transcription and RNA processing of the mutant DNA lead to the mRNA shown in Figure; note that the sixth codon, which normally reads GAG, now reads GUG. Tranlation of this mRNA leads to a substitution of valine for glutamic acid at the sixth position in the polypeptide (black arrow). The transversion mutation is indeed the actual molecular change responsible for b^s hemoglobin. The b^s mutation also obliterates the *MSt* II restriction site, so *MSt* II will not cleave the b^s DNA at this position. This molecular difference provides a rapid and convenient method for *in vitro* diagnosis of sickle cell anemia.

There is another type of missense mutation, which is called a *transition* mutation because it involves the replacement of one purine (in this case G) with another purine (in this case A) or one pyrimidine (in this case C) with another pyrimidine (in this case T). In the transition mutation a CG base pair in the original molecule is replaced with a TA base pair. The corresponding portion of the mRNA is shown in part (C), and in this case the original GAG codon in position 6 is altered to AAG. Translation of the mutant mRNA produces a β-globin polypeptide in which the normal glutamic acid at position 6 is replaced with a lysine. This particular mutation is known as the β^c mutation. Sine the b^c mutation is relatively common in areas of the world where malaria is prevalent, it is thought that heterozygotes for β^c may have some protection against malaria, as is certainly the case with β^c.

Not all base substations are missense mutation. There is also another type of mutation called a nonsense mutation. A nonsense mutation is one that creates a chain-terminating codon [i.e., uAA (orchre), UAG (amber), or UGA (opal)] in a coding region. The name nonsense mutation may seem odd, but it comes from the fact that chain-terminating codons are times referred to as nonsense codons. The transvertion to leads to a UAG terminating codon at the sixth position in the mRNA. Thus, translation of this mRNA leads to a truncated polypeptide because translation terminates at the UAG codon.

In prokaryotes such as *E. coli* and in lower eukaryotes such as yeast, certain mutations have the ability to suppress nonsense mutations. These nonsense-suppressing mutations are called nonsense suppressors, and they involve mutations in genes for transfer RNA (tRNA) that reduce the faithfulness of translation so that chain-terminating codons are sometimes misread.

Such a mutant tRNA will correctly translate serine codons, but, on occasion, will misread a chain-terminating codon (in this case UAG) as a serine codon. Consequently, in some fraction of attempts at translation of the mutant mRNA, the UAG codon will be misread as serine (stipulated arrow), and translation will be able to produce normally beyond this point. If the serine-containing polypeptide is able to function, the phenotypic effects of the original nonsense mutation will have been suppressed by the nonsense suppressor.

Of course, nonsense suppressors can occur in tRNA genes other than serine tRNA. For example, if a UAG suppressor occurs in a gene for glycine tRNA, a UGA codon will sometimes be misread as a glycine codon, and glycine will be inserted into the polypeptide at this position. It should also be emphasized that a particular nonsense suppressor will suppress only one of the chain-terminating codons; that is, an *amber suppressor* will suppress only the UAG (amber) terminator, or *ochre suppressor* will suppress only the UAA (ochre) terminator, and an *opal suppressor* will suppress only the UGA (opal) terminator. Of course, a nonsense suppressor will occasionally misread normal termination codons, too; for example, an amber suppressor will occasionally misread a UAG codon at the end of a normal gene and the result is an abnormally long polypeptide. However, organisms that carry nonsense suppressor are able to survive in spite of the occasional misreading of normal genes.

Not all base-substitution mutations lead to missense or nonsense codons. Because the genetic code contains many synonymous codons, some base substitutions leave the amino acid sequence of the corresponding polypeptide unaltered. Such mutations are said to be silent. In this case a C/G-to-T/A transition at the third position in codon 6 of the β-globin gene leads to an mRNA that has GAA as its sixth codon. However, GAA is a synonymous codon for glutamic acid, so the amino acid sequence of the resulting polypeptide will be completely normal. As a brief inspection of the genetic code will reveal most silent substitutions in coding regions would be expected to involve transitions in the third position of a code. The situation may be very different for intervening sequence, however. Since a particular length and base sequence of an intervening sequence does not seem to be essential for proper gene function, many mutations in intervening sequences—transitions, transversions, even inversions, deletions, or insertions–may be silent.

A final example of a type of mutation that has relatively simple molecular basis is known as a *frameshift* mutation. A frameshift mutation alters the reading frame of an mRNA during translation and thereby greatly alters the amino acid sequence of the corresponding polypeptide. Frameshift mutations are caused by deletions of additions of a small number of nucleotides in a coding region; since the genetic code consists of triplets of nucleotides, any deletion or addition of a number of nucleotides other than an exact multiple of three will cause a shift in reading frame. An example of a frameshift mutation associated with a single nucleotide deletion is Part (a) is again a portion of the first coding region of the human β-globin gene, and the left side of the figure illustrates the normal course of transcription and translation of this sequence. As in previous figure involving this gene, small gaps are introduce in part (c) to show the normal triplet reading frame. The right side of the figure shows the consequence of a deletion of the indicated A/T nucleotide pair. When this deletion-bearing sequence is examined in terms of its triplet reading frame, it can be seen that all triplets beyond the second one are different from those in normal sequence. This lack of correspondence occurs, of course, because the single-nucleotide deletion causes the triplet reading frame beyond the deletion to be shifted one nucleotide to the left. The extensive lack of correspondence due to the frameshift mutation is also evident in the mRNA, and the resulting amino acid sequence in the polypeptide is hardly recognizable as a β-globin sequence. Such out-of-frame translation will continue along the mutant mRNA until a termination codon is encountered.

Mutations Resulting from Unequal Crossing-Over

It has been stated that unequal crossing-over involving duplications can lead to an increase or decrease in the number of copies of the region. When this process occurs at the molecular level, it creates new types of DNA sequences that qualify as mutations. An example of unequal crossing-over creating new mutations is found in the human b-globin gene. Part (a) shows part of the DNA sequence coding for amino acids 83 through 98, which is 18 nucleotides upstream from the second intervening sequence. The boxes indicate two nearby regions that have a perfect eight-nucleotide homology that can act as a duplication. Part (b) shows the coding sequence for amino acids 88 through 103; it has been shifted to the left relative to part (a) to show how the regions of homology can mispair (shaded boxes). As before, the gaps in the DNA sequence indicate the translational reading frame.

Sequence (c) combines the left part of (a) with the right part of (b). This sequence is deleted for the nucleotides that code for amino acids 91 through 95, and it is of some interest that this deletion, called the *Gun Hill deletion*, is found in certain rare individuals. Sequence (d) is the complementary crossover product to sequence (c); this sequence carries a duplication of the nucleotides that code for amino acids 91 through 95.

Mutagenic agents and the Mechanisms of Mutation

Mutations occur in the somatic and germ cell of all organisms. In organisms like Drosophila, mutations are detected for specific loci in one in about 10^5 to 10^6 gametes. Therefore the spontaneous mutation rate for most genes is about 10^{-5} to 10^{-6} mutations per locus per generation. First, agents that increase the mutation rate above the spontaneous level will be discussed because they help explain the spontaneous rate.

The two basic kinds of mutagenic agents are (1) physical ones including radiation (s) and (2) chemical ones, consisting of both "natural" and "unnatural" substances.

The Spontaneous Mutation Rate

An organism carefully protected from known mutagenic chemicals and from known artificially made radioactive sources will still produce gametes or cells with new gene mutations, What causes these spontaneous mutations? A number of factors can be enumerated and may be divided into two categories: environmental and genetic.

Environmental Factors

Background radiation comes to mind as a factor almost immediately, because we have been sensitized since the 1950s to the hazards of ionizing radiation from atomic bombs, industrial and medical applications of nuclear energy, and so forth. However, the average background measured in roentgens per generation is insufficient to account for more than a small fraction of spontaneously occurring sex-linked recessive lethals in Drosophila. But, course, if the background radiation is raised, this could cause additional mutations that might become an important factor in the future of humans and all other organisms.

Ultraviolet light may not be of much significance in the production of mutations since the ultraviolet in sunlight that reaches the surface of the earth has wavelengths only above 300 nm. This range of wavelengths is not effective in inducing mutations to any marked extent except somatic ones in skin. In the higher animals overlying tissue protects the germ cells as already noted.

Chemical in the environment may certainly be a factor, but their effect is difficult to measure for several reasons: (1) a mutagenic chemical may be metabolized to a nonmutagenic state before it reaches the germ cells, or (2) a nonmutagenic chemical may be metabolized to a mutagenic state, and so on. Despite these difficulties, it is now apparent that close attention should be paid to the present and future effects of chemicals on the mutation rate. Our environment is becoming more and more polluted with many different kinds of natural and synthetic chemicals used as insecticides, herbicides, food preservers, medicines, beautifiers, tranquilizers, and hallucinogens. What effect do these how have on the human mutation rate? We do not know, but it is important that we find out, because an increase in the rate could have diastrous results.

For the present, we can assume that the presence of chemicals in the environment may have some effect on the mutation rate of all organisms, but it certainly cannot account for all the spontaneous mutation rate. Metabolic products, such as formaldehyde and hydrogen peroxide produced within cells, may also be involved, but here again it is doubtful that their effects are highly significant.

Genetic Factors

Organisms evolved in a changing environment to which they had to continually adapt as they evolved. Mutations were necessary for this evolution to occur, just as they are necessary for future evolution. But the mutation rate must not be so high that it results in the production of many deleterious mutation that would act as "genetic load" on the population.

It is highly probable that the main component of the spontaneous mutation rate is genetically controlled. As we have described in previous pages, enzymes are involved in the replication of DNA. Mistakes of different types may occur during replication that may lead to mutation, especially if the replicating enzymes themselves are deficient. These mistakes may be replicated or they may be repaired by the repair systems that we known to exist. Thus, two opposing systems, genetically controlled, may be postulated to exist: the replicative synthetic one and the repair system. The spontaneous rate of mutation may be determined by an equilibrium point between them.

It has long been known that different populations of *Drosophila melanogaster* collection from widely scattered areas in different part of the world show significantly different mutation rates (Table 5.1). Furthermore, genes have been identified that have an influence on the mutation rate A second chromosome gene, *hi*, and a third chromosome gene, *mu*, have been located in *melangoaster*. Both of these increase the "spontaneous" mutation rate when homozygous, although *mu* appears to act in this way only in females. Extensive work has been done in E. coli with mutant strains that show a high rate of A:T → C:G transversions. The result is that the mutation rates as specific loci are raised several thousand fold!

Genes that raise the mutation rate are called *mutator* genes. A number of these have been identified in yeast, *Drsosophia*, and E. coli. They have yet to be directly implicated in the DNA replicative and repair processes, but it is difficult to believe that they are not.

In addition to mutator genes, there are also numerous examples of mutable loci in a variety of organisms. Certain alleles of some genes exhibit an unstable condition, so that they mutate from the wild type to mutant condition at a high rate in either the germ cells, or somatic cells, or both.

Table 5.1. Spontaneous mutation rates found in different populations of *D. melanogaster*, sex-linked and second chromosome recessive lethals

	X chromosome			*Second chromosome*		
Stock	*Number tested*	*Number lethals*	*Percent lethals*	*Number tested*	*Number lethals*	*Percent lethals*
Florida inbred	2108	23	1.09	—	—	—
Wooster	1266	8	0.63	—	—	—
Oregon R	3049	2	0.07	—	—	—
Florida No. 10	916	10	1.09	516	9	1.74
Lausanne	955	2	0.21	436	3	0.69
Leningrad	8614	14	0.16	—	—	—
Sukhami	2309	24	1.04	—	—	—

Phenotypic Effects of Mutations

Mutations must normally cause some detectable *phenotypic change* for their presence to be recognized. The effects of mutations on phenotype range from alterations so minor that they can be detected only by special genetic or biochemical techniques to gross modifications of morphology to lethals. A gene is a specific sequence of nucleotide pairs coding for a particular polypeptide. Any mutation occurring within a given gene will thus produce a new form or *new allele* of that gene. Because of the degeneracy of the genetic code, some base pair changes do not change the protein products coded for by the genes in any way. Genes containing mutations with small effects that can be recognized only by special techniques are called "isoalleles." Other mutations result in total loss of gene-product activity. If mutations of the latter type occur in essential genes (genes required for viability), they will, of course, be lethal.

Mutations may be either recessive or dominant. In haploid (or, more accurately, monoploid) organisms like viruses and bacteria, both recessive and dominant mutations can be recognized by their effects on the phenotype of the organism in which they originated. The dominance or recessiveness of mutations in bacteria can be determined only by studying partial diploids. In diploid (or polyploid) organisms, recessive mutations will be recognized only when present in the homozygous condition. Most recessive mutations in diploids will not be recognized at the time of their occurrence, since they will be present in the heterozygous state. Sex-linked recessive mutations are an exception, since they will be expressed in the hemizygous state in the heterogametic sex (males in humans and fruit flies; females in birds). Sex-linked recessive lethal mutations will alter the sex ratio, sinçe hemizygous individuals carrying the lethal will not survive.

The most useful mutations for the genetic analyses of many biological processes are conditional lethal mutations. These are mutations that are (1) lethal in one environment, the so-called restrictive conditions, but are (2) viable in a second environment, the permissive conditions. Such mutations allow geneticists to identify and study mutations in essential genes that result in complete loss of gene-product activity even in haploid organisms. Mutants carrying conditional lethal can be propagated under permissive conditions, and information about the functions of the gene products can be deduced by studying the consequences of their absence under the restrictive conditions. Conditional lethal mutations also provide valuable selective mutations also provide valuable selective sieve for genetic fine structure analysis.

The three major classes of mutants with conditional lethal mutations are (1) *auxotrophic mutants*, (2) *temperature-sensitive mutants*, and (3) *suppressor-sensitive mutants*. Auxotrophic mutants (as opposed to prototrophic "wild-types") are mutants that are unable to synthesize an essential metabolite (amino acid, purine, pyrimidine, vitamin, etc.) That is synthesized de novo by wild-type individuals of the species. Such auxotrophic mutants will grow and reproduce when the metabolite is supplied in the medium (the permissive condition); they will not grow when the essential metabolite is absent (the restrictive condition). Temperature-sensitive mutants will grow at one temperature but not at another temperature. Most temperature-sensitive mutants are heat-sensitive some, however, are cold-sensitive. The temperature sensitivity usually results from increases heat or cold lability of the mutant gene product, for example, an enzyme which is active at low temperature but partially or totally inactive at higher temperatures. Occasionally, only the synthesis of the gene product is sensitive to temperature, and once synthesized, the mutant gene product may be as stable as the wild-type gene product, Suppressor-sensitive mutants are viable when a second genetic factor, a suppressor, is present, but are nonviable in the absence of the suppressor. The suppressor gene may correct or compensate for the defect in phenotype that is caused by the suppressor-sensitive mutation, or it may render the gene product, altered by the mutation, nonessential.

Most of the thousands of mutations that have been identified and studied by geneticists have been found to be deleterious and recessive.

This is to be expected, considering what we know about the genetic control of metabolism and the techniques available for identifying mutations. Metabolism occurs by sequences of chemical reactions, each step of which is catalyzed by a specific enzyme coded for by one or more genes. Mutations in these genes frequently produce blocks in metabolic pathway. These blocks occur because changes in the base-pair sequences of genes often (but not always) cause changes in the amino acid sequences of polypeptides, which may, in turn, result in loss of function. This, in fact has been the most commonly observed effect of easily detected mutations. Given a wild-type allele coding for an active enzyme and mutant alleles coding for less active or totally inactive enzymes, it is apparent why most of the observed mutations might be recessive, as is observed. If the enzyme is metabolically important, such mutations will also be deleterious. If the enzyme catalyzes an essential reaction, the mutations causing total loss of activity will be recessive lethals.

But why should most mutations with phenotypically recognizable effects result in decreased gene produce activity or no gene-produce activity? This result can be predicted if one accepts the effectiveness of natural selection and if one accepts the effectiveness of natural selection and if one assumes the existence of a semiconstant environment during the recent (on the evolutionary scale) evolution of life forms on earth. A "wild-type" allele of a gene coding for a "wild-type" enzyme or structural protein will have been selected for optimal activity for many generations. Mutations resulting in amino acid changes that increased the efficiencies of enzymes in carrying out particular functions will have been preserved by natural selection, and, as the most "fit," they will have become the new "wild-types". Given a sufficient period of time in semiconstant environment, most, if not all, sequences of amino acids (at least all sequences of amino acids that differ form the "wild-type" by a single mutation) will have been tried, and natural selection will have preserved the most efficient one. It will now be the wild-type. Mutations, which produce changes in these very specific sequences of amino acids, will usually result in less activity or no activity at all. As such, they will most frequently be recessive and deleterious.

An analogy can be made with any complex, carefully engineered machine. If you randomly modify any one essential component (e.g., of a watch or an automobile), it seldom performs as well as it did prior to the random change.

Radiation-Induced Mutation

That portion of the electromagnetic spectrum containing wavelengths that are shorter and of higher energy than visible light (wavelengths below about 0.1 μm) can be subdivided into ionizing radiation (X rays, gamma rays, and cosmic rays) and nonionizing radiation (ultraviolet light). Ionizing radiations such as X rays (about 0.1 to 1 nm) are of high energy and thus are useful for medical diagnosis because they can penetrate living tissues. In the process of penetrating matter, these high-energy rays collide with atoms and cause there lease of electrons, leaving positively charged free radicals or ions. These ions, in turn, collide with other molecules, causing the release of further electrons. The net result is that a "core" of ions is formed along the track of each high-energy ray as it passes through matter of living tissues. This process of ionization (thus the name ionizing radiation) is induced by machine-produced X rays, protons, and neutrons, as well as by the alpha, beta, and gamma rays released by radioactive isotopes of the elements (e.g., ^{32}P, ^{35}S, radium, cobalt-90 etc.). Ultraviolet rays having lower energy, penetrate only the surface layer of cells in higher plants and animals and do not induce ionization. Ultraviolet rays dissipate their energy to atoms that they encounter, raising the electrons in the outer orbitals to higher energy levels, a state referred to as excitation. Molecules containing atoms in either ionic forms or excited states are chemically more reactive than those containing atoms in their normal stable states. The increased reactivity of atoms present in DNA molecules is the basis of the mutagenic effects of ultraviolet light and ionizing radiation.

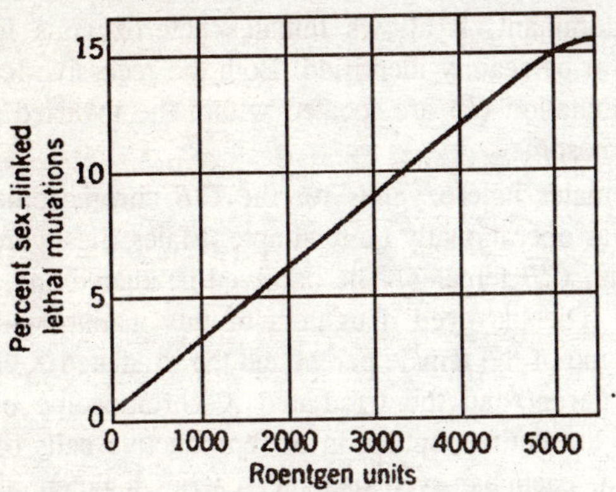

Fig. 5.5. Relationship between the frequency of sex-linked lethal mutations induced in Drosophila sperm and ionizing irradiation dosage.

Ionizing Radiation

In 1927, H. J. Muller first demonstrated that mutation could be induced by an external factor. Muller demonstrated that X-ray treatment markedly increased the frequency of sex-linked recessive lethal mutations in D. melanogaster. Muler's unambiguous demonstration of the mutagenicity of X rays became possible by his development of a technique facilitating the simple and accurate identification of lethal mutations in the X chromosome of Drosophila. This technique, called the ClB method, involves the use of females heterozygous for a normal X chromosome and X chromosome (The ClB chromosome) specifically constructed for Muller's experiment.

The ClB chromosome has three essential components. (1) The C (for crossover "suppressor") refers to the presence of a long inversion, that prevents recombination of genetic markers on the *ClB* chromosome and alleles on the normal X chromosome. The inversion does not actually prevent crossing over, but causes gametes containing X chromosomes produced by crossing over between the *ClB* chromosome and the normal X chromosome to be inviable. Chromosomes resulting from crossing over between a chromosome containing an inversion (an inverted segment of the chromosome) and a normal chromosome will contain duplication and deficiencies (repeated and missing sets of genes). The inversion is required in Muller's experiment to assure that the markers on the *ClB* chromosome stay together through meioses. (2) The *l* refers to a recessive lethal in the *ClB* chromosome. (3) The *B* refers to the presence of the partially dominant mutation that causes the bar eye phenotype, which is a narrow, slit-shaped eye. Because it is partially dominant, it allows females heterozygous for the *ClB* chromosome to be readily identified. Both the recessive lethal (*l*) and the bar eye mutation (*B*) are located within the inverted segment of the *ClB* chromosome.

Given females heterozygous for the *ClB* chromosome. Muller's experiment was operationally quite simple. Males flies were irradiated and mated with *ClB* females. The bar-eyed irradiated and mated with ClB females. The bar-eyed daughters of this mating will carry the *ClB* chromosome of the female parent and the irradiated X chromosome of the male parent and the irradiated X chromosome of the male parent. Since the entire population of reproductive cells of the males was irradiated, each bar-eyed daughter carries a potentially mutated X chromosome. That is, each male is likely to produce some sperm that contain X chromosomes carrying new lethal mutations and some sperm without X-linked lethal mutations. The bar-eyed daughters were then mated individually (in separate bottles) with wild-type males. If the irradiated X chromosomes carried by a bar-eyed daughter contains a sex-linked lethal, all of the progeny of the mating will be female. Since males are hemizygous for the X chromosome, those receiving the *ClB* chromosome will die due to the recessive lethal (*l*) that it carries. Those receiving the irradiated X chromosome will also die if a recessive lethal has been induced in it. Matings of bar-eyed daughters carrying an irradiated X chromosome, in which no lethal mutation has bee induced, with wild-type males will produce female and male progeny in a ratio of 2:1 (only the males with the *ClB* chromosome will die). Scoring for the presence of recessive sex-linked lethals is thus unambiguous and error free using the *ClB* technique—simply scoring for the presence or absence of male progeny. By using this technique. Muller was able to demonstrate an increase in mutation rate of up to 150-fold after X-ray treatment.

Another technique that facilitates the detection of mutations in Drosophila, in this case sex-linked mutations with visible effects on phenotype (often called "visible mutations"), involves the use of *attached-X-chromosomes*. Attached-X-chromosomes undergo compulsory nondisjunction (failure of homologous chromosomes or chromatids to disjoin or separate during anaphase) since the two X chromosomes are joined to single centromere. If females with two attached-X chromosomes plus a Y chromosome (XXY) are mated to normal males, any mutation that occurs on the X chromosome of the male will be expressed in the surviving male progeny. In such attached-X matings, the male progeny receive their X chromosome from their male parent, rather that from their female parent as in a normal mating. If the male parent is treated with a mutagenic agent such as X rays, the increased frequency of recessive visible mutations can be easily assessed by screening the male progeny of the attached-X mating.

X rays and most other forms of ionizing radiation are quantitated in roentgen units (r units, pronounced "runtgen"), which are measured in terms of the number of ionizations per unit volume under a standard set of conditions. More specifically, one roentgen unit is the quantity of ionizing radiation that produces one electrostatic unit of charge in a one cm^3 volume. Note that the dosage of irradiation in roentgen units does not involve a time scale. The same dosage may be obtained by a low intensity of irradiation over a long period of time or a high intensity of irradiation for a short period of time. This is very important because in most studies the frequency of induced point mutations is directly proportional to the dosage of irradiation. In Drosophila sperm, for example, there is an increase of approximately 3 percent in the mutation rate for each 1000 r in increase in irradiation dosage. This linear relationship between mutation frequency and dosage in indicative of so called "single-hit- kinetics." That is, only one event (ionization?) or one "hit" is required to cause a mutation. Or, stated differently, every ionization has some fixed (under a specific set of conditions) probability of inducing a mutation. If the cumulative effects of many ionizations were required to induce a mutation, would plot as a curve that was concave upward.

The linear relationship between mutation rate and radiation dosage is important because it speaks directly to the frequency asked question of What is a safe level of irradiation? Even very low levels of irradiation have certain low, but very real, probabilities of inducing mutations. The question is thus meaningless. There is, no such thing as a safe level. In *Drosophila* sperm, for example, very low levels of irradiation over long periods of time (chronic irradiation) are as effective in inducing mutations as the same total dosage of irradiation administered at high intensity for short periods of time (acute irradiation). This clearly has major practical significance in evaluating the effects of the increased exposure of living organisms to radiation that results from the testing and use of nuclear weapons and nuclear reactors in generators, spaceships, and so on.

In mice, chronic irradiation has been found to induce somewhat fewer mutations than the same dosage of acute irradiation. Moreover, when mice were treated with intermittene doses of irradiation, the mutation frequency was slightly lower than when they were treated with the same total amount of irradiation in a continuous dose. It should be emphasized that all of these irrdiation treatments were mutagenic, albeit, to different degrees, to both Drosophila and mice. The different responses of fruit flies and mice to chronic irradiation may result from differences in their ability to repair damaged DNA. Repair mechanisms may exist in the spermatogonia and oocytes of mice that do not exist in Drosophila sperm.

The single-hit theory implies that one ionization can produce one mutation, but it does not imply anything about the efficiency with which this happens. Several factors have been shown to affect the efficiency with which irradiation induces mutations. The receptivity of a cell at different stages of its metabolic cycle is an important factor in determining rates for induced mutations. A. H. Sparrow has shown marked variations in numbers of chromosome fragments, assumed to be directly related to mutational changes, at different stages in meiosis and cleavage in the plant genus Trillium. Chromosome aberrations were induced about 60 times more frequently at metaphase than it Interphase. Nondividing cells in Trillium showed little radiation damage, whereas rapidly dividing cells were very sensitive.

Oxygen tension and temperature change, when associated with irradiation, also may significantly alter the frequency of mutations. Low-oxygen tension decreases mutations. Oxygen can magnify the effect of radiation, but only if it is present during the irradiation. Oxygen has less effect with intense conditions than with moderate conditions of ionization. Environmental agents that protect germ cells from radiation damage often do so by lowering the oxygen concentration of tissue, and those that enhance the effectiveness of radiation and oxygen.

Ionizing radiation also induces various kinds of gross changes in chromosome structure (chromosome aberrations) such as deletions, duplications, inversions, and translocations. These changes in chromosomes structure result from breaks in chromosomes caused by ionizing radiation. Since they require two breaks, the kinetics of induction are two "hit" as expected, rather than single "hit."

Ultraviolet Radiation

Ultraviolet rays do not possess sufficient energy to induce ionizations. They are, however, readily absorbed by certain substances such as purines and pyrimidines, which then enter a more reactive or excited state. Because of their lower energy, they penetrate tissues only slightly, usually only the surface layer of cells in multicullular organisms. Nevertheless, ultraviolet light (UV) is at a wavelength of 254 nm. Maximum mutagenicity also occurs at 254 nm, suggesting that the UV-induced mutation process is mediated directly by the absorption of UV by purines and pyrimidines. In vitro studies show that the pyrimidines (especially thymine) absorb strongly at 254 nm and, as a result, become very reactive.

The two major products of UV absorption by pyrimidines appear to be pyrimidine hydrate and pyrimidine dimers. Several lines of evidence indicate that thymine dimerization is probably the major mutagenic effect of UV. Thymine dimers appear to cause mutations indirectly in two ways. (1) Dimers apparently perturb the DNA double helix and interfere with accurate DNA replication. (2) Occasional errors are made during the processes that cells possess for the repair of "damaged" DNA, such as DNA containing thymine dimers. The relationship between mutation rate and UV dosage is highly variable, depending on the type of mutation, the organism, and the conditions employed. "single-hit kinetics" are only occasionally observed, in contrast to ionizing radiation.

Mutation Frequency

Mutation is necessary to provide the genetic variability required for the evolutionary adaptation of species to environmental changes. On the other hand, most are deleterious. Thus, if mutation were to become too frequent in a species, it would create a sizable "genetic load" of deleterious effects. Clearly, if this "genetic load" became too large, the species would face extinction. Human technology has already contaminated the earth with increased levels of radiation and chemicals that are known to be mutagenic. While the consequences of the present increased level of mutagenes in the environment cannot yet be accurately assessed, most scientists agree that further increases in the levels of mutagens in the environment should be avoided. Yet significant quantities of hundreds of new chemicals are introduced into the environment each year, most of them with insufficient, if any, mutagenicity tests. Whether they will cause harmful increases in mutation rate (and/or cancer incidence) is a question of utmost concern to everyone.

Each gene probably has its own characteristic mutational behaviour. Some genes mutational behaviour. Some genes undergo mutations more frequently than others in the some organism. Those with unusually high mutation rates are called unstable or mutable, but a wide range of mutation rates exists among genes that are considered stable. The mutation rate per gene in bacteria is of the order of 1 in 100,000 to 1 in 10 million (10^{-5} to 10^{-7}) per cell generation. For fruit flies, the average for mutation in a particular gene is in the order of 1 in 100,000, with a range from 1 in 200,000 to 1 in 100,000, with a range from 1 in 20,000 to 1 in 200,000 gametes. Since most of the data on mutation rates in fruit flies have been obtained from experiments with males, questions have arisen concerning a possible sex difference in overall mutation rates. B. Wallace has shown through extensive experiments that mutation rates are not significantly different in the two sexes of D. Melanogaster. However, mutation rates do differ in different strains.

Estimates of mutation rates humans indicate a somewhat greater frequency than those cited for stable gens in most other organisms.

Samples collected thus far have been small, and the methods used were indirect and subject to large errors. Genes associated with such human traits as intestinal polyposis and muscular dystrophy have been estimated to mutate once in 104 to 105 people. A human generation is equal to about 50 cell generations. By expressing any mutation rate as probability of mutation per cell per generation, a mutation rate is defined independently of exact physiological conditions and stage in life cycle. This definition is based on a time unit proportional to a cell's division time. When expressed in terms of cell generations, rates for fruit and humans are generally comparable with those for bacteria.

ESTIMATING MUTATION RATES

Mutation rates are usually defined in terms of mutation events per generations or, less frequently mutations per unit time. The direct experimental determination of mutation rates is complex. Because mutation is must be sampled to get accurate estimates of mutation rates. In addition, recurrent sampling is required unless the initial population is sufficiently small that its probability of containing a mutant organism is negligible. Simply enumerating the number of mutant individuals in a population does not indicate how many mutational events have occurred since the mutants will usually undergo exponential growth like the nonmutants, although often at a slower rate than the latter.

Consider a "wild-type" allele a^+ mutating to a mutant allele a with a mutation rate μ (defined as a constant a probability of a mutational event per cell duplication or per organism duplication), that is,

$$a^+ \to a$$

clearly, the number of mutation events will depend on the number of a+ alleles in the population. For simplicity, consider a haploid organism. If one starts with population in which the number of mutant organisms equals *M* and the total population size equals *N*, then

$$dM = \left(\mu + \frac{M}{M}\right) dN$$

This *assumes* that (1) M is very small relative to N, so that the difference between N and N - M is negligible; (2) the mutant and nonmutant organisms duplicate at the same rate; (3) reversion of a → a+ does not occur or is negligible. If one starts with a population in which M is zero, and makes all the same assumptions, then the relationship simplifies to

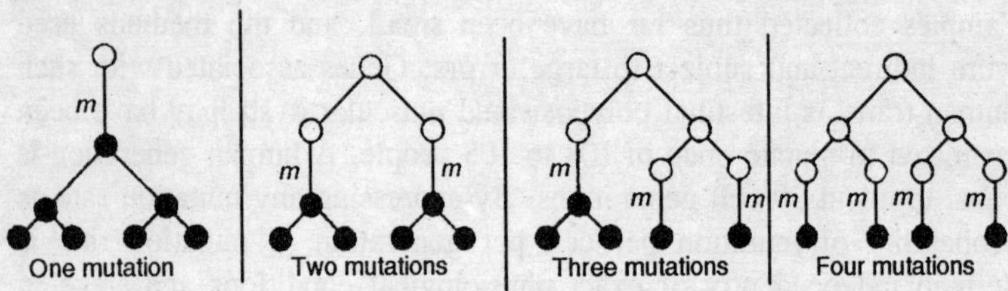

Fig. 5.6. A given number of mutant organisms in a population can result from various numbers of mutational events, depending on when the mutation occurs during the exponential growth of the population.

$$dM = \mu dN$$

These equations can be integerated and used to obtain estimates of mutation rates in experimental populations. More frequently, however, populations are analyzed in terms of allele frequencies, or changes in allele frequencies due to mutation, selection, or migration. Moreover, mutation is not a unidirectional process; reverse mutations occur. A more realistic picture is thus

$$a^{+} = \Leftrightarrow a$$

in which the "wild-type" allele a+ "wild-type" gene to a mutant form (usually a nonfunctional or only slightly functional form) is investigated, what is actually measured is the summation of many different mutation events at many different sites in the gene. The "average gene" is usually considered to be a DNA sequence of about 1000 nucleotide pairs coding for polypeptide that is about 333 amino acids long. Amino acid changes at many different positions in the polypeptide (resulting from base-pair substitutions at many different positions in the structural gene) may be expected to result in an inactive gene product. Thus, when one measures the mutation rate of a "wild-type" gene to the nonfunctional mutant from, one will actually be measuring the sum of many different mutations, each occurring at its own specific rate.

Mutation "Load" Versus Genome Size

In the preceding section, an average mutation rate of 1 per 100,000 (or 10^{-5}) per gene per generation was cited for Drosophila. Estimates of the total mutation rate—the summation of mutation rates of the entire genome - in Drosophila are of the order of 5 percent (or 5×10^{-2}) per generation (per haploid complement, or per gamete). These values can be used to estimate the number of genes in the Drosophila genome (haploid complement). If the mutation rate per gene in m and the mutation rate per total genome is M, then the number of genes in the genome (N), will equal M/m. Using the above estimates, N = 5 × 10–2/10–5 = 5000. This estimate agrees well with estimates of the number of genes in Drosophila obtained by other genetic criteria. On the other hand, one biochemical estimate suggests that Drosophila has about three times that many genes.

In the Drosophila genome (haploid) contains only 5000–15,000 genes, there is a major difference in the organization of the Drosophila genome and the genomes of the extensively studied bacteria and viruses. This difference, in fact, appears to be a basic difference between prokaryotes and eukaryotes. In prokaryotes, essentially all of the DNA of the genome appears to represent structural genes. Estimates of the number of genes from mutation studies (in some case, identification of all or most of the genes of the organs) agree well with estimates obtained by dividing the total DNA content in nucleotide pairs) of the genome by 1000 (103 nucleotide pairs per "average" gene). The phage lambda chromosome consists of 45 × 10^3 nucleotide pairs and about 40 known genes. The phage T4 chromosome contains about 200 × 10^3 nucleotide pairs. Over 100 genes have been mapped in phage T4, and a significant number remain to be identified. The E. coli genome contains about 4000 × 10^3 nucleotide pairs, while mutation analyses suggest that it contains about 3000 genes. In these prokaryotic organisms, then, the genomes are composed almost entirely of structural genes. The Drosophila melanogaster haploid chromosome complement contains about 120,000 × 10^3 nucleotide pairs. If all of the DNA represented structural genes of average size (103 nucleotide pairs), then Drosophila should have about 120,000 genes, rather than 5000 to 15,000 genes as indicated by the calculation above.

It now seems clear the most of the DNA of Drosohila and other eukaryotes, including humans, does not represent structural genes. What is the function of this "excess" or noncoding DNA? Many geneticists believe that this noncoding DNA is regulatory, playing important roles in the regulaion of gene expression. Others believe that it has an important role in some aspect of chromosome structure. Still others believe that much of the noncoding DNA is just "junk" DNA with no important function, possibly a reservoir of nucleotide-pair sequences available for the evolution of new genes. In any case, this DNA does not code for proteins. This "noncoding" DNA is of two types: DNA that is transcribed, but for which the transcripts never leave the nuclei, and (2) nontranscribed DNA. These two types of DNA may have very different functions. The first type of noncoding DNA clearly includes many of the noncoding intervening sequences or introns of eukaryotic genes.

An estimate of the maximum number of essential genes per genome can also be made from considerations of average mutation rates and the maximum mutation "loads" that a species could be expected to survive. For example, consider an average mutation rate of 4×10^{-6} recessive lethal mutation per essential gene per generation (per gamete). J. L. King, and others, have emphasized that it is unlikely that a species could tolerate more than 0.8 new lethal mutation per zygote (or 0.4 per gamete). Dividing 0.4 new recessive lethal mutation per gamete by 4×10^{-6} new recessive lethal mutation per essential gene per gamete gives a maximum number of 10^5 essential genes per gamete. At 10^3 nucleotide pairs per gene, the 105 essential genes would represent a total for 10^8 nucleotide pairs. The haploid chromosome complement of mammals, including humans, contains about 3×10^8 nucleotide pairs. Thus, according to thee estimates, essential genes can represent a maximum of a only about 3 percent of the total DNA in the genome of mammals. The actual number of essential genes in mammals is probably closer to $1–2 \times 10^4$, or less than 1 percent of the total DNA.

Mutable Genes

Most gene are relatively stable and mutate infrequently, but in some organisms a few genes mutate spontaneously so often that individuals carrying them are mosaics of mutated and unmutated genes. The R locus, which is involved with anthocyanin pigment synthesis in maize, for example, was found by R.A. Emerson to undergo alteration much more frequently than other loci. The change from Rr to rr, for example, was found to occur, at the rate of 1 per 20,000 gametes. These "mutable" genes are either more unstable than others or they are influenced by other factors in the genetic environment. McClintock's conclusion from extensive studies on maize was that a mutable gene is not an autonomous entity, but is derived from an agent that is integrated at the site of the mutable gene to cause instability.

Examples of highly mutable genes have been found in both plants and animals, but they appear to be more common in plants. Mutations in highly mutable genes occur frequently in somatic tissue and occasionally in germ cells. Somatic mutations may show their effects as colour variegations (mosaics) in such plant part as endosperm, leaves, and petals. Many common plants—including the larkspur, snapdragon, sweet pea, four-o'clock, and morning glory—have colour variegations suggesting unstable or mutable genes.

The classical investigations of M. Demerec (1941) on a mutable gene called miniature-alpha in Drosophila virilis provided the first substantial data on the genetic properties of mutable gens in animals. These propertie are: (1) mutation occurs primarily before meiosis; (2) mutation occurs in both females and males, implying that meiotic crossing over is not involved (crossing over occurs only rearely in Drosophila males) and (3) mutation is strongly influenced by neighbouring genes. More recently, M.M. Green and others have described several mutable gene systems involving the white locus (eye colour) and other genes in D. melanogaster. White-crimson (w^c), for example, mutates to wild-type and phenotypes other than white-crimosn at a frequency of 10–3. This mutable system and several others at the w locus are associated with chromosome alterations, particularly deficiencies. This led to the hypothesis that a "*controlling element*" from the cytoplasm is integrated into the chromosome at the site of mutability, and that this agent is responsible for chromosome aberrations.

Many of these "controlling elements" have now been identified by G. Rubin, P. M. Bingham, and colleagues, and shown to be transposable elements with structures similar to the TN elements of bacteria. The unstable allele ww^c mentioned above has been shown to result from the insertion of an approximately 10,000 nucleotide-pair-long transposable element belonging to the class called FB (for "foldback") elements.

Another unstable allele, w^{a1} has been shown to have a transposable element called *copia* (because its sequence is present in RNA in *copious* amounts) inserted at the site of mutation. Copia has been isolated and sequenced. It is about 5,000 nucleotide pairs long with 276 nucleotide-pair perfect direct repeats at its ends. These 276 nucleotide-pair repeats, in turn, have 17 nucleotide-pair inverse repeats at each end. Note the similarity of *copia* to E. coli elements Tn 9 and Tn 10.

Mutator and Antimutator Genes

Genes in maize were shown by McClintock to influence the stability of other genes. These have been called mutator genes. A striking example of the action of a mutator gene with a specific effect on a basic colour gene in maize was described by M. M. Rhoades. The colour of maize leaves and other plant parts is dependent on a complex of three complementary genes, symbolized A, C and R. A colour other than green is produced only when the dominant alleles (A, C, and R) of all three genes are present. The colour may be purple if gene P is also present, or red (pp), or some other colour depending on what other genes are included along with A-C-R. Plant soft genotype aa, cc, or rr have green leaves regardless of alleles present at the other loci. Plants with the genotype aaC-R- would be expected to be green, but in the presence of a mutator gene called Dt, they are variegated. Light-coloured corn kernels have purple spots at locations where somatic mutations have occurred. The Dt gene produces its effect by influencing one or more of the a alleles of the aa genotype to mutate to A. Patches of cells scattered throughout the plant carry A alleles resulting from these mutate to A. Patches of cells scattered throughout the plant carry A alleles resulting from the mutations. On the leaves and kernels, these patches give a speckled appearance. The size of the spots depends on the stage of development at which the mutations occurred. Green cells contain unmutated genes; that is, they are of genotype aa. In this case, the allele (Dt or dt) present at one locus thus influences the mutation rate of an allele (a) present at another locus.

J. F. Speyer and others have found broad-spectrum mutator genes in bacteriophage T4. Temperature sensitive (ts) mutants of gene 43, for example, alter the mutation rates of point mutations in other genes. Some mutator genes exert their mutagenic effect during DNA replication by altering polymerase activity, producing mutagenic base analogs, or modifying DNA bases, thus influencing the mutation rates of other genes. One mutator in phage T4 gene 43 (ts L88) produces a DNA polymerase that utilizes incorrect nucleotides at a higher frequency than does the wild-type enzyme.

Several mutations in gene 43 exhibit powerful negative or antimutator activities, particularly during the formation of A : T $\rightarrow$ C : G substitutions. Experiments have shown that DNA polymerase (gene 43 products) that are isolated from E. coli that were infected with mutator, antimutator and wild-type strains for T4 bacteriophage discriminate between adenine and 2-aminopurine to different degrees during DNA synthesis in vitro. Significantly larger amounts of 2-aminopurine are incorporated into DNA by wild-type and mutator than by antimutator enzymes. This indicates that organisms have the potential to evolve mechanism that result in mutation rates that are lower than those presently existing in wild-type strains of somatic organisms.

When the mutator and antimutator DNA polyemerase of phage T4 were studied in vitro, they were found to have altered ratios of polymerase activity to 3'→5' exonuclease activity. Mutator have increased polymerase/exonuclease activity. Antimutators exhibit decreased polymerase/exonuclease activity. This suggests that the mutation rate is, at least in part, controlled by the relative rates of polymerization and proofreading rates decrease mutation rates (antimutator). Decreased proofreading efficiencies (or increased polymerization rates) increase mutation frequencies (mutator).

Several mutator genes have also been identified and characterized in E. coli.Some increase the frequency of only certain types of base-pair substitutions. Others cause an increase in all types of point mutations. R. C. von Borstell and others measured spontaneous mutation rates in yeast and isolated many mutator genes with different kinds of mutator and antimutator activity.

Decreases as well as increases in mutation rates, as compared with wild-type, can thus result from new mutations. With both mutators and antimutaors operating in the same system, particular spontaneous mutation rates may be optimized through natural selection. Mutators and antimutators thus become relative terms, because no standards are available for optimum rates. Results of studies on yeast have borne out the conclusions of those on viruses and bacteria; that spontaneous mutation rates are under the genetic control of the cell itself.

Insertion Sequence ("Is Elements") As Mutators

A significant proportion of the mutations that occur in bacteria and some bacteriophages are now known to be effects of small sequences of DNA called IS elements or insertion sequences. These DNA sequences, from about 800 to about 1400 nucleotide paris in length are present in E. coli chromosomes, for example, in several copies. The first five IS elements to be characterized were: ISI (768 nucleotide pairs long) and IS2, IS3, IS4, and IS5 (all between 1200 and 1400 nucleotide pairs in length). The IS elements are transposable; that is, they have the ability to move from one position in the chromosome or to other chromosomes in the same cell. They can move, for example, from the E. coli chromosome to bacteriophage chromosomes or to plasmids (small circular molecules of DNA, or "minichromosomes," often present in bacteria.

The chromosome of the strains of E. coli that have been studied contain about eight copies of ISI elements are physically and covalently inserted as linear sequences into the chromosomes when an IS element is inserted (or insert itself?) into a gene, it destroys the function of, mutates, that gene. While insertion of IS elements occur at many sites in the E. coli chromosomes, it is not completely at random. Preferred sites of insertion have been demonstrated in some case.

Detectable IS-induced mutations occur with a frequency of about 10–6 to 10–7 per cell generation in E. coli. IS-induced mutations are revertible. They can be distinguished from all mutations with similar effects in that their reversion frequency (usually in the range 10–6 to 10–7) can not be enhanced with mutagens. The fact that IS-induced mutations are revertible means that IS elements can excise themselves from the chromosome with precision to the single nucleotide pair. Reversion to the functional state requires restoration of the original nuclotide-pair sequence of the gene; no base-pairs can be added or deleted during the insertion-excision process. The failure of mutagens to hence reversion is to be recombination mechanism, not a mutation process.

The "controlling elements" and mutable loci (or unstable genes) in higher plants and animals are probably caused by sequences of DNA analogous to the IS elements in prokaryotes.

Induced Mutations

Mutations can be induced by either physical or chemical means: such mutations are called induced mutations. Irradiation is an example of a physical mutagen, with x-rays, gamma-rays, and ultraviolet light being the most common examples used. One consequence of x- or gamma-ray irradiation is the breakage of chromosomes, which may result in chromosomal rearrangements, or the events may be lethal to the cell.

5-Bromouracil

5-Bromouracil (5-BU) is a base analog; that is, its structure closely resembles one of the bases normally found in DNA. 5-BU can exist in two states. In its usual keto state it exhibits properties similar to those of thymine and thus will pair with adenine in DNA. Rarely it switches to the enol state, and in this form it will pair specifically with guanine.

Mutations can be induced by 5-BU (and in general by base-analong muta-gens) in two ways. The first involves the incorporation of the usual form of 5-BU into DNA during replication. If 5-BU shifts to its rare enol state during the next round of replication, the result will be a transition mutation from AT to GC.

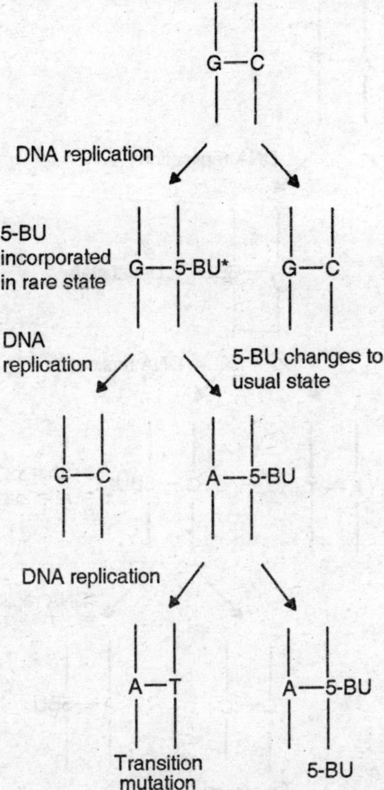

Fig. 5.7. Mutagenic action of 5-BU when it incorporates into DNA in its usual keto state and then shifts to its rare state during the next round of replication.

The second way that 5-BU can induce mutations is if the base analog g is incorporated in the DNA while it is the rare enol state. This dictates insertion opposite a G on the complementary strand, and subsequent replication with a shift of the 5-BU to the usual keto state will result in a transition mutation from GC to AT.

Thus 5-BU can induce either AT to GC or CG-to-AT transition mutations. (In the jargon of this area, 5-BU is said to induce two-way transition mutations.) Therefore it is possible to correct a 5-BU induced transition mutation by treating with 5-BU for a second time. This is called *reversion* of the mutation.

2-Aminopurine

2-Aminopurine (2-AP) is also a base analog and, like 5-BU, it can exist in two states. In its usual state it behaves like adenine and will

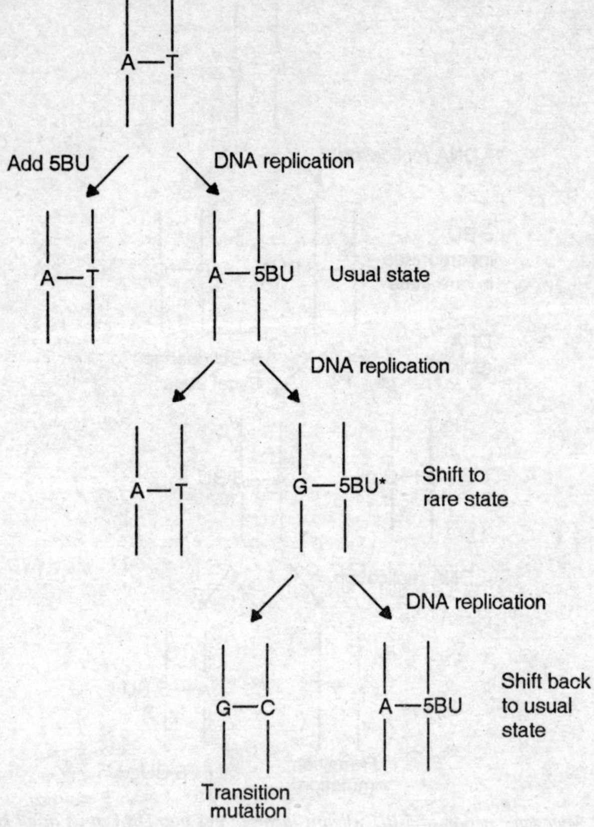

Fig. 5.8. Mutagenic action of 5-BU when it incorporates into DNA in the rate state and then shifts to the usual keto state during the next round of replication.

form two hydrogen bonds with thymine. In its rare amino state 2-AP behaves like guanine and forms two hydrogen bonds with cytosine. Thus 2-AP can induce transition mutations both from AT-to-GC and from GC-to-AT. 2-AP induced mutations can therefore be reverted by 2-AP treatment.

Nitrous Acid

Nitrous acid (NA : HNO_2) is a deminating, agent; it acts by removing amino groups (NH_2) from the bases. In some but not all instances this alters their base-pairing abilities and hence induces mutations. The three bases that have amino groups are adenine, guanine, and cytosine. When adenine is treated with NA, it is changed to hypoxanthine, which will pair with cytosine. This results in an AT-to-a

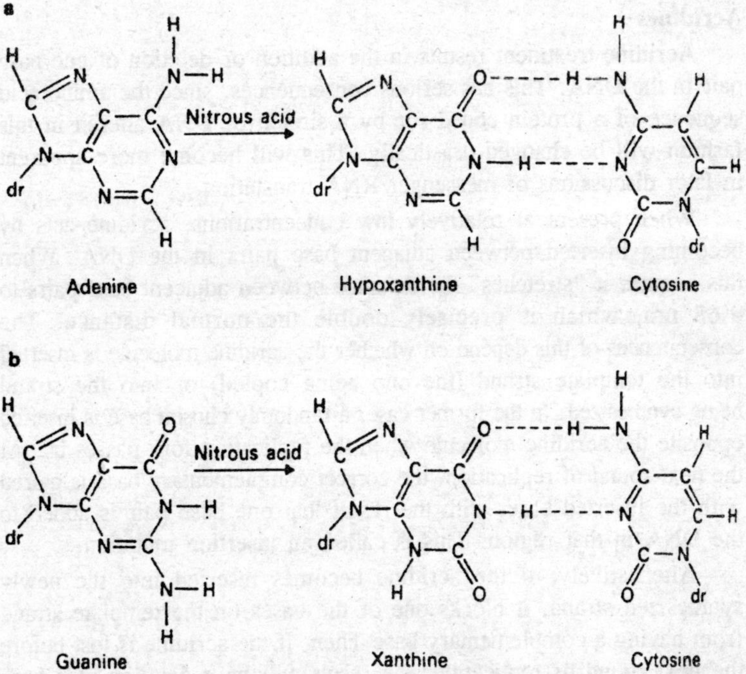

Nitrous acid

Adenine Hypoxanthine Cytosine

b

Nitrous acid

Guanine Xanthine Cytosine

Fig. 5.9. Mutagenic action of nitrous acid. (a) Deamination of adenine by nitrous acid treatment produces hypoxanthine, which pairs with cytosine, (b) Deamination of guanine produces xanthine, which pairs with cytosine.

GC transition mutation. Treatment of guanine with NA removes the amino group from the 2-carbon position and produces xanthine. However, since both guanine and xanthine pair with cytosine, no base-pair mutation results.

Deamination of cytosine by NA produces uracil, which, or course, pairs with adenine. This results in a GC-to-AT transition mutation–the opposite of NA's effect on adenine. Thus mutations induced by nitrous acid can be reverted by nitrous acid treatment: in other words, nitrous acid induces two-way transition mutations.

Hydroxylamine

Hydroxylamine (NH_2OH) reacts only with cytosine, hydroxylating it so that it can then only with adenine . Thus hydroxlamine induces one-way transition mutation from GC to AT. Because of this, mutations induced by hydroxylamine cannot be reverted to revert by 5-BU. 2-AP, or NA treatment since these mutagens can bring about a GC-to-AT transition.

Acridines

Acridine treatment results in the addition or deletion of one base pair in the DNA. This has serious consequences, since the amino acid sequence of a protein coded for by a stretch of DNA altered in this fashion will be changed drastically. This will become more apparent in later discussions of messenger RNA translation.

When present at relatively low concentrations, acridine acts by becoming inserted between adjacent base pairs in the DNA. When this occurs, it "stretches" the distance between adjacent base pairs to 0.68 nm, which is precisely double the normal distance. The consequences of this depend on whether the acridine molecule is inserted into the template strand (the one being copied) or into the strand being synthesized. In the former case, a randomly chosen base is inserted opposite the acridine molecule when the replication fork passes by. At the next round of replication, the correct complementary base is paired with the inserted base, with the result that one base pair is added to the DNA in that region. This is called an insertion mutation.

Alternatively, if the acridine becomes inserted into the newly synthesized strand, it blocks one of the bases on the template strand from having a complementary base Then, if the acridine is lost before the next round of replication, the result will be a deletion of a base pair. Therefore it is possible to revert an acridine-induced mutation by a second treatment with acridine.

In summary, the mutagens described in this section have different modes of action and cause different mutation changes.

Table 5.2. Summary of the method of action of various chemical mutagens.

Mutagen	*Base-pair changes*
5-Bromouracil	AT ↔ GC two-way transitions
2-Aminopurine	AT ↔ GC two-way transitions
Nitrous acid	AT ↔ GC two-way transitions
Hydroxylamine	GC → AT one-way transitions
Acridines	+1 or −1 insertion or deletion

Some of the chemical mutagens described in this section are commonly used in the laboratory. Many other chemicals appear to cause mutations, and public awareness of this is increasing as industrial effluents, cosmetic ingredients, food additives, etc. are examined carefully for any mutagenic activity in test organisms.

6

Gene Structure

This chapter provides some basic background in molecular biology that is required for studying evolutionary processes at the DNA level. The genomic structure of prokaryotic and eukaryotic genes and genetic codes are described, as are some basic properties of amino acids and proteins. Furthermore, various types of mutation and the pattern of point mutation inferred from nucleotide substitutions in noncoding sequences are discussed. Additional molecular biology background will be provided in the relevant chapters.

Nucleotide Sequences

The hereditary information of all living organisms, with the exception of some viruses, is carried by *deoxyribonucleic acid* (DNA) molecules. DNA usually consists of two complementary chains twisted around each other to form a right-handed helix. Each chain is a linear polynucleotide consisting of four nucleotides. There are two *purines*: *adenine* (A) and *Guanine* (G), and two *pyrimidines*: *cytosine* (C) and *thymine* (T). The two chains are joined together by hydrogen bonds between pairs of nucleotides. Adenine pairs with thymine by means of two hydrogen bonds, also referred to as the *weak bond*, and guanine pairs with cytosine by means of three hydrogen bonds, the *strong bond*.

Each nucleotide in a DNA sequence contains a pentose sugar (deoxyribose), a phosphate group, and a purine or a pyrimidine base. The backbone of the DNA molecule consists of sugar and phosphate moieties, which are covalently linked together by asymmetrical 5'—3' phosphodiester bonds. Consequently, the DNA molecule is polarized, one end having a phosphoryl radical (—P) on the 5' carbon of the terminal nucleotide, the other possessing a free hydroxyl (—OH) on

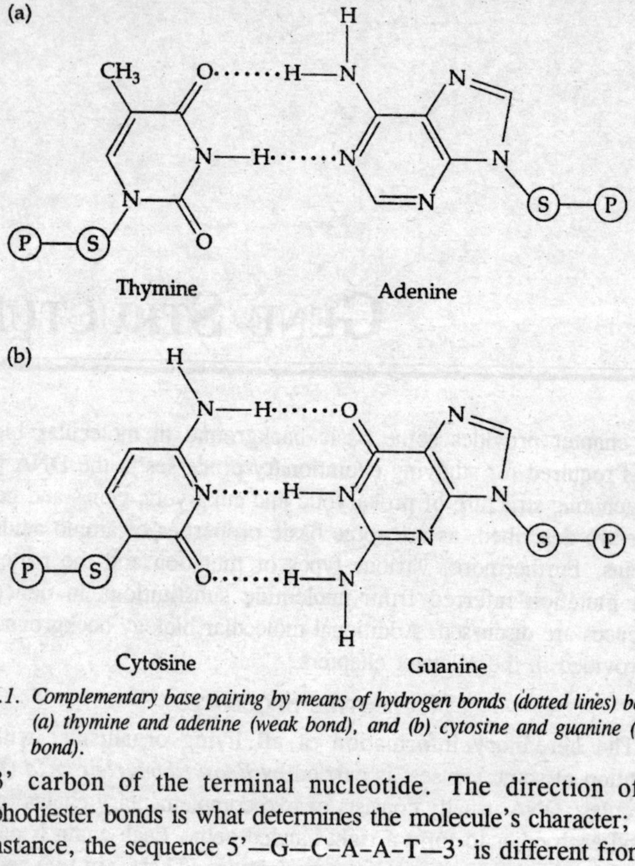

(a)

Thymine Adenine

(b)

Cytosine Guanine

Fig. 6.1. Complementary base pairing by means of hydrogen bonds (dotted lines) between (a) thymine and adenine (weak bond), and (b) cytosine and guanine (strong bond).

the 3' carbon of the terminal nucleotide. The direction of the phosphodiester bonds is what determines the molecule's character; thus, for instance, the sequence 5'—G—C-A-A-T—3' is different from the sequence 3'—G-C-A-A-T—5'. By convention, DNA sequences are written in the order they are transcribed, i.e., from the 5' end to the 3' end, also referred to as the *upstream* and *downstream* directions, respectively. The double helical form of DNA has two strands in antiparallel array.

Ribonucleic acid (RNA) is found as either a double- or a single-stranded molecule. RNA differs from DNA by having ribose as its backbone sugar moiety, instead of deoxyribose, and by using the nucleotide uracil (U) instead of thymine (T). In RNA, in addition to the canonical complementary base pairs G:C and A:U, the G:U pair is also stable; in DNA, stable pairing does not occur between G and T.

Adenine, cytosine, guanine, and thymine/uracil are referred to as the standard nucleotides. Some functional RNA molecules, most notably

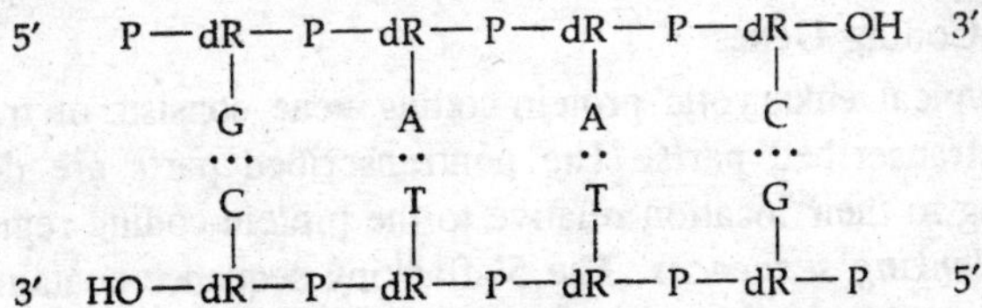

Fig. 6.2. Schematic representation of the antiparallel structure of double-stranded DNA. P, phosphate; dR, deoxyribose; OH, hydroxyl; A, adenine; G, guanine; C, cytosine; T, thymine; —, covalent bond; ··, weak bond; ···, strong bond.

tRNAs, contain nonstandard nucleotides (i.e., chemical modifications of standard nucleotides) that have been introduced into the RNA after transcription.

Gene Structure

Traditionally, a gene was defined as a segment of DNA that codes for a polypeptide chain or specifies a functional RNA molecule. Recent molecular studies, however, have altered our perception of genes, making a somewhat vaguer definition necessary. Accordingly, a gene is a sequence of genomic DNA or RNA that performs a specific function. Performing the function may not require the gene to be translated or even transcribed.

At present, three types of genes are recognized: (1) *protein-coding genes*, which are transcribed into RNA and subsequently translated into proteins, (2) *RNA-specifying genes*, which are only transcribed, and (3) *regulatory genes*. According to a narrow definition, the third category includes only untranscribed sequences. Transcribed regulatory genes essentially belong to one of the first to categories. Protein-coding genes and RNA-specifying genes are also referred to as *structural genes*. Note that some authors restrict the definition of structural genes to include only protein-coding genes.

The transcription of both protein-coding genes and RNA-specifying genes in bacteria is carried out by one type of DNA-dependent RNA polymerase. In contrast, in nuclear genomes of eukaryotes, three types of RNA polymerases are employed. Ribosomal RNA (rRNA) genes are transcribed by RNA polymerase I (Pol I), protein-coding genes by RNA polymerase II (Pol II), and small cytoplasmic RNA (scRNA) genes, such as genes specifying transfer RNAs (tRNAs), by RNA polymerase III (Pol III). Some small nuclear RNA (snRNA) genes (e.g., U2) are transcribed by Pol II, others (e.g., U6) by Pol III. The snRNA gene U3 is transcribed by Pol II in vertebrates and lower eukaryotes, but by Pol III in plants.

Protein-Coding Genes

A typical eukaryotic protein-coding gene consists of transcribed and nontranscribed parts. The nontranscribed parts are designated according to their location relative to the protein-coding regions as 5' and 3' *flanking sequences*. The 5' flanking sequence contains several signals (specific sequences) that determine the initiation, tempo, and timing (or tissue specificity) of the transcription process. Because these regulatory sequences promotes the transcription process, they are also referred to as *promoters*, and the region in which they reside is called the *promoter region*. The promoter region usually consists of the following signals: the *TATA box* located 19-27 base pairs (bp) upstream of the starting point of transcription, the *CAAT box* farther upstream, and one or more copies of the *GC box*, consisting of the sequence GGGCGG or its variants and surrounding the CAAT box. The TATA box, which is usually surrounded by GC-rich sequences, controls the choice of the starting point of transcription. The CAAT and GC boxes, which may function in either orientation, control the initial binding of RNA polymerase II to DNA. Note, however, that none of the above signals is uniquely essential for promoter function. Some genes do not possess a TATA box and thus may not have a unique starting point of transcription; for example, many housekeeping-enzyme genes such as the hypoxanthine phosphoribosyl transferase and dihydrofolate reductase genes have a GC-rich region instead of the TATA box. Other genes possess neither a CATT box nor any GC box, and their transcription

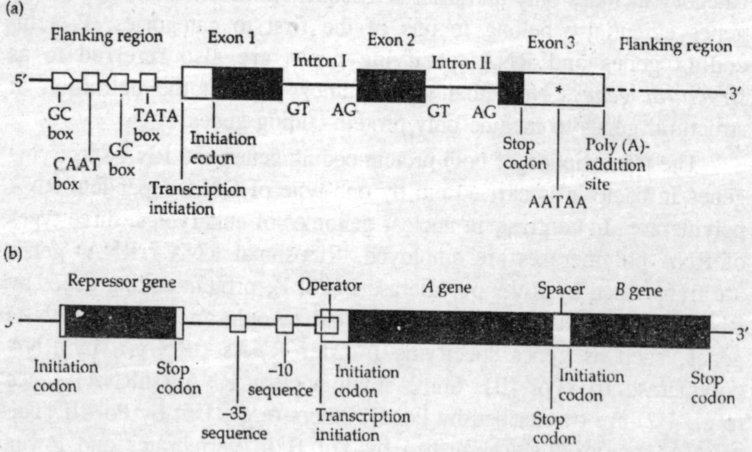

Fig. 6.3. (a) Schematic structure of a typical eukaryotic protein-coding gene. (b) Schematic structure of an induced prokaryotic operon.

initiation is controlled by other elements in the 5' flanking region. The 3' flanking sequence contains signals for the termination of the transcription process and poly(A)-addition. Presently, it is still difficult to delineate with precision the points at which a gene begins and ends.

In addition to the signals in the promoter region, there is another type of signal, called the enhancer, that is located outside the promoter region but can enhance the initiation of transcription. An enhancer may consist of several modular elements, may function in either direction, and can be in the 5' flanking region, in an intron, or in the 3' flanking region. Such elements may be targets for tissue-specific or temporal regulation.

The transcription of protein-coding genes in eukaryotes starts at the *transcription initiation site* or (the *cap site* in the RNA transcript), and ends at the *termination site*, which may or may not be close to the *polyadenylation* or *poly(A) addition site* of the mature messenger RNA (mRNA) molecule. In other words, termination of transcription may occur farther downstream from the poly(A) addition site. The transcribed RNA, also referred to as *pre-messenger RNA* (*pre-mRNA*), contains 5' and 3' *untranslated regions*, *exons*, and *introns*. Introns, or intervening sequences, are those transcribed sequences that are excised during the processing of the pre-mRNA molecule. All genomic sequences that remain in the mature mRNA following splicing are referred to as exons. Exons or parts of exons that are translated are referred to as protein-coding exons or *coding regions*. The DNA strand from which the pre-mRNA is transcribed is called the antisense strand. The untranscribed complementary strand, the sequence of which is identical to that of the pre-mRNA is transcribed is called the antisense strand. The untranscribed complementary strand, the sequence of which is identical to that of the pre-mRNA, is called the sense strand.

There are several types of introns, characterized by the specific mechanism with which the intron is cleaved out of the primary RNA transcript. Here we are concerned only with the introns in the nuclear genes that are transcribed by RNA polymerase II. These introns are called *spliceosomal* introns and are enzymatically cleaved out of the pre-mRNA during the maturation of the molecule. They are different from group I and group II introns, which have self-splicing activities and are found in mitochondrial and chloroplast DNA. The *splicing site* or *junctions* of spliceosomal introns are probably determined by nucleotides at the 5' and 3' ends of each intron, known as the *donor* and *acceptor* sites, respectively. For instance, all eukaryotic nuclear introns begin with GT and end in AG (the *GT-AG rule*), and these sequences have been shown to be essential for correct excision splicing, though there are exceptional cases. Exon sequences adjacent to introns may contribute to the determination of the splicing site. In addition, each intron contains a specific sequence, called the *TACTAAC box*, located about 30 bp upstream of the 3' end of the intron. This sequence is well conserved in yeast, though it is more variable in higher eukaryotes. The splicing involves the cleavage of the 5' splice junction and the creation of a phosphodiester bond between the G at the 5' end of the intron and the A in the sixth position of the TACTAAC box. Subsequently, the 3' splice junction is cleaved and the two exons are ligated.

The number of introns varies from gene to gene. Some genes possess dozens of introns, and some introns may be thousands of nucleotides long. Others (e.g., most histone genes) are devoid of introns altogether. Exons are not distributed evenly over the length of the gene. Some exons are clustered; others are located at great distances from neighbouring exons. The majority of nuclear protein-coding genes in higher eukaryotes consist mostly of introns. Not all introns interrupt coding regions. Some occur in untranslated regions, mainly in the region between the transcription initiation site and the initiation codon.

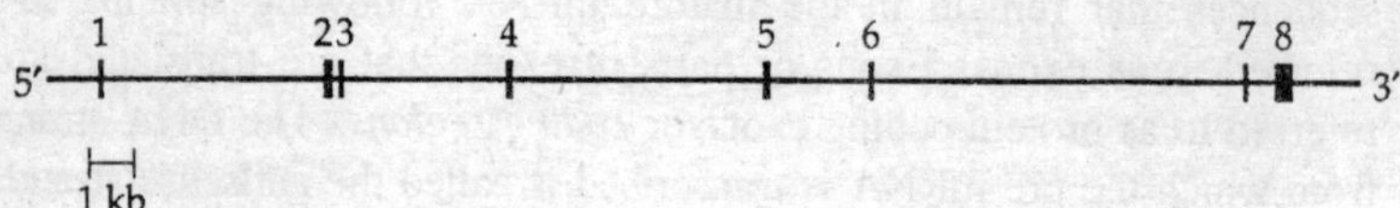

Fig. 6.4. The localization of the eight exons in the human factor-IX gene.

Protein-coding genes in eubacteria are different from those in eukaryotes in several respects. Most importantly, they do not contain introns. Promoters in eubacteria contain a –10 sequence and –35 sequence, so named because they are located, respectively, 10 bp and 35 bp upstream of the initiation site of transcription. The former, also known as the *Pribnow box*, has the sequence TATAAT or its variant, while the latter has the sequence TTGACA or its variant. A prokaryotic promoter may also contain other specific sequences farther upstream from the –35 sequence.

Several structural genes in prokaryotes may be arranged consecutively to form a unit of gene expression (a polycistron) that is transcribed into one molecule of mRNA and subsequently translated into different proteins. Such a unit usually contains genetic elements that control the coordinated expression of the genes belonging to the unit. The entire arrangement of genes is called an *operon*.

RNA Specifying Genes

The structure of RNA-specifying genes is usually similar in eukaryotes and prokaryotes. These genes generally contain no introns. However, in some organisms, such as ciliates, slime molds, and some bacteria, RNA-specifying genes may contain introns that must be spliced out before the RNA molecule becomes functional. Sequence elements involved in the regulation of transcription of some RNA-specifying genes are sometimes included within the sequence specifying the functional end product. In particular, all eukaryotic tRNA specifying genes contain an internal transcriptional start recognized by RNA polymerase III.

Many RNA molecules are modified following transcription. Such modifications include the incorporation of standard and nonstandard ones, and the enzymatic addition of terminal sequences of ribonucleotides to either the 5' or the 3' end.

Regulatory Genes

Our knowledge of regulatory genes is much less advanced than that of the other types of genes. Traditionally, these genes would be called regulatory signals or sequences because they are not transcribed. Several such types of genes or families of genes have been tentatively identified. (1) *Replicator genes* specify the sites for initiation or termination of DNA replication. For example, this term may apply to the specific sequences at the origin of each replicon, which is a unit of DNA replication in a genome. These sequences may serve as binding sites for specific initiator or repressor molecules. (2) *Telomeres* are short repetitive sequences at the end of eukaryotic chromosomes; e.g., the repeat sequence is CCCTAA in humans and CCCTAAA in the plant *Arabidopsis*. The telomere provides a protective "cap" for the end of a chromosome, so that it is not "sticky" and does not react with the end of another chromosome and is not susceptible to exonucleolytic degradation. (3) *Segregator genes* provide specific sites for attachment of the chromosomes to the segregation machinery during meiosis and mitosis. This type of gene includes centromeric sequences, which have been found to consist of short repetitive sequences similar to those of telomeres. (4) *Recombinator genes* provide specific recognition sites for recombination enzymes during meiosis. Many more regulatory genes may exist, some of which could be independent of structural genes in both function and location, and could be involved in complex regulatory functions, such as the ontogenetic development of multicellular organisms.

Amino Acids

Amino acids are the basic structural units of proteins. All proteins in all organisms, from bacteria to humans, are constructed from the 20 amino acids, for rare and posttranslationally modified amino acids of proteins.

Each amino acid has an NH_2 (amine) group and a COOH (carboxylic) group on either side of a central carbon atom, which is called the α carbon or C_α.. Also attached to the α carbon are a hydrogen atom and a distinct *R group*, which is also referred to as a *side chain*. The side chain varies in size, shape, charge, hydrogen-bonding capacity, and chemical reactivity, and it is the side chain that makes one amino acid different from another. In fact, classification of amino acids is usually made on the basis of their R groups.

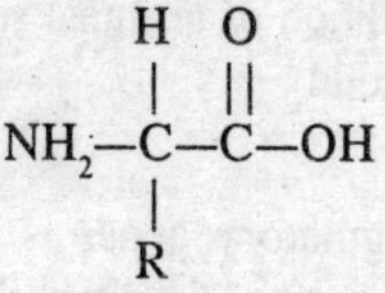

```
     H   O
     |   ||
NH2—C—C—OH
     |
     R
```

Fig. 6.5. Structure of an amino acid.

Based on the polarity of their R groups at pH 6.0 to 7.0 (the zone of intracellular pH), the 20 amino acids can be classified into four main classes: (1) *nonpolar* or *hydrophobic* groups; (2) *neutral* (*unchanged*) *polar* groups; (3) *positively charged* groups; and (4) *negative charged* groups. However, as well shall see, the four groups are not clear-cut and certain amino acids (e.g., glycine and tyrosine) straddle the boundary between categories. The following description is largely from Lehninger (1982), Dickerson and Geis (1983), and Stryer (1988).

Amino Acids with Nonpolar (Hydrophobic) R Groups

There are eight amino acids in this category. Alanine has a methyl R group; leucine, isoleucine, valine, and proline have an aliphatic (i.e., having an open-chain structure but no aromatic ring) hydrocarbon group; phenylalanine and tryptophan have an aromatic (benzene-like) ring; and methionine contains a sulfur within the hydrocarbon chain that acts like a CH_2 group. In general, these amino acids are less soluble in water than the amino acids with polar R groups.

Hydrophobic ("water-hatching") amino acids prefer to be in the internal part of a protein molecule, away form the aqueous environment of the cell. Thus, they are also called *internal* amino acids. Alanine is the least hydrophobic member of the nonpolar class and is near the borderline between nonpolar amino acids and those with uncharged polar R groups. Thus, it does not have a strong preference to be internal or *external*. Such an amino acid is said to be *ambivalent* or *indifferent*.

Proline differs from all the other standard amino acids in that its side chain loops back to make a second connection to its amine nitrogen. This ring forces a contorted bend on the polypeptide chain. Thus, proline is especially useful in reversing the direction of a peptide backbone.

Amino Acids with Uncharged Polar R Groups

This class includes seven amino acids whose R groups contain neutral (uncharged) polar functional groups that can hydrogen-bond with water. Glycine, the borderline member of this group, is sometimes classified as a non-polar amino acid, but its R group, a single hydrogen atom, is too small to influence the high degree of polarity of the α-amino and α-carboxyl groups. It is indifferent with respect to preference to be internal or external.

Serine and threonine have an aliphatic hydroxyl group. They are neither so hydrophobic nor so polar as to have a strong preference to be internal or external; they, too, are indifferent.

Tyrosine has an aromatic hydroxyl group and cysteine has an aliphatic sulfhydryl (SH) group. These R groups tend to lose protons by ionization far more readily than the R groups of other amino acids of this class. However, since they are only slightly ionized at pH 7.0, cysteine and tyrosine also belong to the indifferent group. Indeed, tyrosine is placed in the hydrophobic class by many authors. In extracellular proteins cysteine usually occurs in its oxidized form cystine, in which the thiol groups of two molecules of cysteine have been oxidized to a disulfide group to provide a covalent cross-linkage between them. In intracellular proteins, the redox potential prevents disulfide bonds from forming.

Asparagine and glutamine are derived, respectively, from aspartic acid and glutamic acid by replacing the OH group with NH_2 i.e., they are the amides of aspartic acid and glutamic acid. They are easily hydrolyzed by acid or base to aspartic acid and glutamic acid, if they are not buried inside a polypeptide. Since they have very polar side chains, they are generally found on the outside of the molecule.

Amino Acids with Negatively Charged (Acidic) R Groups

The two members of this class are aspartic acid and glutamic acid, each with a carboxyl R group that is fully ionized and thus negatively charged at pH 6 to 7. The two amino acids are hydrophilic and have a strong preference to be on the external part of a protein.

Amino Acids with Positively Charged (Basic) R Groups

Lysine has an amine-containing R group, whose nitrogen electron pair can attract a proton and acquire a positive charge. Arginine has a guanidino R group (CH_4N_2), which can acquire a proton and become positively charged. Both amino acids have a net positive charge at pH

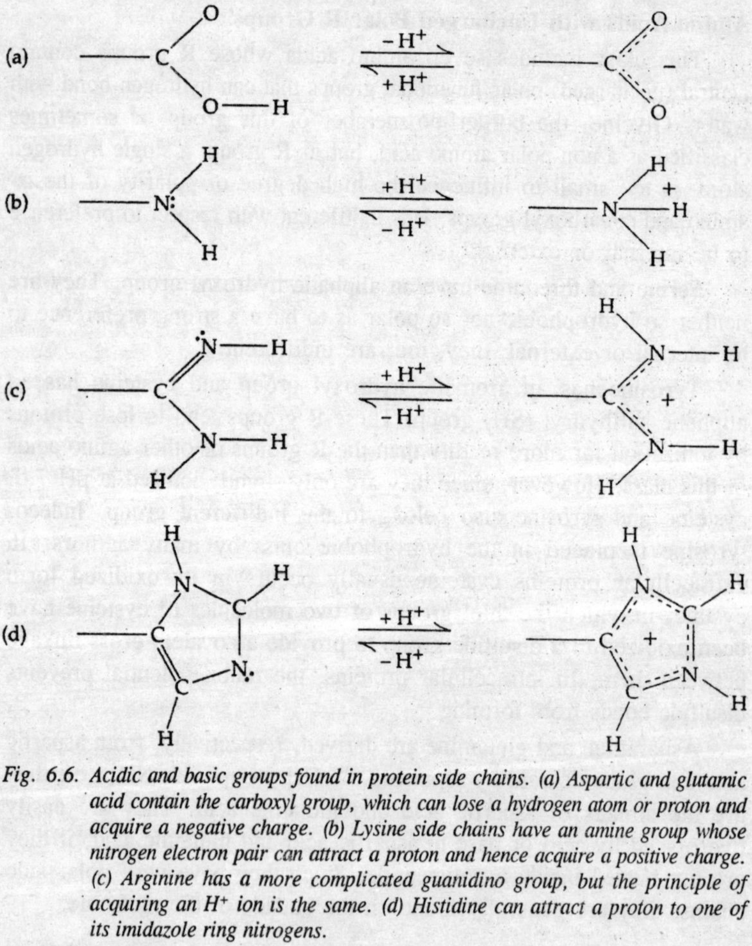

Fig. 6.6. Acidic and basic groups found in protein side chains. (a) Aspartic and glutamic acid contain the carboxyl group, which can lose a hydrogen atom or proton and acquire a negative charge. (b) Lysine side chains have an amine group whose nitrogen electron pair can attract a proton and hence acquire a positive charge. (c) Arginine has a more complicated guanidino group, but the principle of acquiring an H^+ ion is the same. (d) Histidine can attract a proton to one of its imidazole ring nitrogens.

7.0. Histidine can attract a proton to one of its imidazole ($C_3H_3N_2$) ring nitrogens. However, whether histidine is positively charged or neutral depends on its local environment. At pH 6.0, somewhat over 50 percent of histidine molecules possess a protonated, positively charged R group, but at pH 7.0, less than 10 percent have a positive charge. Histidine as well as lysine and arginine are hydrophilic and almost commonly external.

Among the 20 amino acids, glycine has the smallest molecular weight (75) because its R group has only a single hydrogen atom. This makes it particularly useful in tight corners in the interior of a protein where there is no room for a bulky group. In addition, like proline, it is an α-helix breaker. The largest amino acid is tryptophan, which has a molecular weight of 204. The average molecular weight of amino acids in proteins is around 100-110.

Two amino acids are said to be similar if they have similar physicochemical properties. These properties include polarity and volume (size). It has been suggested that composition is also an important property for determining similarity. The "composition," c, of an amino acid is defined as the atomic weight ratio of noncarbon elements in end groups or rings to carbons in the side chain. For example, for the serine side chain —COH, c = 17/12 (atomic weight of hydroxyl over that of carbon) and for lysine —$CCCCNH_2$, c = 16/48 (amino over four carbons). An exchange between two amino acids are said to be conservative if the two amino acids have similar physicochemical properties (e.g., Leu and Ile) but nonconservative if otherwise (e.g., Leu and Ser).

PROTEIN STRUCTURE

A protein consists of one or more polypeptide chains, each of which has a unique, precisely defined amino acid sequence. In a polypeptide chain, the α-carboxyl group of one amino acid is joined to the α amino by a peptide bond. For example, the formation of a dipeptide (glycylalanine) from two amino acids (glycine and alanine) with the loss of a water molecule. Note that in this dipeptide, glycine acts as the carboxylic acid, while alanine acts as the amine. If the two roles are reversed, then the dipeptide is alanylglycine and is different from the dipeptide, glycylalanine. Thus, a polypeptide has a direction; the standard presentation of a sequence is form left to right, usually in terms of the one-letter code. For instance, glycylalanine can be presented as G.A or GA and alanylglycine as A.G or AG. The end containing the free amino group (i.e., the "left" end) of a polypeptide

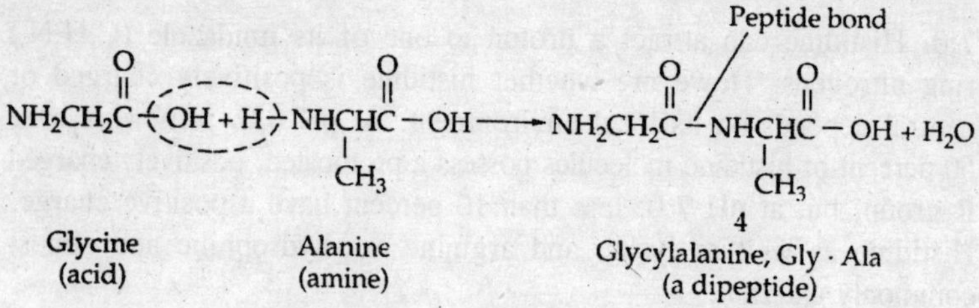

Peptide bond

$$\underset{\text{Glycine (acid)}}{NH_2CH_2\overset{\overset{O}{\|}}{C}-OH} + \underset{\text{Alanine (amine)}}{H-NH\underset{\underset{CH_3}{|}}{C}H\overset{\overset{O}{\|}}{C}-OH} \longrightarrow \underset{\text{Glycylalanine, Gly}\cdot\text{Ala (a dipeptide)}}{NH_2CH_2\overset{\overset{O}{\|}}{C}-NH\underset{\underset{CH_3}{|}}{C}H\overset{\overset{O}{\|}}{C}-OH} + H_2O$$

4

Fig. 6.7. Formation of a dipeptide from two amino acids.

is termed the amino terminal or N-terminal end, while the end containing the free carboxyl group is termed the carboxyl terminal or C-terminal end. The amino acid sequence of a polypeptide or protein is called the *primary structure*. Polypeptides vary greatly in size, usually from 50 to 1000 amino acids long. Each amino acid unit in a polypeptide is called a residue.

A polypeptide chain rarely exists in a randomly flexed, stringlike manner. Rather, noncovalent forces (e.g., those associated with hydrogen bonds) often force regions of a polypeptide into some particular flexed conformations or structures. This level of protein structure is called the *secondary structure*. Two periodic structures, the α helix and the β-pleated sheet (or simply β sheet), are of particular interest.

In the α helix, which is a rodlike structure, the main chain coils as a right-handed screw with all the side chains sticking outward in a helical array. A hydrogen bond extends from the oxygen of the CO group of each amino acid to the hydrogen of the NH group of the amino acid that is situated four residues ahead in the linear sequence. Although these hydrogen bonds are individually weak force of attraction, they collectively add up and lead to a periodic structure. In this structure each residue is related to the next residue by a translation of 1.5 Å along the helix axis and a rotation of 100 degrees. Because this rotation gives 3.6 amino acid residues per turn of helix, an interesting consequence is that amino acids spaced three or four apart in the linear sequence are spatially quite close to each other, whereas those spaced two apart are situated on opposite sides of the helix and so do not make contact.

In contrast to the tightly coiled structure in the α helix, the polypeptide chain in the β-pleated sheet is almost fully extended. In fact, the axial distance between adjacent amino acids is 3.5 Å, more than twice the distance (1.5 Å) in the α helix. The β sheet is formed when two or more polypeptide chains are laid down next to one another, either parallel or antiparallel, and then cross-connected by CO···HN hydrogen bonds at right angles to the chain direction. Note that the hydrogen bonds are now between different polypeptide chain (or different stretches of the same chain) rather than CO and NH groups located close together in the same chain. In β sheets the side chains are pushed outward as far as possible away from the sheet surface.

Some proteins such as hemoglobin, myoglobin, and collagen have a very high α helix content, whereas others such as trypsin and chymotrypsin are virtually devoid of α helix. Two or more α helices or their variants can entwine around each other to form a coiled cable. Such cables are found in several proteins: keratin in hair, myosin in muscle, epidermis in skin, and fibrin in blood clots. They serve a mechanical role in forming stiff bundles of fibres. The β sheet is found in many enzymes and is the basic structure of silk fibers. In silk fibers the protein chains lie parallel to the fiber direction and make the fibers tough and hard to break. Also, the ability of stacks of sheets to slide over one another like pages in a book makes silk fibers flexible and supple.

The three-dimensional structure of a protein is termed its *tertiary structure*. This structure is formed from the packing of local features such as helices and sheets in a polypeptide into a three-dimensional molecule by noncovalent forces, mostly hydrogen bonds. In the process of folding, amino acid residues with hydrophobic side chains tend to be buried on the inside of the protein while those with hydrophilic side chains tend to be on the outside. This tendency is particularly strong in globular proteins, which are usually in an aqueous environment. For example, in the tertiary structure of myoglobin, the polypeptide has managed to place virtually all hydrophobic residues inside and has left hydrophilic residues outside.

Another force that can affect tertiary structure is the attraction between a positively charged and a negatively charged residue. As mentioned above, aspartic and glutamic acid residues each carry a negative charge, $—COO^-$, and arginine and lysine (and sometimes histidine) residues each carry a positive charge, $—NH_3^+$. These oppositely charged residues may attract each other if they are spatially close to each other. Such a force of attraction in a protein is called a salt bridge.

Once a polypeptide chain is formed, the molecule automatically twists itself in response to forces of potential hydrogen bonds, salt bridges, and water-avoidance or water-attraction properties into the secondary and tertiary structural features. Thus, the primary sequence

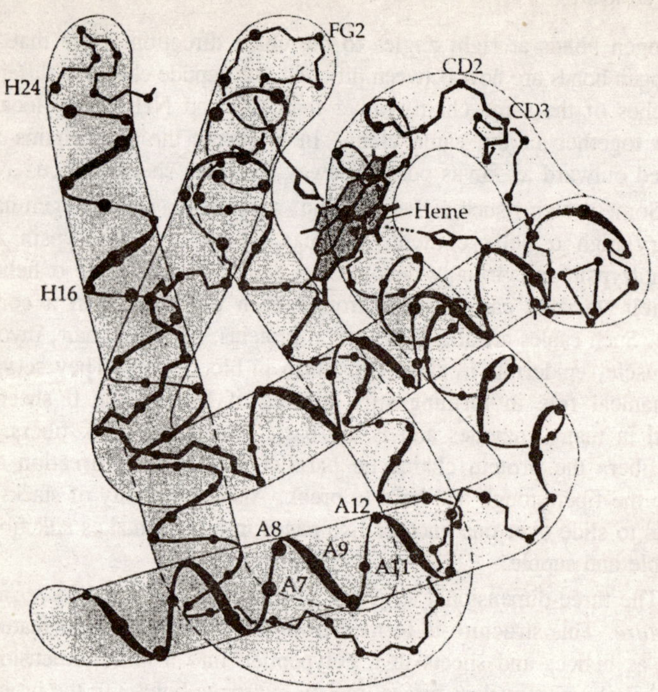

Fig. 6.8. Myoglobin, the oxygen carrier in muscles.

of a protein determines its secondary and tertiary structures. However, our knowledge of protein folding is not yet good enough to predict accurately the secondary and tertiary structures of a protein from its primary structure. This is a challenging problem in protein chemistry today.

A protein may consist of more than one polypeptide. For example, hemoglobin, the oxygen carrier in the blood, is formed from four polypeptides: two α globins, and two β globins. The final level of complexity that arises from the particular collection of polypeptides making up a protein and from the way in which the independent subunits are arranged is called the *quarternary structure*. Forces between oppositely charged sites and hydrogen bonds are involved in holding subunits together to stabilize the quaternary structure.

Genetic Codes

The synthesis of proteins involves a process of decoding, whereby the genetic information carried by the mRNA is translated sequentially into amino acids through the use of transfer RN (tRNA) mediators.

Translation starts at the translation initiation site and proceeds to a stop signal. Translation involves the sequential recognition of triplets of adjacent nucleotides called *codons*. The phase in which a sequence is translated is determined by the initiation codon and is referred to as the *reading frame*. In the translational machinery, at the interface between the ribosome and the mRNA molecule, each codon is translated into a specific amino acid, which is subsequently added to the elongating polypeptide. The correspondence between the codons and the amino acids is determined by a set of rules called the *genetic code*. With few exceptions, the genetic code for nuclear protein-coding genes is *universal*, i.e., the same code is used for the translation of almost all eukaryotic nuclear genes and prokaryotic genes.

Since a codon consists of three nucleotides and since there are four different nucleotides, there are $4^3 = 64$ possible codons. Of these, 61 code for specific amino acids and are called *sense codons*, while the remaining 3 are *nonsense* or *stop* codons that act as signals for the termination of the translation process. Since there are 61 sense codons and only 20 primary amino acids in proteins, most amino acids are encoded by more than one codon. Such a code is referred to as a *degenerate codons*. Synonymous codons that differ from each other at the third position only are referred to as a *codon family*. For example, the four codons for valine form a four-codon family. In contrast, the six codons for serine are divided into a four-codon family (UCU, UCC, UCA, and UCG) and a two-codon family (AGU and AGC).

The first amino acid in most eukaryotic proteins is a methionine encoded by the *initiation codon* AUG. This amino acid is usually removed in the mature protein. Most prokaryotic genes also use the AUG codon for initiation, but the amino acid initiating the translation process is a methionine derivative, namely formylmethionine.

The universal genetic code is also used in the translation process employed by the genomes of plastids, such as the chloroplasts of vascular plants. In contrast, most animal mitochondrial genomes and a few nuclear ones (e.g., *Mycoplasma* and *Tetrahymena*) use codes that are different from the universal genetic code. However, there are usually only minor differences between these codes and the universal genetic code. For example, in the vertebrate mitochondrial code, the two codons AGA and AGG (denoted as AGR), which specify arginine in the universal code, are used as termination codons, the codon UGA is used for tryptophan rather than for termination, and the codon AUG codes for methionine instead of isoleucine.

Table 6.1. The universal genetic code

Codon	*Amino acid*	*Codon*	*Amino acid*	*Codon*	*Amino acid*	*Codon*	*Amino acid*
UUU	Phe	UCU	Ser	UAU	Tyr	UGU	Cys
UUC	Phe	UCC	Ser	UAC	Tyr	UGC	Cys
UUA	Leu	UCA	Ser	UAA	Stop	UGA	Stop
UUG	Leu	UCG	Ser	UAG	Stop	UGG	Trp
CUU	Leu	CCU	Pro	CAU	His	CGU	Arg
CUC	Leu	CCC	Pro	CAC	His	CGC	Arg
CUA	Leu	CCA	Pro	CAA	Gln	CGA	Arg
CUG	Leu	CCG	Pro	CAG	Gln	CGG	Arg
AUU	Ile	ACU	Thr	AAU	Asn	AGU	Ser
AUC	Ile	ACC	Thr	AAC	Asn	AGC	Ser
AUA	Ile	ACA	Thr	AAA	Lys	AGA	Arg
AUG	Met	ACG	Thr	AAG	Lys	AGG	Arg
GUU	Val	GCU	Ala	GAU	Asp	GGU	Gly
GUC	Val	GCC	Ala	GAC	Asp	GGC	Gly
GUA	Val	GCA	Ala	GAA	Glu	GGA	Gly
GUG	Val	GCG	Ala	GAG	Glu	GGG	Gly

As many deviations from the universal code have been observed, the genetic code is evolving, not frozen. For example, the codon AUG, which codes for isoleucine in the universal code, has changed to code for methionine in yeast and metazoa, but has reverted back to code for isoleucine in echinoderms. To explain changes in the code, Jukes and Osawa (1991) proposed the following hypothesis: The changes are preceded by disappearance of a codon from coding sequences in mRNA of an organism or organelle. The function of the codon that disappears is assumed by other synonymous codons (e.g., the codon mutase to a synonymous codon), so that there is no change in amino acid sequences of proteins. The deleted codon then reappears, but it now codes for a different amino acid or serves as a stop codon.

In the mitochondrial genome of plants, the triplet CGG was once thought to code for tryptophan instead of arginine because CGG codons were often found in plant mitochondrial genes at positions corresponding to those encoding conserved tryptophans in other organisms. However, it was later found that the C residues in these codons for the conserved tryptophans were converted to U residues in the corresponding mRNA sequences, so that the CGG codons were changed to the UGG codons in the mRNA. Since UGG indeed codes for tryptophan in the universal code, it appears that plant mitochondrial genomes also use the universal code. The conversion of C to U, known as a type of *RNA editing*, also occurs at some other C residues, but the specificity and structural determinants of conversion are not known at the present. For example, it is not clear why in the cytochrome c oxidase II gene in wheat the TCA codon at position 56 is converted to UUA in the mRNA, whereas the TCA codon at position 5 is not. Because of the lack of complete knowledge, direct translation of plant mitochondrial genes into polypeptides cannot be done at the present time. There is evidence that RNA editing occurs in the mitochondria of all major groups of land plants except the Bryophyta. How such an editing system has evolved is a mystery.

Mutation

Mutations are errors in DNA replication or errors in DNA repair. They maybe classified by the length of the DNA sequence affected by the mutational event. For instance, a mutation may affect a single nucleotide (a *point mutation*), or several adjacent nucleotides. Mutations can also be classified by the type of change caused by the mutational event into (1) *substitutions*, the replacement of one nucleotide by another; (2) *recombination*, which includes crossing-over and gene conversion; (3) *deletions*, the removal of one or more nucleotides from the DNA;

(a) AAGGCAAACCTACTGGTCTTATGT

(b) AAGGCAAATCTACTGGTCTTATGT

(c) AAGGCAAACCTACTGCTCTTATGT

(d) AAGGCAACTGGTCTTATGT (ACCTA)

(e) AAGGCAAACCTACTAAAGCGGTCTTATGT

(f) AAGGTTTGCCTACTGGTCTTATGT

Fig. 6.9. Types of mutations. (a) Original sequence; (b) a transition from C to T; (c) a transversion from G to C; (d) a deletion of the sequence ACCTA; (e) an insertion of the sequence AAAGC; (f) an inversion of 5'–GCAAAC–3' to 5'–GTTTGC–3'.

(4) *insertions*, addition of one or more nucleotides to the sequence; and (5) *inversions*, the rotation by 180° of a double-stranded DNA segment consisting of two or more base pairs.

Substitutional Mutation

Nucleotide substitutions can be classified into *transitions* and *transversions*. Transitions are substitutions between A and G (purines) or between C and T (pyrimidines). Transversions are substitutions between a purine and a pyrimidine; e.g., A→C and A→T are transversions. There are four types of transitions and eight types of transversions.

Nucleotide substitutions occurring in protein-coding regions can also be characterized by their effect on the product of translation, the protein. A substitution is said to be *synonymous* or *silent* if it causes no amino acid change. Otherwise, it is *nonsynonymous*. Nonsynonymous, or amino acid-altering, mutations are further classified into *missense* and *nonsense* mutations. A missense mutation changes the affected codon into a codon that specifies a different amino acid from the one previously encoded. A nonsense mutation changes a codon into one of the termination codons, thus prematurely ending the translation process and ultimately resulting in the production of a truncated protein.

```
(a) Ile  Cys  Ile  Lys  Ala  Leu  Val  Leu  Leu  Thr
    ATA  TGT  ATA  AAG  GCA  CTG  GTC  CTG  TTA  ACA
                                    ↓
    ATA  TGT  ATA  AAG  GCA  CTG  GTA  CTG  TTA  ACA
    Ile  Cys  Ile  Lys  Ala  Leu  Val  Leu  Leu  Thr

(b) Ile  Cys  Ile  Lys  Ala  Asn  Val  Leu  Leu  Thr
    ATA  TGT  ATA  AAG  GCA  AAC  GTC  CTG  TTA  ACA
                                  ↓
    ATA  TGT  ATA  AAG  GCA  AAC  TTC  CTG  TTA  ACA
    Ile  Cys  Ile  Lys  Ala  Asn  Phe  Leu  Leu  Thr

(c) Ile  Cys  Ile  Lys  Ala  Asn  Val  Leu  Leu  Thr
    ATA  TGT  ATA  AAG  GCA  AAC  GTC  CTG  TTA  ACA
                   ↓
    ATA  TGT  ATA  TAG  GCAAACGTCCTGTTAACA
    Ile  Cys  Ile  Ter
```

Fig. 6.10. Types of point mutations in a coding region: (a) synonymous, (b) missense, and (c) nonsense.

Each of the sense codons can mature to nine other codons by means of a single nucleotide substitution. For example, CCU (Pro) can experience six nonsynonymous substitutions, to UCU (Ser), ACU (Thr), GCU (Ala), CUU (Leu), CAU (His), or CGU (Arg), and three synonymous substitutions, to CCC, CCA, or CCG. Since the universal genetic code consists of 61 sense codons, there are $61 \times 9 = 549$ possible nucleotide substitutions. If we assume that nucleotide substitutions occur at random and that all codons are equally frequent in coding regions, we can complete the expected proportion of the different types of nucleotide substitutions from the genetic code. Because of the structure of the genetic code, synonymous mutations occur mainly at the third position of codons. Indeed, almost 70% of all the possible nucleotide changes at the third position are synonymous. In contrast, all the mutations at the second position of codons are nonsynonymous, as are the vast majority (96%) of nucleotide changes at the first position.

Crossing-Over and Gene Conversion

There are two types of homologous recombination: *crossing over* and *gene conversion*. Figure 6.11 shows a typical experiment in yeast for distinguishing between them. A diploid is constructed by crossing one haploid strain with the linked wild type alleles *A*, *B*, and *C* to a haploid containing mutant alleles *a*, *b*, and *c*. This diploid is shown at the top of the figure. The left side of Figure 6.11 shows a crossover between the first two linked loci; the segregation of the genetic markers at the two loci following the crossover will be 1*AB*:1*Ab*:1*aB*:1*ab* (tetratype) instead of 2*AB*:2*ab* (the normal parental ditype). Crossing-over is also termed *reciprocal recombination* because the tetrad that has the recombinant spore *Ab* also has the reciprocal recombinant product *aB*. The right side of Figure 6.11 shows a conversion event at the *B* locus; the segregation at the *B* locus will be 3*B*:1*b* instead of 2*B*:2*b*, although segregation at each of the other loci will be 2:2. Since it appears that one allele at the *B* locus has been converted by the other, the event is called *gene conversion*. If the *A* and *B* genes are examined together in the converted tetrad, it is noted that one spore contains the recombinant product *aB* but the other three spores contain parental combinations of markers. For this reason, gene conversion is also called *nonreciprocal recombination*. Gene conversion and crossing-over are often associated events; in yeast the frequency of association can be 50% or higher.

Figure 6.11 represents recombination at meiosis. Recombination can also occur at mitosis, but it is usually much less frequent than

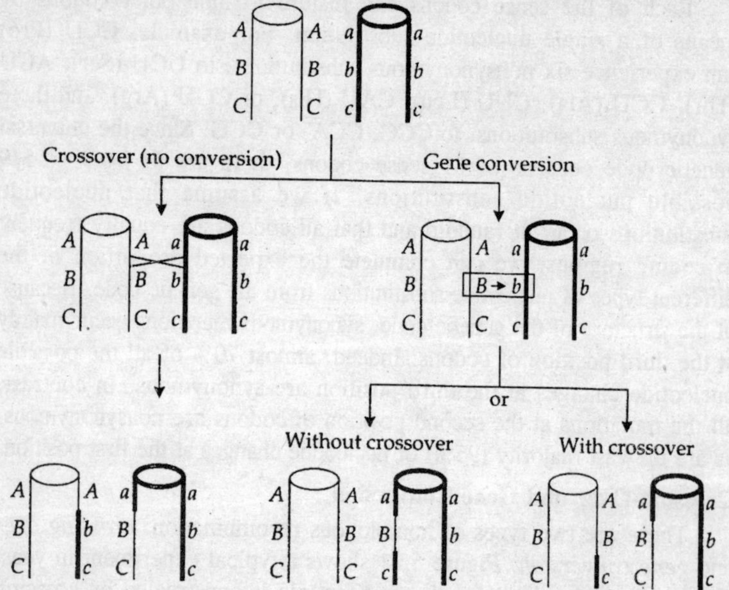

Fig. 6.11. Meiotic crossing-over and gene conversion.

meiotic recombination; in yeast, for example, the former is usually 1000-fold less frequent than the latter.

Deletions and Insertions

Deletions and insertions can occur by several mechanisms. One mechanisms is *unequal crossing over*. Figure 6.12a shows a simple model in which an unequal crossing-over between two chromosomes results in the deletion of a DNA segment in one chromosome and a reciprocal addition in the other. The chance of unequal crossing-over is much increased if a DNA segment is duplicated in tandem, because of the increased chance of misalignment.

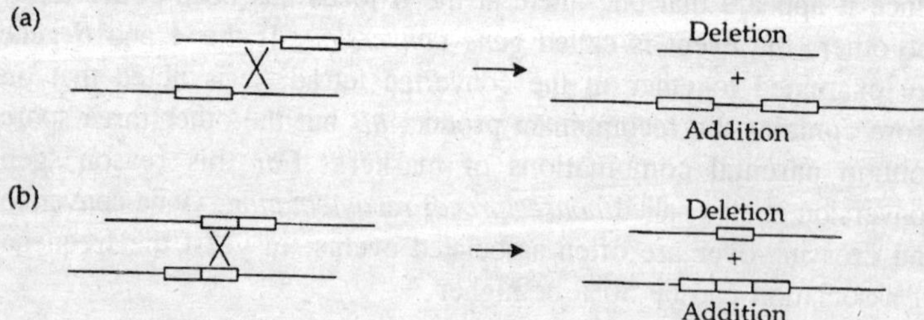

Fig. 6.12. Unequal crossing over. A box denotes a particular stretch of DNA. When a DNA segment is duplicated in tandem as in (b), the chance of misalignment increases and so does the chance of unequal crossing-over.

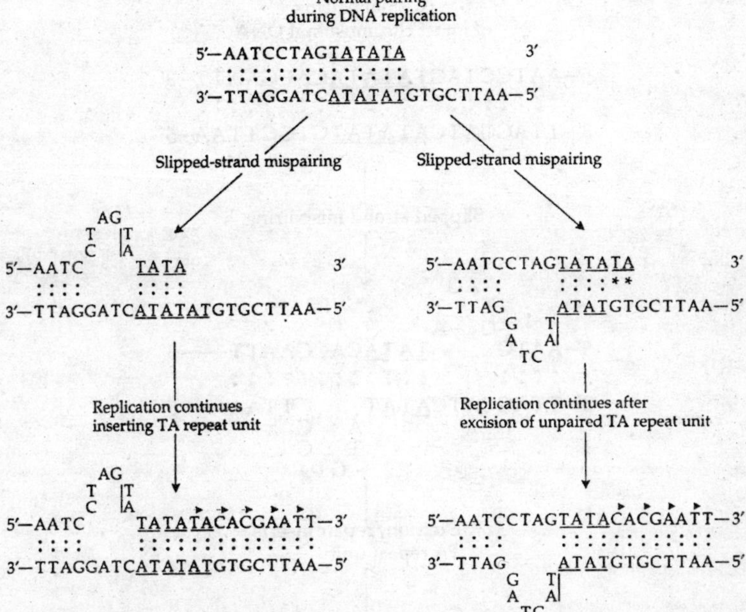

Fig. 6.13. A two-base slippage in an TA repeat during DNA replication.

Another mechanism is *replication slippage* or *slipped-strand mispairing*. This type of event occurs in DNA regions containing contiguous short repeats. During DNA replication, slippage can occur because of mispairing between neighbouring repeats and that slippage can result in either deletion or duplication of a DNA segment, depending on whether the slippage occurs in the 5'→3' direction or in the opposite direction. Slipped-strand mispairing can also occur in nonreplicating DNA.

A third mechanism responsible for insertion or deletion of DNA sequences is DNA transposition.

Deletions and insertions are collectively referred to as *gaps* (or indels), because when a sequence involving either an insertion or a deletion is compared with the original sequence, a gap will appear in one of the two sequences. The number of nucleotides involved in a gap event ranges from one or a few nucleotides to contiguous stretches involving thousands of nucleotides. The length of the gaps essentially exhibit a bimodal type of frequency distribution, with short gaps (up to 20-30 nucleotides) being mostly caused by errors in the process of DNA replication, such as the slipped-strand mispairing discussed above,

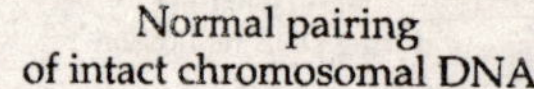

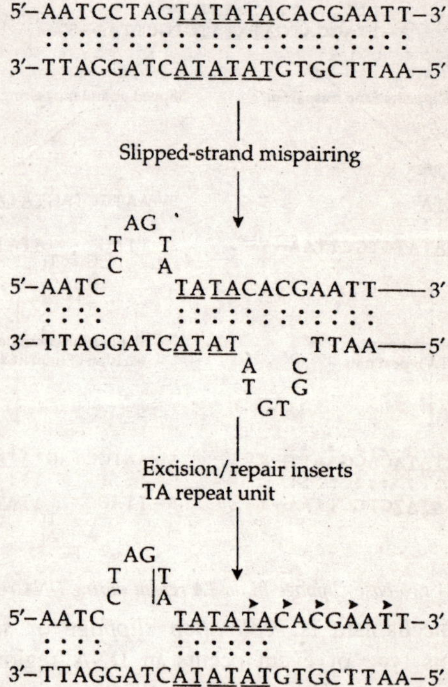

Fig. 6.14. A two-base slippage in an TA repeat in nonreplicating DNA.

and with long insertions and deletions occurring mainly because of unequal crossing-over or DNA transposition.

In a coding region, a gap event involving a number of nucleotides that is not a multiple of three will cause a shift in the reading frame, so that the coding sequence downstream of the gap will be read in the wrong phase. Such a mutation is known as a *frameshift mutation*. Consequently, a gap not only introduces numerous amino acid changes, but may also obliterate the termination codon or bring into phase a new stop codon, thus resulting in a protein of abnormal length.

Spatial Distribution of Mutations

Mutations do not occur randomly throughout the genome. Some regions are more prone to mutate than others, and they are called *hotspots* of mutation. One such hotspot of mutation is the dinucleotide 5'—CG—3' (often called CpG), in which the cytosine is frequently methylated in higher eukaryotes and replicated with error, changing it

(a) Lys Ala Leu Val Leu Leu Thr Ile Cys Ile Ter
AAG GCA CTG GTC CTG TTA ACA ATA TGT ATA TAA TACCATCGCAATAGGG
↓
G

AAG GCA CTG TCC TGT TAA CAATATGTATATAATACCATCGCAATAGGG
Lys Ala Leu Phe Cys Ter

(b) Lys Ala Asn Val Leu Leu Thr Ile Cys Ile Ter
AAG GCA AAC GTC CTG TTA ACA ATA TGT ATA TAA TACCATCGCAATAGGG
↑
G

AAG GCA AAC GGT CCT GTT AAC AAT ATG TAT ATA. ATA CCA TCG CAA TAG GG
Lys Ala Asn Gly Pro Val Asn Asn Met Tyr Ile Ile Pro Ser Gln Ter

Fig. 6.15. Examples of frameshifts in reading frames caused by deletion or insertion. (a) A deletion of a G causes premature termination and (b) an insertion of G obliterates a stop codon.

to 5'—TG—3'. The dinucleotide 5'—TT—3' is a hotspot of mutation in prokaryotes but not in eukaryotes. In bacteria, regions within the DNA containing short *palindromes* (i.e., sequences that read the same on the complementary strands, such as 5'—GCCGGC—3' 5'—GGCGCC—3', and 5'—GGGCCC—3') were found to be more prone to mutate than other regions. In eukaryotic genomes, short tandem repeats are often hotspots for deletions and insertions, probably as a result of slipped-strand mispairing.

Pattern of Nucleotide Substitution in Noncoding Sequences

Since point (substitution) mutation is one of the most important factors in the evolution of DNA sequences, molecular evolutionists have long been interested in determining the *pattern of spontaneous mutation*. This pattern provides empirical data for modeling DNA sequence evolution and can serve as a standard for inferring how far the observed frequencies of interchange between nucleotides in any given DNA sequence have deviated from the values expected under no selection, i.e., under *selective neutrality*.

One way to study the pattern of point mutation is to examine the pattern of substitution in regions of DNA that are subject to no selective constraints. Pseudogenes are particularly useful in this respect. Since they are devoid of function, all mutations occurring in pseudogenes are selectively neutral and become fixed in the population with equal probability. Thus, the pattern of nucleotide substitution in pseudogenes should reflect the pattern of spontaneous point mutation. As the pseudogene data that have been used were from mammalian nuclear pseudogenes, the pattern inferred should be taken as an estimate of the mutation pattern in mammalian nuclear DNA. We shall also discuss the pattern inferred from the control (D-loop) region of human mitochondrial DNA (mtDNA), which may be taken as an estimate of the mutation pattern in mammalian mtDNA.

Substitution Pattern in Pseudogenes

Figure 6.16 shows a simple method for inferring the nucleotide substitutions in a pseudogene sequence. Sequence 1 is a pseudogene, sequence 2 is its functional counterpart from the same species, and sequence 3 is a functional sequence that has diverged before the emergence of the pseudogene. Suppose that, at a certain nucleotide site, sequences 1 and 2 have A and G, respectively. Then we can assume that the nucleotide in the pseudogene sequence has changed from G to A if sequence 3 has G, but that the nucleotide in sequence 2 has changed from A to G if sequence 3 has A. However, if sequence 3 has T or C, we cannot decide the direction of change, and in this case the site is excluded from the comparison. Since the rate of substitution is usually much higher in pseudogenes than in the homologous functional gene, differences in the nucleotide sequence between a gena and a pseudogene are expected to have occurred in the pseudogene in the majority of cases.

The matrix Table 6.2 represents the combined patter of substitution inferred from 13 mammalian pseudogene sequences. Each entry, f_{ij}, in the matrix represents the expected number of base change from i to j among every 100 substitutions in a random sequence (i.e, in a sequence

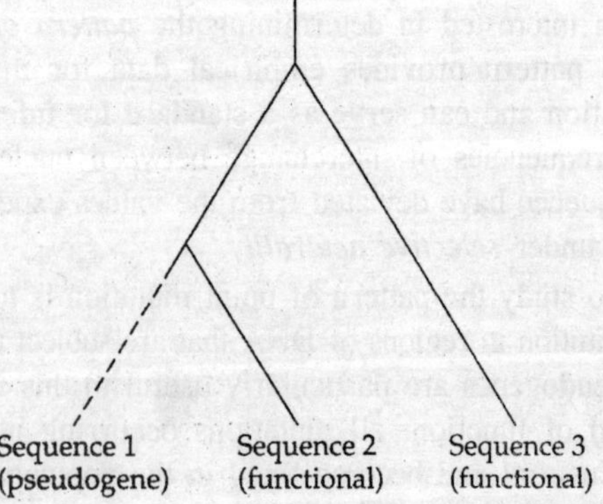

Fig. 6.16. A tree for inferring the pattern of nucleotide substitution in a pseudogene sequence. The dashed line implies "nonfunctional."

in which the four bases are equally frequent). For example, f_{AT} = 4.7, that is 4.7% of all substitutions are form A to T.

Table 6.2. Pattern of nucleotide substitution in pseudogenes

	To				
From	*A*	*T*	*C*	*G*	*Row total*
A	—	4.7±1.3	5.0±0.7	9.4±1.3	19.1
	—	(5.3±1.4)	(5.6±0.8)	(10.3±1.4)	(21.2)
T	4.4±1.1	—	8.2±1.3	3.3±1.2	15.9
	(4.8±1.1)	—	(9.2±1.3)	(3.6±1.3)	(17.6)
C	6.5±1.1	21.0±2.1	—	4.2±0.5	31.7
	(7.1±1.3)	(18.2±2.3)	—	(4.2±0.6)	(29.5)
G	20.7±2.2	7.2±1.1	5.3±1.0	—	33.2
	(18.6±1.9)	(7.7±1.3)	(5.5±1.3)	—	33.2
Column total	31.6	32.9	18.5	16.9	
	(30.5)	(31.2)	(20.3)	(18.1)	

The direction of mutation is clearly nonrandom. For example, A changes more often to G than to either T or C. The four elements from the upper-right corner to the lower-left corner are the f_{ij} values for transitions, while the other eight elements represent transversions. All transitions, and in particular C→T and G→A, occur more often than transversions. The sum of the relative frequencies of transitions is 59.3% (56.3% if CG dinucleotides are excluded). Note that under random mutation the expected proportion of transitions is only 33%, for there are only four types of transitions but eight types of transversions. The observed proportion is almost twice the value expected under random mutation.

Note also that some nucleotides are more mutable than others. The last column of Table 6.2 lists the relative frequencies of all mutations from A, T, C, and G to any other nucleotide. Were all four nucleotides equally mutable, we would expect a value of 25% in each of the column's elements. However, Table 6.2 shows the G mutates with a relative frequency of 33.2% (i.e., G is a highly mutable nucleotide), while T mutates with a relative frequency of 15.9 (i.e., it is less mutable than G). The bottom row of Table 6.2 shows the relative frequencies of all mutations that result in A, T, C or G, respectively. Note that 64.5% of all mutations result in either A or T, while the random expectation is 50%. Since there is a tendency for C and G to change frequently to A to T, and since A and T are not as mutable as C and G, pseudogenes are expected to become rich in A and T. This should also be true for other noncoding regions that are subject to no functional constraint. Indeed, noncoding regions are generally found to be AT-rich. The result in Table 6.2 are expressed in terms of the sense strand, i.e., the untranscribed strand. Thus, a change from G to A actually means that a G:C pair is replaced by an A:T pair. This change can occur as a result of either a G mutating to A in the sense strand or a C to T can occur as a result of either a C mutating to T in one strand or a G mutating to A in the other. If there is no difference in the pattern of mutation between the two strands, we should have $f_{GA} = f_{CT}$. Similarly, we should obtain $f_{AG}=f_{TC}$, $f_{AT}=f_{AT}$, $f_{AC}=f_{TG}$, $f_{CA}=f_{GT}$, and $f_{CG}=f_{GC}$. These equalities hold only approximately, and there may be minor asymmetries in the mutation pattern between the two strands. However, Bulmer's (1991) study of the substitution pattern in the β-globin region did not reveal any significant deviations from symmetry.

It is known that, in addition to base mispairing, the transition from C to T can also arise from conversion of methylated C residues to T residues upon deamination. The effect will elevate the frequencies of C:G→T:A and G:C→A:T; i.e., f_{CT} and f_{GA}. Since about 90% of methylated C residues in vertebrate DNA occur at 5'—CG—3' dinucleotides, this effect should be expressed mainly as changes of the CG dinucleotide to TG or CA. After a gene becomes a pseudogene, such changes would no longer be subject to any functional constraint and can therefore contribute significantly to C→T and G→A transitions if the frequency of CG is relatively high before silencing of the gene (i.e., loss of function) occurs. The substitution pattern obtained by excluding all nucleotide sites where the CG dinucleotides appear to have occurred in the ancestral sequences of these pseudogenes is given in these in Table 6.2. This pattern is probably more suitable for predicting the pattern of mutations in a sequence that has not been subject to functional constraints for a long time (e.g., some parts of an intron), because in such a sequence there would exist few CG dinucleotides. The pattern with CG dinucleotides excluded is somewhat different from that obtained otherwise. In particular, the differences among the relative frequencies of the four transitions become somewhat less conspicuous, and the relative frequencies of the transversions become slightly higher, except for G→C and C→G.

Substitution Patterns in the Control Region of mRNA

The discovery of a high ratio of transitional to transversional substitution in primate mtDNA has attracted much attention, for the

Table 6.3. Pattern of nucleotide substitution in the control region of mtDNA

	To				
From	*A*	*T*	*C*	*G*	*Row total*
A	—	0.4	1.1	14.1	15.6
T	0.3	—	33.8	0.3	34.4
C	1.1	25.8	—	0.5	27.4
G	20.0	1.1	1.6	—	27.7
Column totals	21.4	27.3	36.5	14.9	

ratio was estimated to be close to 20, much higher than the ratio of ~1 in nuclear DNA. Although the original observation was from coding sequences and so might not reflect the spontaneous mutation pattern, the high ratio was also observed in the control region, which, except for the central part, appears to be subject to very weak selective constraints and has evolved very rapidly. Recently, Kocher and Wilson (1991) and Tamura and Nei (1993) have studied the substitution pattern in the control region using extensive sequence data from human and chimpanzee. The estimated pattern shown in Table 6.3 was based on 95 human sequences; the conservative central portion was excluded from the analysis. Obviously, all eight types of transversion occurred with very low frequencies and the average transition/transversion ratio is 15.7, similar to the value (15.0) estimated by Vigilant et al. (1991). Moreover, the transitional rate between pyrimidines (C and T) is higher than that between purine (A and G), and f_{TC} and f_{GA} are, respectively, higher than f_{CT} and f_{AG}, which suggests asymmetry in forward and backward mutation.

7

MOLECULAR AND MENDELIAN GENETICS

The chapter provides an introduction to the genetics that you need to understand the evolutionary biology contained in this book. It begins with the molecular mechanism of inheritance, and moves on to the Mendelian principles. It then considers how Darwin's theory almost required heredity to be Mendelian, because natural selection cannot operate with a blending mechanism of inheritance.

The molecule called DNA (deoxyribose nucleic acid) provides the physical mechanism of heredity in all living creatures. DNA carries the information used to build a new body, and to differentiate its various parts. DNA molecules exist inside almost all of the cells of a body, and in all the reproductive cells (or gametes). Its precise location in the cell depends on cell type.

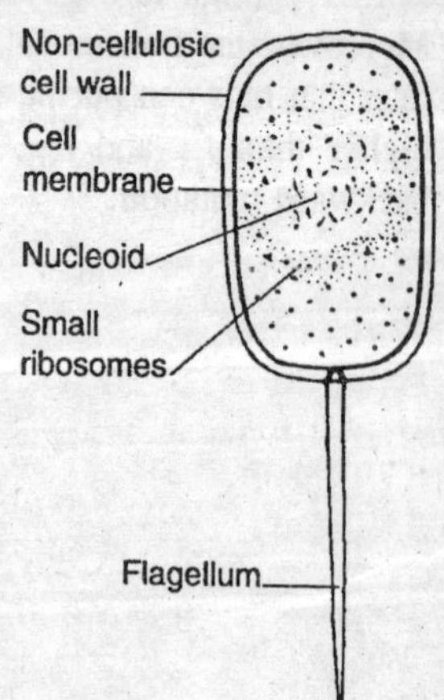

Fig. 7.1. Fine structure of prokaryote cell.

There are two main types of cells: *eukaryotic* and *prokaryotic*. Eukaryotic cells have a complex internal structure, including internal organelles and a distinct region, surrounded by a membrane, called the nucleus; eukaryotic DNA exists within the nucleus. Prokaryotic cells are simpler and have no nucleus; prokaryotic DNA lies within the cell, but in no particular region. All complex multicellular organisms,

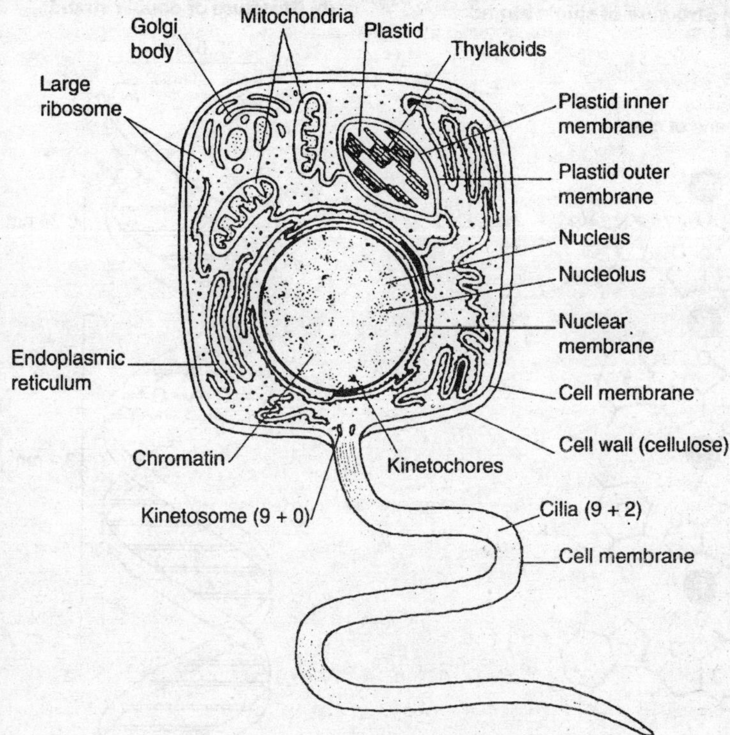

Fig. 7.2. Fine structure of eukaryote cell.

including all plants and animals, possess eukaryotic cells, as do fungi (such as yeast) and protozoans (such as amebas). Prokaryotic cells are found in other unicellular organisms, particularly in bacteria and spirochetes.

Within a eukaryotic cell nucleus, the DNA physically is carried in structures called *chromosomes*; chromosomes can be seen through a light microscope at certain stages in the cell cycle. Individuals of different species characteristically have different numbers of chromosomes—each individual human has 46, for example, whereas a fruitfly *Drosophila melanogaster* has eight chromosomes, and other species have other numbers. The finer structure of the DNA is too small to be seen directly, but it can be inferred by the method of x-ray diffraction. The molecular structure of DNA was unraveled by Watson and Crick in 1953.

The DNA molecule consists of a sequence of units; each unit, called a *nucleotide*, consists of a phosphate and a sugar group with a

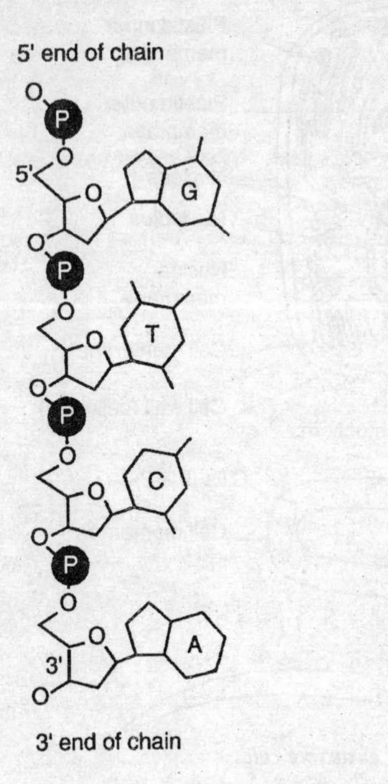

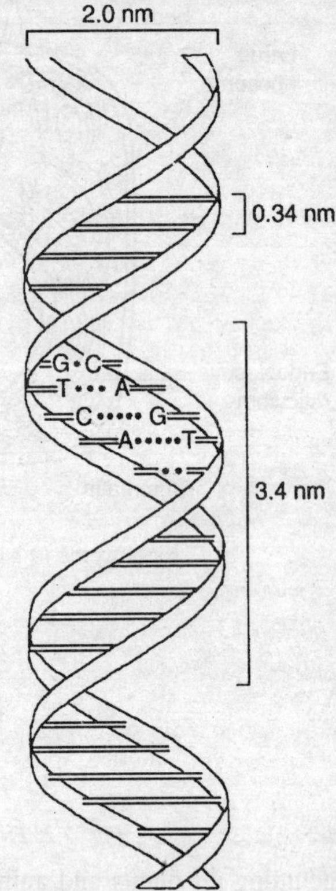

Fig. 7.3. The structure of DNA. (a) Each strand of DNA is made up of a squence of nucleotide units. (b) The full DNA molecule has two complementary strands, arranged in a double helix.

base attached. The alternating sugar and phosphate groups of successive nucleotides form thc backbone of the DNΛ molecule. The full DNA molecule consists of two, paired complementary strands, each consisting of sequences of nucleotides. The nucleotides of opposite strands are chemically bonded together. The two strands exist as a double helix.

DNA Encoding Body's Proteins

How does the DNA encode the information to build a body? The DNA in an individual human cell contains about 3×10^9 nucleotide units. This total length can be divided into *genes* and other kinds of DNA. Some genes lie immediately next to neighoring genes; others are separated by regions of varying length. Each gene contains the information that codes for a particular protein.

A crude but workable way to describe the role of proteins is to say that bodies are built from proteins. Different parts of the body develop their distinct characteristics because of the kinds of proteins of which they are made. Skin, for example, mainly consists of a protein called keratin; oxygen is carried in red blood cells by a protein called hemoglobin; eyes are made sensitive to light by pigment proteins such as rhodopsin (rhodopsin itself consists of a protein called *opsin* combined with a derivative of vitamin A); and metabolic processes are catalyzed by a whole battery of proteins called *enzymes* (cytochrome c, for instance, is a respiratory enzyme, while alcohol dehydrogenase is a digestive enzyme). Coding for proteins by DNA ultimately enables the egg cell to develop into an adult body.

Proteins are made up of particular sequences of amino acids. Twenty different amino acids are found in most kinds of living things; each amino acid behaves chemically in distinct ways, such that different sequences of amino acids result in proteins with very different properties. A protein's exact sequence of amino acids determines its nature. Hemoglobin, for example, consists of two α-globin molecules, which are 141 amino acids long in humans, and two β-globins, which are 144 amino acids long; insulin is created by a sequence of 51 amino acids. Hemoglobin binds oxygen in the blood, whereas insulin stimulates cells (particularly muscle cells, and others) to absorb glucose from the blood. The different behaviours of hemoglobin and insulin result from the chemical properties of their different amino acids, arranged in their characteristic sequence.

The idea that one gene encodes for one protein is a simplification. A more accurate statement would be that one cistron encodes for one polypeptide. A *cistron* is defined as the length of DNA that codes for a polypeptide. A polypeptide is a chain of amino acids—the polypeptides in a body are of various lengths. The description involving cistrons is more exact because some proteins, such as hemoglobin, are assembled from more than one polypeptide, and the different polypeptides are encoded in separate (rather than contiguous) regions of the genome. Hemoglobin is created from four polypeptides, and the cistrons for two of those polypeptides lie far from the cistrons for the other two. (Thus, all proteins are polypeptides, but not all polypeptides are proteins. Each of the four polypeptides that make up hemoglobin represents only a part of a protein.) No single "gene" encodes hemoglobin. However, for many purposes it is not grievously wrong to describe DNA as made up of genes (and noncoding regions), and genes as coding for proteins.

How exactly do genes in the DNA encode for proteins? The answer is that the sequence of nucleotides in a gene specifies the sequence of amino acids in the protein. DNA contains four types of nucleotides that differ only in the base part of the nucleotide unit; the sugar and phosphate groups remain the same in all four types. The four types of nucleotides are called adenine (A), cytosine (C), guanine (G), and thymine (T). Adenine and guanine belong to the chemical group called purines, while cytosine and thymine are pyrimidines. In the double helix, an A nucleotide in one strand always pairs with a T nucleotide in the other; and a C always pairs with a G. If the nucleotide sequence in one strand was . . . AGGCTCCTA .. ., then the complementary strand would be. . . TCCGAGGAT. . . . Because the sugar and phosphate are constant, it is often more convenient to imagine a DNA strand as a sequence of bases, like the . . . AGGCTCCTA . . . sequence given above.

Transcription and Translation

There are four types of nucleotides, but 20 different amino acids. A one-to-one code of nucleotide encoding amino acid would therefore be impossible. In fact, a triplet of bases encodes one amino acid; the nucleotide triplet for an amino acid is called a *codon*. The four nucleotides can be arranged in 64 ($4 \times 4 \times 4$) different triplets, each of which codes for a single amino acid. The relation between triplet and amino acid is called the *genetic code*.

The mechanism by which the amino acid sequence is dictated by the nucleotide sequence of the DNA is understood in molecular detail. The full detail is unnecessary for our purposes, but we should distinguish two main stages. At the first, an intermediate molecule called *messenger RNA* (mRNA) is transcribed from the DNA; the process is called *transcription*. mRNA has a structure similar to that of DNA, except that mRNA is single-stranded and uses a base called uracil (U) rather than thymine (T). The DNA sequence AGGCTCCTA would therefore have an mRNA with the sequence UCCGAGGAU transcribed from it. The genetic code is usually expressed in terms of the codons in the mRNA. The mRNA sequence UCCGAGGAU, for example, codes for three amino acids: serine-glutamic acid-aspartic acid. The beginning and end of a gene are signaled by distinct base sequences, which (in a sense) punctuate the DNA message.

Transcription takes place in the nucleus. After the mRNA molecule has been assembled on the gene, it then leaves the nucleus and travels to one of the structures in the cytoplasm called *ribosomes*; ribosomes

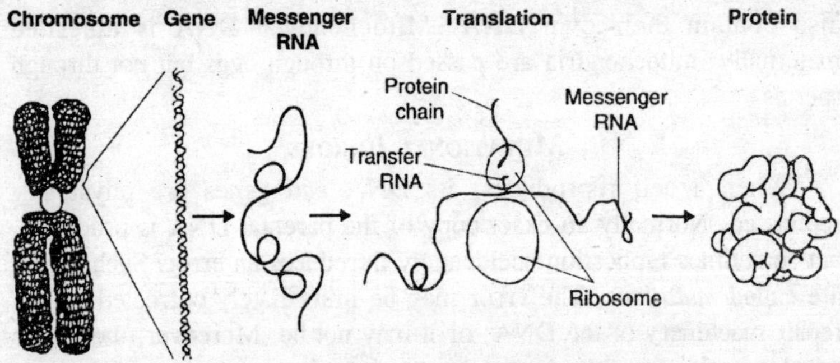

Fig. 7.4. The transfer of information in a cell.

are made of another kind of RNA called *ribosomal RNA* (rRNA). The ribosome is the site of the second main stage in protein production, in which the amino acid sequence is read off from the mRNA sequence and the protein is assembled. The process is called *translation*. The actual translation is achieved by yet another kind of RNA, called *transfer RNA* (tRNA). tRNA has a base triplet recognition site, which binds to the complementary triplet in the mRNA, and has the appropriate amino acid attached at the other end. There are 61 possible types of tRNA, one for each coding triplet; all 61 are not found in any one cell because some tRNA molecules can bind to more than one triplet in the mRNA. (This phenomenon is called *wobble*, and the "wobble rules" govern which tRNA molecules bind which combinations of codon.) Thus, protein assembly consists of tRNA molecules lining up on the mRNA at a ribosome. Other molecules are also needed to supply energy and attach the RNAs correctly.

Different proteins contain different numbers of amino acids, but a very approximate average figure is 100-300 amino acids. Insulin (51 amino acids) is a short protein; long proteins can include thousands of amino acids. The human genome may contain approximately 100,000 (10^5) genes (and therefore a human body makes use in its life of about 100,000 kinds of protein). A protein consisting of 300 amino acids needs roughly 1000 (10^3) nucleotides, which leads to the estimate of 10^8 coding bases in the human genome. Because the total length of an individual's DNA is about 3×10^9 bases, about 90% of the DNA in the genome is of uncertain function.

In addition to the DNA on the chromosomes in the nucleus, certain organelles in the cytoplasm possess much smaller quantities of DNA. Mitochondria—the organelles that control respiration—have some DNA, and in plants the organelles called chloroplasts that control photosynthesis also contain their own DNA. Mitochondrial DNA is inherited maternally: mitochondria are passed on through eggs but not through sperms.

Mutational Errors

When a cell reproduces, its DNA and genes are physically replicated. Normally an exact copy of the parental DNA is produced, but sometimes replication accidentally introduces an error. Such errors are called *mutations*. The error may be immediately corrected by the repair machinery of the DNA, or it may not be. Moreover, the repair process itself can make a mistake and introduce a mutation. The new sequence of DNA that results from a mutation may code for a form of protein with different properties from the original. Mutations can happen in any cell, but the most important mutations, for the theory of evolution, are those occurring in the production of the gametes. These mutations are passed on to the offspring, who may differ from their parents because of the mutation.

Various kinds of mutations can occur. In point mutation, a base in the DNA sequence changes to another base. The effect of a point mutation depends on the kind of base change. Synonymous, or silent, mutations are mutations between two triplets that code for the same amino acid, and have no effect on the protein sequence; non-synonymous, or meaningful, point mutations do change the amino acid. Because of the structure of the genetic code, most synonymous mutations occur in the third base position of the codon. Approximately 70% of changes in the third position are synonymous, whereas all changes in the second and most (96%) at the first position are meaningful. Another distinction

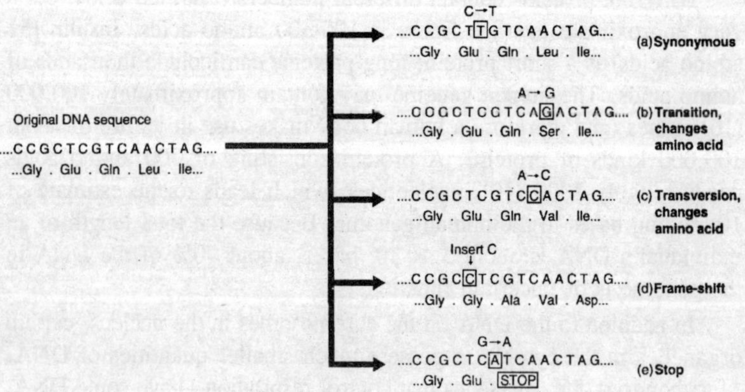

Fig. 7.5. Different types of mutations.

for point mutations relates to transitions and transversions. Transitions are changes from one pyrimidine to the other, or from one purine to the other—that is, between A and G, and between C and T. Transversions replace a purine base by a pyrimidine, or vice versa—that is, from A or G to T or C (and from C or T to A or G). The distinction is interesting because transitional changes occur much more frequently in evolution than transversions.

Successive amino acids are read from consecutive base triplets. If, therefore, a mutation inserts a base pair into the DNA, it can alter the meaning of every base "downstream" from the mutation. Such *frame-shift mutations* will usually produce a completely nonsensical, functionless protein. Another kind of mutation is for a previously coding triplet to mutate to a "stop" codon; the resulting protein fragments will probably be functionless. Mutations may also involve whole chunks of chromosomes rather than single bases. A length of chromosome may be translocated to another chromosome, or to another place on the same chromosome, or be inverted. Whole chromosomes may fuse, as has happened in human evolution; chimps and gorillas (our closest living relatives) have 24 pairs of chromosomes, whereas humans have 23. Some or all of the chromosomes may be duplicated.

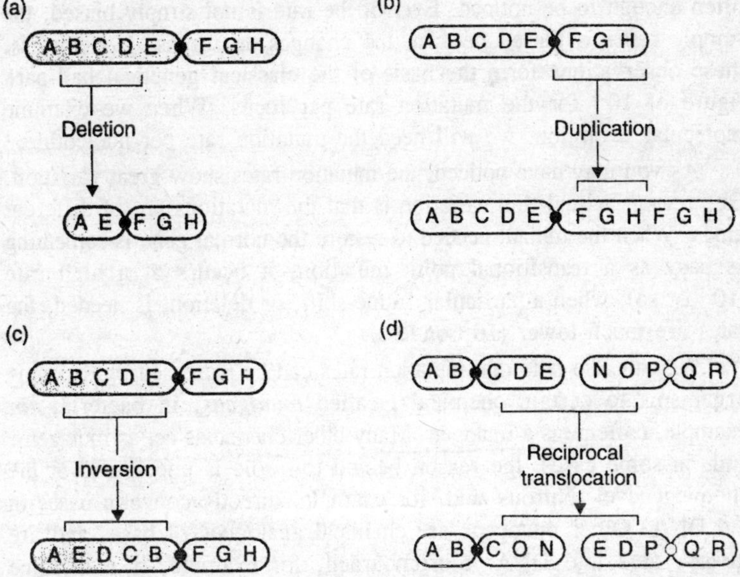

Fig. 7.6. Chromosomes can mutate by (a) deletion; (b) duplication of a part; (c) inversion; or (d) translocation.

It is more difficult to generalize about the phenotypic effects of these chromosomal mutations. If the break-points of the mutation divide a protein, that protein will be lost in the mutant organism. If the break occurs between two proteins, however, any effect will reflect whether the expression of a gene depends on its position in the genome. In theory, it might not matter whether a protein is transcribed from one chromosome or another; in practice, gene expression is probably at least partly regulated by relations between neighbouring genes and a chromosomal mutation will then have phenotypic consequences.

Rates of Mutation

What is the rate of mutation? A direct method of finding out is to observe the rate at which a visible mutant phenotype arises in a laboratory population of a species. The method has been used with many laboratory species, such as bacteria, yeast, *Drosophila*, and mice. Many are mutations from a rare, often dysfunctional, form back to the normal form of the gene. A memorable figure is that mutation rates per gene are approximately 10^{-6}; however, these estimates, from direct observation, may be biased because the genetic changes are ones that are easily noticed. They may well have higher-than-average mutation rates, because changes with low mutation rates simply may not happen often enough to be noticed. Even if the rate is not simply biased, the sample remains biased because the changes are gross. Nonetheless, these observations form the basis of the classical genetical ball-park figure of 10^{-6} for the mutation rate per locus. When we examine molecular evolution, we will need the mutation rate per nucleotide.

As you may have noticed, the mutation rates show great variation. One reason behind this variation is that the mutations are of different kinds. When the mutant needed to restore the normal gene is something as easy as a transitional point mutation, it occurs at a high rate (10^{-3} or so); when a particular frame-shift, or deletion, is needed, the rates are much lower (10^{-9} or less).

We can also note that mutation rates can be increased by exposing organisms to certain chemicals, called *mutagens*; in bacteria, for example, caffeine is a mutagen. Many other chemicals act as mutagens, and, in some cases, the reason behind this role is understood at the chemical level. Nitrous acid, for example, directly converts bases in the DNA. Other mutagens are chemical analogues of bases and are incorporated into DNA: 5-bromouracil, for example, is a thymine analogue and increases the mutation rate when it is substituted for thymine bases in the DNA. X-rays also increase mutation rates, particularly the rates of breaks and rearrangements in chromosomes. These artificial influences on mutation rates are environmentally important, but may not have been of major importance in natural evolution.

Double Set of Genes

DNA is physically carried on chromosomes. Humans, as noted earlier, have 46 chromosomes, which consist of two sets of 23 distinct chromosomes. (More correctly, an individual has a pair of sex chromosomes—which are similar [XX] in females, but noticeably different [XY] in males—plus a double set of 22 non-sex chromosomes, called *autosomes*.) The condition of having two sets of chromosomes (and therefore two sets of the genes carried on them) is called *diploidy*. The estimate of 3×10^9 nucleotides in the human genome is for only one of the sets of 23 chromosomes; the total DNA library of a human cell has about 6×10^9 nucleotides.

Diploidy is important in reproduction. An adult individual possesses two sets of chromosomes. Its gametes (eggs in the female, sperm or pollen in the male) have only one set—a human egg, for example, has only 23 chromosomes before it is fertilized. Gametes are said to be *haploid*. They are formed by a special kind of cell division, called *meiosis*; in meiosis, the double set of chromosomes is reduced to result in a gamete with only one set. When male and female gametes fuse, at fertilization, the resulting *zygote* (the first cell of the new organism) has the double chromosome set restored) and it develops to produce a diploid adult. The cycle of genesis can then repeat itself.

Because each individual possesses a double set of chromosomes, it also has a double set of each of its genes. Any one gene is located at a particular place on a chromosome, called its *genetic locus*; an individual is therefore said to have two genes at each genetic locus in its DNA. One gene comes from the organism's father and the other from its mother. The combination of the two genes at a locus is called a *genotype*. The two copies of a gene in an individual may be the same, or slightly different (i.e., the amino acid sequences of the proteins encoded by the two copies may be identical or have one or two differences). If they are the same, the genotype is a *homozygote*; if they differ, it is a *heterozygote*. The different forms of the gene that can be present at a locus are called *alleles*.

Genes and genotypes are usually symbolized by alphabetic letters. For in- stance, if two alleles at the genetic locus under consideration are designated as *A* and *a*, an individual can then have one of three genotypes: *AA*, *Aa*, or *aa*. The genotype at a locus should be distinguished from the *phenotype* it produces. If two alleles occur at a locus in a population, the two can combine into three possible genotypes: *AA*, *Aa*, and *aa*. (If there are more than two alleles, more than three genotypes are possible.) The genes will influence some property of the organism, and the property mayor may not be easily visible. Suppose the genes influence colour. The *A* gene might encode a black pigment and *AA* individuals would be black; *aa* individuals, lacking the pigment, might be white. The colouration is then determined by the phenotype controlled by the genotype at that locus: an individual's phenotype consists of its body and behaviour as we observe them.

If we consider only the *AA* and *aa* genotypes and phenotypes in this example, then a one-to-one relationship exists between genotype and phenotype. However, a different relationship is possible, as can be illustrated by considering two possibilities for the phenotype of the *Aa* genotype. One possibility is that the colour of *Aa* individuals is intermediate between the two homozygotes—that is, they are gray. In this case, three phenotypes exist for the three genotypes, in a one-to-one relationship. The second possibility is that the *Aa* heterozygotes resemble one of the homozygotes; they might be black, for instance. The *A* allele is then called *dominant* and the *a* allele *recessive*. (An allele is dominant if the phenotype of the heterozygote looks like the homozygote of that allele; the other allele in the heterozygote is then recessive.) If dominance occurs, only two phenotypes will be possible for the three genotypes, and there will no longer be a one-to-one relation between them. If you know only that an organism has a black phenotype, you do not know its genotype.

Any degree of dominance could appear at different genetic loci. Full dominance, in which the heterozygote resembles one of the homozygotes, and no dominance, in which it is intermediate between the homozygotes, are extreme cases. The phenotype of the heterozygote could lie anywhere between the two homozygotes. Instead of being either black or gray, it could exhibit any degree of grayness. Dominance is only one of a number of factors that complicate the relationship between genotype and phenotype. The most important such factor is the environment in which an individual grows up.

Genes and Mendelian Ratios

Mendelian ratios express the proportions of different genotypes in the offspring of parents of particular combinations of genotypes. The easiest case is a cross between an *AA* male and an *AA* female. After meiosis, all males gametes and all female gametes contain the *A* gene. They combine to produce *AA* offspring. The Mendelian ratio is therefore 100% *AA* offspring.

Now consider a mating between an *AA* homozygote and an *Aa* heterozygote. Again, all of the *AA* individual's gametes contain a single *A* gene. When a heterozygote reproduces, half of its gametes contain an *A* gene, and half an *a* gene. The pair will produce *AA*:*Aa* offspring in a 50:50 ratio.

Finally, consider a cross between two heterozygotes. Both male and female produce half a gametes and half *A* gametes. If we consider the female gametes (eggs, or ovules), half of them are *a*; half of these *a* female gametes will be fertilized by *a* sperm, and half by *A* sperm. The other half of the female gametes are *A*, of which half will be fertilized by *a* sperm and half by *A* sperm. The resulting ratio of offspring is $^1/_4$ *AA* : $^1/_2$ *Aa* : $^1/_4$ *aa*.

The separation of an individual's two genes at a locus into its offspring is called *segregation*. The ratios of offspring types produced by different kinds of matings are examples of Mendelian ratios. They were discovered by Gregor Mendel in about 1856-1863. Mendel was a monk, later Abbe, in St. Thomas's Augustinian monastery in what was then Brunn in Austro-Hungary.

Mendelian ratios can also exist for more than one genetic locus. If the alleles at one locus are *A* and *a*, and at a second *B* and *b*, then an individual will have a double genotype, such as *Ab*/*Ab* (double homozygote) or *Ab*/*ab* (single heterozygote). It has a double set of genes at each locus, one set from each parent. The segregation ratios now depend on whether the genetic loci lie on the same or different chromosomes. Recall that an individual human has a haploid number of 23 chromosomes and about 100,000 genes. Thus, there must be on average about 5000 genes per chromosome. Different genes on the same chromosome are described as being *linked*. Genes that are very close together are tightly linked, while those farther apart are loosely linked. Genes on separate chromosomes are unlinked.

The easy case involves two unlinked loci; the genes at the two loci then segregate independently. Imagine a cross in which only one of the loci is heterozygous, such as a cross between an *Ab*/*Ab* male and an *Ab*/*ab* female. All the genes at the B locus are the same, while at the A locus the male is *AA* and the female is *Aa*. The ratio of offspring will be $^1/_2$ *AAbb* and $^1/_2$ *Aabb*, a simple extension of the one-locus case.

Now suppose both loci are heterozygous. The ratios of B locus genotypes associated with each A locus genotype are those predicted by applying Mendel's principles independently to each locus. For instance, a cross between two *Bb* heterozygotes produces a ratio of offspring of $^1/_4$ *BB*: $^1/_2$ *Bb*: $^1/_4$ *bb*, and this ratio will be the same within each A locus genotype. A cross between a male *AB*/*Ab* and a female *AB*/*ab* will generate $^1/_2$ *AA* and $^1/_2$ *Aa* offspring. Of the $^1/_2$ that are *AA*, $^1/_4$ are *AB*/*AB*, $^1/_2$ are *AB*/*Ab*, $^1/_4$ are *Ab*/*Ab*. The same concept applies to the $^1/_2$ *Aa* genotypes. Add the two *A* genotypes and the total offspring ratios are:

AB/*AB*	*AB*/*Ab*	*Ab*/*Ab*	*AB*/*aB*	*AB*/*ab*	*Ab*/*ab*
$^1/_8$	$^1/_4$	$^1/_8$	$^1/_8$	$^1/_4$	$^1/_8$

The segregation of unlinked genotypes is called *independent segregation*.

When the loci are linked on the same chromosome, they do not segregate independently. At meiosis, when haploid gametes are formed from a diploid adult, an additional process called *recombination* occurs. The pair of chromosomes physically line up and, at certain places, their strands join together and recombine. Recombination shuffles the combinations of genes. If an individual inherited *AB* from its mother and *ab* from its father, and recombination occurs between the two loci, it will produce *Ab* and *aB* gene combinations in its gametes.

Recombination is a random process; it may or may not "hit" any point in the DNA. It occurs with a given probability, usually symbolized by r, between any two points on a chromosome. The probability r can be defined between nucleotide sites or genes. Thus, if the A locus and the B locus are linked, the chance of a recombination between them in an individual is r and the chance of no recombination is $(1 - r)$. In any one individual, recombination either does or does not happen, but the chance of recombination determines the frequencies of genotypes in the gametes produced by a population. If we consider a large number of *AB*/*ab* individuals, they will produce gametes in the following proportions:

gamete	AB	Ab	aB	ab
proportion	$^1/_2\,(1 - r)$	$^1/_2\,r$	$^1/_2\,r$	$^1/_2\,(1 - r)$

These fractions can be used to calculate the Mendelian ratio for a cross involving an *AB*/*ab* individual. The principle is logically easy to understand, but the ratios can be laborious to work out in practice. The case of independent segregation corresponds to $r = {}^1/_2$. That is, when the A and B loci lie on separate chromosomes, $r = (1 - r) =$

$^1/_2$ and the *Ab*/*aB* parent produces *Ab*, *aB*, *AB*, and *ab* gametes in the ratio 1:1:1:1.

For any two genes, recombination can "hit" more than once between them in an individual (a process called "multiple hits"). If two hits occur between a pair of loci, they cancel one another and the chromosome has the same combination of genes as if no recombination occurred. Thus, it is more accurate to say that the probability of recombination *r* equals the probability that an odd number of hits will occur, and the probability (1 – *r*) equals the chance of no hits plus the chance of an even number of hits.

The Mendelian ratios, in which paired diploid genes segregate into haploid gametes and the gametes of different individuals then combine at random, are the basis of all the theory of population genetics.

Blending Mechanism of Heredity

Mendel's theory of heredity plugged a dangerous leak in Darwin's original theory, and the two theories together eventually came to form the synthetic theory of evolution, or neo-Darwinism. The problem was Darwin's lack of a sound theory of heredity; indeed, it had been shown even in Darwin's time that natural selection would not work if heredity was controlled in the way thought by most biologists before Mendel. Before Mendel, most theories of heredity were *blending* theories. We can see the distinction in much the same terms as have just been used for Mendelism. Suppose gene *A* causes its bearers to grow up dark blue in colour, while gene *a* causes its bearers to grow up white. We can imagine that, as in the real world of Mendelism, individuals in our imaginary world of blending heredity are diploid and have two copies of each "gene." An individual could then either inherit an *AA* genotype from its parents and have a dark blue phenotype, inherit an *aa* genotype and have a white phenotype, or inherit an *Aa* genotype and have a light blue phenotype. (Thus, in the Mendelian version of the system, we would say that no dominance exists between the *A* and *a* genes.)

The interesting individuals for this argument are the ones that have inherited an *Aa* genotype and have grown up to be light blue. They could have been produced in a cross of a dark blue and a white parent: the offspring will then be light blue whether inheritance is Mendelian (with no dominance) or blending. But consider the next generation. Under Mendelian heredity, the light blue *Aa* heterozygote passes on intact to its offspring the *A* and *a* genes it had inherited

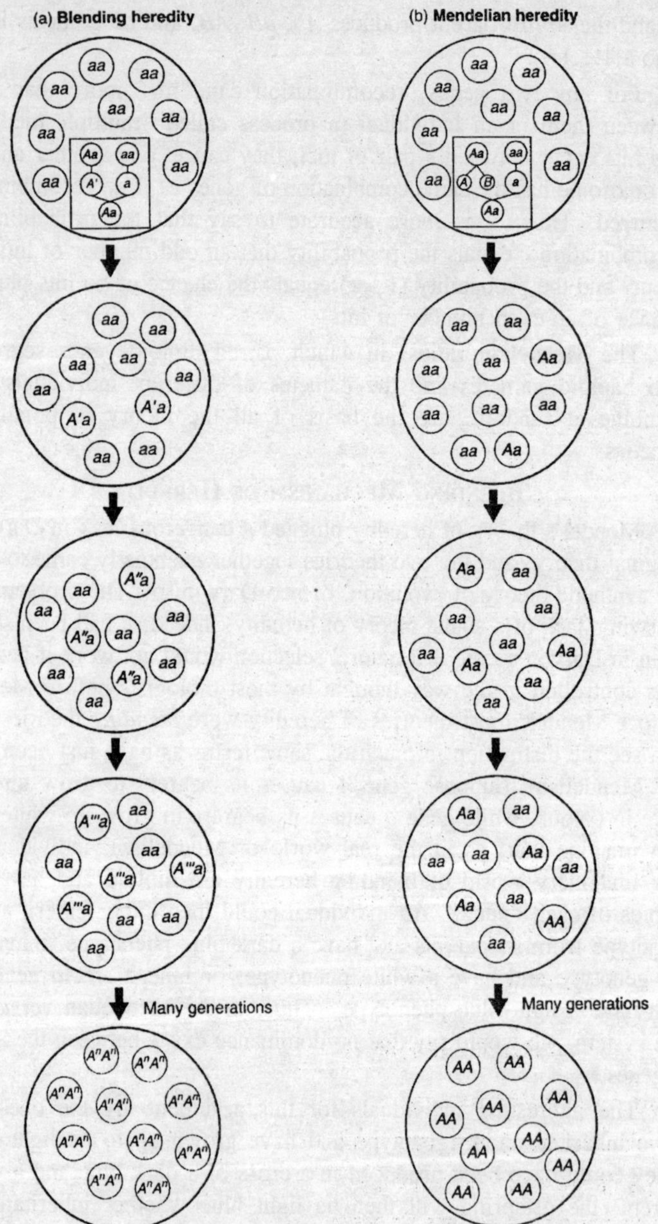

Fig. 7.7. Two population with 10 individuals each one with blending heredity and the other with Mendelian heredity

from its father and mother. Under blending heredity, an individual does not pass on the same genes as it inherited. If an individual inherited an *A* and an *a* gene, the two would physically blend in some way to form a new sort of gene (let us call it *A*') that causes light blue colouration. Instead of producing $^1/_2$ *A* gametes and $^1/_2$ *a*, the individual would then produce all *A*' gametes. This would affect the second generation. Whereas in Mendelian heredity the dark blue and white colours segregate out again in a cross between two heterozygotes, in the analogous cross with blending heredity they do not—all the grandchildren are light blue.

Mendelism is an atomistic theory of heredity. Not only do discrete genes encode discrete proteins, but the genes are also preserved during development and passed on unaltered to the next generation. In a blending mechanism, the "genes" are not preserved. The genes that an individual inherits from its parents are physically lost, as the two parental sets are blended together. In Mendelism, it is perfectly possible for the *phenotypes* of the parents to be blended in the offspring (as they are in the initial *AA* × *aa* cross), but the genes do not blend. Indeed, the phenotypes of real mothers and fathers often do blend in their offspring, which led most students of heredity before Mendel to assume that inheritance must be controlled by some blending mechanism. However, the case of heterozygotes that are intermediate between the two homozygotes (i.e., no dominance) shows that the blending of phenotypes need not mean blending of genotypes. In fact, the underlying genes are preserved.

One way of expressing the importance of Mendelism for Darwin's theory is to say that it efficiently preserves genetic variation. In blending inheritance, variation is rapidly lost as extreme types mate together and their various "genes" are blended out of existence into some general mean form; in Mendelian inheritance, variation is preserved because the extreme genetic types (even if disguised in heterozygotes) are passed down from generation to generation.

Why does this preservation of genes matter for Darwinism? Armed with only the elementary account of natural selection, however, it is still possible to understand why Darwin "needed" Mendel.

Suppose that a population of individuals is white in colour and has the *aa* genotype, and heredity blends. For some reason, it is advantageous for individuals in this population to be dark blue in colour: dark blue individuals would survive better and leave more offspring. Moreover, it is better to be somewhat blue (i.e., light blue) than to be white.

Suppose now that a single new light blue individual somehow crops up by mutation, and it has an *Aa* genotype. This *Aa* individual will survive better than its *aa* fellow members of the population and produce more offspring. However, the advantageous gene cannot last long with blending. In the first generation, it produces *A*' gametes; these combine with *a* gametes (because all other members of the population are white) and produce *A'a* offspring. We can suppose that these individuals are slightly lighter blue in colour than the original *Aa* mutant; they still have an advantage, but it is less marked.

The *A'a* individual's genes blend, such that all its gametes will have an *A*" gene. When that unites with an *a* gamete (because still almost everyone else is white) an *A"a* offspring results, which is even lighter blue in colour. After a matter of time, the original favourable mutation will be blended almost out of existence. The best result possible would be a population that was very slightly less white than the initial population. A population of dark blue individuals could not possibly result from the mutation, because the original gene, which potentially could produce dark blue individuals, will cease to exist after only a single generation. The most famous objection was made in Darwin's time by a Scottish engineer, Fleeming Jenkin. Though Darwin had known about it since his early thinking in the 1840s, he never did find a wholly satisfactory way around it.

Mendelism was what Darwin needed. In the example just given, the original light blue mutation will be in an *Aa* heterozygote, and fully half its offspring will be light blue—because they are also *Aa* heterozygotes. Ample time exists for natural selection to increase the proportion of light blue individuals, and eventually enough will be produced so that two could potentially mate and produce some *AA* homozygotes among their offspring. It is now theoretically possible for a population of dark blue *AA* individuals to result. Thus, natural selection is a powerful process when combined with Mendelian heredity, because Mendelian genes are preserved over time; in contrast, natural selection is a weak process when combined with blending inheritance, because potentially favourable genes are diluted before they can be established.

8

Genotype to Phenotype

The classical genetic approach to the analysis of evolution was based on a mathematics that reduced the entire process to the study of changes in gene frequencies within populations. In effect, the developmental interactions responsible for producing evolutionary changes were ignored. In Neo-Darwinian theory, selection, coupled sometimes with neutral drift, was regarded as the only mechanism responsible for the spread and establishment of biological novelty. In treating the genome simply as a collection of individual genes, however, this theory fails to come to terms with the implications of the turnover of multigene families implicit in the concept of molecular drive. It is simply not adequate to graft such recent molecular findings onto conventional evolutionary theory. Thus two decades of analyzing the polymorphisms that are evident in single genes has given no insight into the origin of morphological change. The signal failure of this approach has, in turn, led to recourse to 'explanations' in terms of 'regulatory' genes and their role in generating evolutionary change.

Molecular analyses of genome structure, by contrast, have highlighted the rapid flux that characterizes the genomes of eukaryotes. In addition it has drawn attention to the fact that most of these genomes are choked with DNA sequences that represent nothing more than evolutionary debris. These sequences are the natural outcome of processes involving the replication, recombination, amplification, insertion, excision and conversion of the DNA molecules within a genome. Many of these sequences are mobile, and most of them make no significant contribution to developmental pro- grammes and so do not impinge on phenotype.

Molecular analyses of the variation in multigene and repetitive DNA families, both within and between species, as well as the experimental addition of specific cloned DNA sequences to a genome, emphasize that many DNA sequences are naturally apostolic and missionary oriented since they continuously convert each other. These findings have provided the basis for the theory of molecular drive which has added a new dimension to the spread of variation through a population and has directed attention to the need to examine the developmental role of multigene families.

Molecular analyses are also now focusing attention on executive genes, as opposed to these which service and maintain universal metabolic back-up systems. One class of these executive genes, typified by the Bithorax complex of *Drosophila*, has been shown to have homologies in all major classes of vertebrates. The structure and mode of expression of such executive genes draws attention to the concept of developmental circuits and the ways in which alterations in circuitry lead to phenotypic changes of evolutionary significance. Many genes affecting early embryogenesis in *Drosophila* have also now been cloned, and the localization of their transcripts and protein products has revealed unexpected distributions.

Finally, the genetic analyses of neural cell lineages in the roundworm *Caenorhabditis elegans*, the monoclonal antibody studies on neuronal pathfinding in the grasshopper, and the recombinant DNA analyses of RNA populations in the vertebrate brain, have revealed both unexpected simplicities and complexities of neuronal developmental lineages.

These developmental insights, stemming from reductionist molecular approaches to the genome, offer for the first time to provide an objective basis for defining and analyzing the modes of morphological change that play a major role in evolution. Recombinant DNA technologies and their spin-offs have revealed that eukaryote genomes are remarkably variable in DNA amounts, chromosomal architecture, gene number and distribution. Enormous variances in amounts of DNA can persist in the face of almost identical morphologies as in frogs, for example, whereas radically different morphologies may stem from what initially appears as a near constancy of genome size as in mammals.

We argue that most of the DNA of eukaryote genomes is irrelevant both to basic metabolism and developmental programmes and hence also to phenotype. It is effectively evolutionary flotsam and jetsam. Most organisms are simply unable to avoid the production and turnover of DNA sequences that have little or no influence on phenotype. Such sequences are by-products of the enzymic processes which replicate and reorganize DNA sequences.

That part of the genome that contributes significantly to developmental decisions, and hence to morphogenesis, consists of a small and exclusive set of genes. These genes are now rapidly being cloned in *Drosophila* and the first of these, Bithorax, has been shown to have homology with genes in amphibia, chicken, mice and men. Such homologies provide us with the first real evidence of any molecular commonality between the development of vertebrates and invertebrates.

We contrast this explosion of knowledge and the new evolutionary vistas it has revealed with the lack of meaningful progress that has been made under the largely mathematical aegis of Neo-Darwinism.

A Genomic Overview

DNA Amounts

Examination of the total amount of nuclear DNA in fish, for example, reveals a staggering range which is certainly not paralleled by an equivalent range of morphological complexity. When one compares related species within a single amphibian genus, the morphological differences between species are slight, yet in these there is a threefold difference in genome size. On a grander scale, *Plethodontid* salamanders which again show little morphological differentiation, vary from 19,000 to 64,000 million base pairs (mbp). The chromosome numbers in both these examples are constant so that it has been wholesale amplification, deletion or movement of DNA sequences which has led to the present day disparity in genome sizes. Whatever the genomic rearrangements have involved, the excess DNA of the larger genomes has clearly made no contribution to morphological variation.

In contrast to these examples of morphological monotony in the face of cataclysmic genome reorganization, mammals illustrate quite the reverse. Here a relatively constant genome size of approximately 3300 mbp, has been no impediment to the evolution of such morphological extremes as bats, whales and humans.

Transcriptive Activity

The gross transcriptive activity of various vertebrate genomes is instructive in this regard. When the total RNA complexity is measured for different tissues or different developmental stages, it is effectively constant irrespective of genome size. Thus, although the two amphibians

Table 8.1. Nuclear genome sizes in different vertebrates, *n* denotes haploid chromosome number.

Vertebrate	*Genome size (million base pairs)*	*n*
Fish		
puffer (*Tetraodon fluviatilis*)	360	
salmon (*Oncorhynchus kisutch*)	2700	
shark (*Squalus acanthias*)	6600	
lungfish (*Protopterus aethiopicus*)	130,000	
Frogs		
Limnodynastes ornatus	900	11
Limnodynastes terraereginae	1900	11
Limnodynastes dumerilii	3100	11
Plethodontid salamanders		
Plethodon c. cinereus	19000	14
Plethodon vehiculum	34000	14
Plethodon vandykei	64000	14

Xenopus laevis and *Triturus cristatus* differ at least sevenfold in amounts of their nuclear DNA, their RNA complexities are quite comparable. Similarly, although the chicken genome is only a third that of a mammal, its somatic RNA complexities (excluding brain tissues) are equivalent. When egg RNA complexities are examined in both vertebrate and invertebrate genomes, whose sizes vary from almost 100 mbp to 10,000 mbp, they show only about twofold variation. These examples reveal that only a small proportion of the total genome is used in a transcriptive sense. The remainder is essentially debris from a phenotypic viewpoint.

Table 8.2. RNA sequence complexities in different vertebrates

Species	*Genome size (million base pair*	*Tissue/stage*	*polyA$^+$ mRNA (million nucleotides)*
Xenopus laevis	3000	egg	35
Triturus cristatus	23000		40
chicken	1200	liver	20
		oviduct	30
		myofibril	32
mouse	3300	embryo	20
		kidney	20
		liver	21

Gross Genomic Organization

Major karyotypic changes have often been considered as being of paramount evolutionary significance in terms of generating novel morphologies. Examination of related species, however, reveals that gross genomic organization can be remarkably fluid in the absence of any meaningful phenotypic change. In related species of muntjak deer the haploid genome can be organized as 3 or 23 chromosomes. Thus a genome that has been truly 'blitzed' in terms of rearrangements remains largely unperturbed in terms of its phenotypic expression. The zebra and the horse, which can hardly be described as anatomically or physiologically very different, have 16 and 32 chromosomes respectively. The muntjak parallel in fungi is provided by two related yeasts which have 3 and 17 chromosomes respectively. Finally, in ciliate protozoans, while *Tetrahymena* has five conventional chromosomes in its macronucleus, *Oxytrichia* has its macronuclear genome shattered into thousands of gene-sized pieces averaging 2200 bp.

Table 8.3. Genomic rearrangements and morphological similarities in related eukaryotes.

Animal	*Haploid chromosome number*
Deer	
Muntiacus muntjak	3
Muntiacus reevesi	23
Equids	
Equus zebra	16
Equus equus	32
Yeasts	
Schizosaccharomyces pombe	3
Saccharomyces cerevisiae	17
Ciliates	
Tetrahymena pyriformis	5
Oxytrichia similis	(thousands of gene-sized pieces 2.2 kb)

The conclusions that flow from these and similar examples are that the genome can be mercilessly modified by gross karyotypic rearrangements and still produce equivalent phenotypes. Indeed, in no case has it been demonstrated that an evolutionarily significant morphological change has been the direct consequence of any chromosomal rearrangement, as opposed to a point mutation in a gene or its controlling sequences, or gene inactivation stemming from the mobile sequences that often abound in a genome. The requirement, then, is to differentiate between chromosomal rearrangements as causes of evolutionary novelty, as opposed to other genomic changes that have gone on in parallel.

Genomic Dissection by Recombinant DNA Technologies

Drosophila melanogaster

Although an invertebrate, this organism is proving the most useful starting point for the analysis of eukaryote genome organization and developmental biology simply because it is the best known eukaryote in terms of genetic and molecular organization. The information it provides offers an objective basis for an evaluation of similar data from vertebrates. The 165 million base pairs of haploid nuclear DNA in the fly consists roughly of, first, 50 mbp of highly repetitive sequences localized predominantly around centromeres, secondly, 20 to 25 mbp of nomadic or mobile DNA sequences together with some dispersed repetitive sequences whose status as nomadic or sedentary has as yet not been finalized, and thirdly, 90 mbp of sequences that are classically defined as unique or single-copy DNA.

The localized highly repetitive sequences are certainly dispensable from both the metabolic and the developmental arena since large deficiencies and duplications of this material produce individuals which, in a somatic sense, are phenotypically indistinguishable from controls and whose early embryogenesis is likewise unperturbed. These sequences in both vertebrates and invertebrates suffer almost every conceivable enzymic insult from point mutation, amplifications, deletions, conversions, insertions and rearrangements without causing overt phenotypic change. Furthermore, some sequences are totally absent from one species, yet present in its closely related relatives. None of this is really surprising since these repetitive sequences are not transcribed into RNA except by default when readthrough occurs from adjacent promoters. They are thus oblique to discussions on phenotypic change during evolution.

The nomadic sequences of *D. melanogaster* consist of about 50 families which are structurally heterogeneous and are evolutionary transients and some are intimately related to vertebrate retroviruses. The nomadics differ not only in frequency between different strains of *D. melanogaster* but also between species of *Drosophila*. When nomadic elements are isolated from *D. melanogaster* and tested against the sibling species *D. simulans*, it is evident that *D. simulans* has only about one-eighth the frequency of *D. melanogaster* nomadics and has not supplemented them in any substantial way with different families. Thus these sibling species differ by about 20 million base pairs of mobile DNA and yet are morphologically nearly identical. These mobile elements can act as internal mutagens since they insert and excise from genie landscapes and cause predictable lesions. They can also mobilize chromosomal segments and move them to other genomic landscapes.

All eukaryotes contain families of nomadic, or formerly nomadic, elements which are heterogeneous in structure and in family membership. The prominent human nomadic family, termed the *Alu* family, exists in roughly 300,000 copies per genome and is found inserted in a variety of genomic locations, genie and non-genic as are some retroviral sequences. As in the case of highly repetitive DNAs, there is no evidence that the middle repetitive fraction, which often consists predominantly of mobile sequences, plays any significant röle *per se* in morphogenesis. The situation was appreciated by Doolittle and Sapienza, as early as 1980 in their critique of the phenotype paradigm. They argued convincingly that no phenotypic or evolutionary role was necessary for middle repetitive sequences and that the obsessive search for functional explanations 'may prove, if not intellectually sterile, ultimately futile'.

Development, Genes and Phenotype

We now turn our attention to the genie compartments of the genome and consider whether any subsection exerts a greater influence than any other in development. Morphological complexity stems ultimately from the correct developmental expression of integrated genie circuits. These genie cascades invariably mean that it is difficult to classify eukaryote genes as 'regulatory' or 'structural' because of the numerous interactions that occur between genes, their products and the intra- and intercellular transport of these products. When, for example, polymorphisms for structural genes have been studied and found to be near identical between chimpanzees and man, it has been a convention to resort to 'regulatory' genes as the prime movers in phenotypic differences. This just defers the issue by semantics. Unlike prokaryotes, where the distinction is relatively straightforward, the concept of distinct 'regulatory' and 'structural' genes is too ill defined to be useful for eukaryotes. It is more pragamatic to classify the available genes in terms of their relative importance as developmental switches.

In terms of 'gene regulation' *per se*, current fashion in recombinant DNA studies focuses heavily on the upstream and downstream landscapes surrounding genes in an effort to elucidate the important sequences used in the control of gene transcription. The existence of various chromatin configurations, enhancers, hypersensitive sites, differential processing at the RNA and protein levels and mRNA stability is relevant only to when, where, and how rapidly, certain proteins are made, modified and dispensed with. They tell us little about developmental regulation, much of which involves interaction between active sites or structural components of proteins and cell-cell interactions. Such interactions are only now being examined by the newer technologies. We focus on them because we believe that it is only by considering cellular phenomena that we will be able to decide which gene pathways are initially worthwhile pursuing in a developmental context.

Cellular Interactions

At gastrulation, entire sheets of cells undergo movement resulting in the particular disposition of the germ layers which allows correct organogenesis to subsequently proceed. What gene programmes underpin these cell determinations and cell movements? If we had some inkling of the genetic pathways involved, we could rationalize how changes in these processes at the cell and tissue level lead to morphological variation. We could then estimate which parts of the pathways were more subject to genetic and environmental stress and hence discover what perturbations are likely to cause significant phenotypic changes during evolution.

Oster and Alberch (1982) provide a striking illustration of a developmental system, epithelial morphogenesis, which brings home the importance of cell-cell interactions and their effects on morphology. The sequence of events that culminates in the formation of feathers, scales, teeth and hair depends initially on cell shape changes involving the epidermis. Here a basal lamina separates an epidermal from a dermal cell layer. The epidermal layer thickens to yield epidermal placodes with cell shape changing from cuboidal to columnar, and cells within the dermal layer then condense beneath the placodes. A binary decision is now made; if *invagination* of the epidermal layer occurs, then either scales (reptiles) or feathers (birds) are formed; if *invagination* is the order of the day, hair or skin glands results (mammals). The critical bifurcation in this system is the inward or outward folding of the epithelial layer. Some relevant questions to be asked in dissecting such a system mechanistically are thus; how do we determine the biochemical events that underlie this process? Is it possible to isolate key genes or key proteins which, in the first instance, allow the determinative events to occur?

So shockingly rudimentary is our knowledge of the control of developmental processes that we are forced to start at any chink in the biological armour. Without any doubt, the most promising avenue is the molecular analysis of development in *Drosophila*. Here the application of recombinant DNA and monoclonal technologies, classical genetics, embryology and gene transfer techniques has already revealed that a number of developmentally significant executive genes have a common 61 amino-acid protein domain, termed the homeo box, which is also found in most vertebrates. It appears that the most profitable approach is to continue to isolate, characterize and perturb *Drosophila* systems and then, by raiding vertebrate genomic libraries, to evaluate if the genes and developmental processes are similar or divergent.

Developmental Genes in Drosophila

Despite 30 years of research, the molecular bases of pattern formation are still obscure. We still do not know how the position-dependent activity of molecules is laid down and interpreted. The situation is now rapidly changing owing to the isolation of three major classes of genes which control spatial organization in the *Drosophila* embryo. There are: (1) maternal effect genes that control spatial co-ordinates; (2) segmentation genes that determine body segment number and polarity by interpreting the positional information supplied by the maternal effect genes; and (3) homoeotic genes that specify segment identity. Not only are these genes being cloned and characterized, but in the case of both the sequentation genes and the homoeotics their sites of transcription within the early embryo are being defined by the use of suitable molecular probes.

Nusslein-Volhard and Wieschaus (1980) undertook a saturation coverage of the genome to search for embryonic lethals which altered the segmentation pattern of the embryo. They uncovered more than a dozen loci which were implicated in various stages of segmentation but none of which affected the overall polarity of the embryo. There were genes that effected pattern duplication in each segment, pattern deletion in alternating segments and deletion of adjacent segments.

Other genes have been isolated in less complete searches. One gene, *fushi tarazu* (lack of enough segments), is of particular interest. In homozygous mutant *fushi tarazu* embryos alternate body segments are lacking and the embryos are lethal. The gene has been cloned and characterized and the distribution of its 1.9 kb poly A^+ RNA transcripts in the wild-type embryo determined. The RNA is expressed only from early blastoderm to the gastrula stage and is first detected during nuclear cleavage. At that time, the transcripts are equally localized between 15 and 65% of the egg length with zero being at the posterior pole. The transcripts then gradually become restricted so that by the blastoderm stage they form seven evenly spaced bands and give the embryo a zebra pattern which anticipates its eventual segmentation.

The currently most fashionable developmental genes are the homoeotics, which specify segment identity. These are genes which, when mutated effectively, transform one cell lineage into another. They are thus switch genes. The reason for the current excitement is that a number of them have been found to contain a conserved 61 amino-acid protein domain which has the molecular characteristics of a DNA binding region. These domains have now been found in developmentally regulated vertebrate genes and the accumulated genetic knowledge of *Drosophila* is now being tested there.

The 61 amino-acid protein domain of the various *Drosophila* homoeotic genes is more than 90% conserved between flies and humans, although most interestingly in humans it is embedded in a totally non-homologous sequence to that in *Drosophila*. It is also highly conserved between flies and mice and between flies and amphibians. In *X. laevis* one homoeo-box containing gene is abundantly transcribed during oogenesis, whereas another is expressed at late gastrulation. All this lends credence to the hypothesis that these protein domains may represent some form of genetic regulatory element that may be involved in the early determination of embryonic cell types in vertebrates.

In *Drosophila*, a sequence from the Bithorax locus has been expressed as a fusion protein in bacteria and then used to raise monoclonal antibodies. The antibodies reveal that the Bithorax protein is localized exclusively in nuclei, and it may well be a DNA binding protein whose specificity of binding regulates other genes. Alternatively, it may be far less spectacular and represent a homing device which instructs the protein to return to the nucleus.

The general significance of all these results is that for the first time genes have been defined and analyzed which are developmentally important and which have been isolated and characterized on the basis of impeccable developmental pedigrees. Whether other genes or domains which regulate these various maternal effects, segment identity, and homoeotic classes will also be found to be present in vertebrates, is an intriguing question.

Finally, the way in which developmental circuits have altered over evolutionary time should at last be amenable to analysis. The finding of such high degrees of amino acid conservation means that the 'active sites' of the homoeo-box domain are not overly prone to accepting alterations. Since the homoeo-box domain has been so well conserved between organisms such as yeast (which is not segmented) and man, and indeed has good homology to DNA binding proteins in prokaryotes, its evolutionary significance in morphogenetic terms is difficult to evaluate. Since these organisms differ considerably in phenotype, there must be modes of escaping from this strict conservatism, if indeed it is the homoeo-box domain which is so important. Alternatively, it may well be that other domains within the same protein in which the homoeo-box is embedded are of developmental significance. Whatever the outcome, the homoeo box has labelled some eukaryote genes which, in the amphibia at least, are expressed in oogenesis and gastrulation, two critical periods in development.

Neural Cell Lineages

The central nervous system has, for obvious reasons, been a much neglected component of evolutionary considerations. However, it is the very first system to be defined developmentally and, as Cans and Northcutt (1983) have pointed out, vertebrates have probably evolved from protochordates principally by expansion of the epidermal nerve plexus. Thus many of the sensory and integrative functions are derived embryologically from the neural crests and the epidermal placodes. In effect, the vertebrate head is an enlarged addition to a protochordate body. Moreover, more than half the transcribed portion of a vertebrate genome is expressed in the brain so, with the majority of gene expression involved in its embryonic architecture and functioning, it is an appropriate place to examine genes and phenotype.

Conventionally, evolution is thought of in terms of new structures with new functions. Thus the changes that have given rise to the lungs of vertebrates, the limbs of tetrapods, the feathers of birds and the inner ear of mammals all enable the organism to perform a new function. The evolution of the nervous system operates on a quite different principle. Here, new functions arise by using the same neuronal components in different and more complex combinations. The basic building blocks of the nervous system, the neurons, have been available since the origin of multicellularity and it has been new and altered neuronal *connectivity* that has been largely responsible for evolutionary changes in the nervous system. It is all well and good to evolve sight, sound, taste, smell and touch and a neuromuscular system as well as mechanisms for monitoring and co-ordinating internal activities, but these systems need to be integrated in multimodal convergence to give rise to sensible behaviour, and such integration is a function of neurons. Since each human neuron synapses on average with 1000 others, the total connections approach 10^{15}. How are these connections laid down faithfully in any nervous system? What genes or gene families control these processes and how are they altered during evolution to yield novel behaviours?

Once again it will be easier to glean rudimentary principles from invertebrates, but should the reader find this a somewhat presumptuous exercise, we emphasize that nearly half of the monoclonal antibodies directed against *Drosophila* nervous tissue cross-hybridize to the human brain.

Studies using the simple roundworm *Caenorhabditis elegans* make it clear that the development of the nervous system operates through binary decision-making processes. Consider, for example, the fate of six ectoblast cells of the lateral ectoderm of this organism. Five of them undergo a common developmental lineage giving rise to hypodermal cells of the cell surface. One cell, V5, produces both hypodermal and neural cells. This lineage also includes programmed cell death as well as producing a lateral sensory structure known as a postdeirid. This structure consists of a dopaminergic neuron, a non-dopaminergic neuron, a sheath cell and a socket cell.

The homoeotic mutation *lin*-22 disrupts this lineage in a specific way. Cells VI, 2, 3 and 4 now express the exact characteristics of V5, so the organism finds itself with superfluous sensory structures and reciprocally lacks hypodermal cells.

There is also a gene, *ced*-3, the mutant form of which rescues cells that normally are destined to undergo programmed cell death. In the V5 lineage described above, a non-dopaminergic neuron and cell death was the normal outcome. In the presence of *ced*-3, however, death does not occur and in- stead a supernumerary dopaminergic neuron is generated.

Caenorhabditis is also in the vanguard for the analysis of heterochronic genes. Changes in the temporal pattern of gene expression, and their concomitant morphological novelties has always ranked high as prime movers of evolution. In this organism, two major classes of mutants can be summarized as follows: in *delayed* (or retarded) cases juvenile characteristics are expressed in the adult, whereas in *early* (or precocious) cases adult characteristics appear in immature individuals. Ambros and Horvitz (1984) have isolated a number of such mutants. In *lin*-14, -28 and -29, for example, certain developmental events occur either earlier or later than normal. Particular *lin*-14 alleles can lead to reiterations where moulting and larval cuticle events are repeated. In other alleles, a hypodermal blast cell does not produce its normal progeny during the L2 stage but generates a lineage produced a larval generation earlier by its grandparental precursor cell.

The important point to note from this simple organism is that genes can be identified which cause heterochronic changes. Their relevance in an evolutionary context will be discernable when they are cloned and their protein products analyzed, and any cross homologies to other organisms evaluated.

Neural Pathfinding Genes and Phenotype

When one examines the setting up of a nerve network, the problems, the genes and the final phenotype, are far from simple. Once again the initial inroads stem from molecular approaches in invertebrates since here the developing neurons consist of a segmental chain which arises from the metameric segmentation of the arthropod body. The studies of Thomas et al. (1984) and Goodman et al. (1984) reveal that in the grasshopper the early growth cones of nerve cells send out filopodia and use cell surface cues to build an initial scaffold along which later developing neurons find their way. Thus a developing G neuron, for example, makes a choice when it encounters the A/P nerve bundle. Which of the four neurons within this bundle will it follow? It makes its choice on the basis of cell-surface molecules on axons and their selective adhesion properties.

Rather surprisingly, when other arthropods are examined, such as a crayfish (*Procambarus*) and *Drosophila*, astounding similarities are seen in the location of identified neurons, in the location of cell bodies, in the axonal scaffolds and in the bifurcations that are made by the respective growth cones. For example, the G neuron in *Drosophila* has near it at least 25 nerve bundles and yet it invariably selects the A/P bundle and within it the P, but not the A, axons, just as does the grasshopper.

Thus, in examining the descriptive biology of how the embryonic nervous system develops, a remarkable early conservatism is seen in arthropods as different as crayfish, grasshopper and *Drosophila*. The important future steps will be first, to describe the network and discover the embryonic rules to see how they may be perturbed; secondly, to determine what causes different neurons to express specific surface labels; thirdly, to determine how growth cones are guided to their final destinations on the basis of selective substrate affinities and fourthly, to determine the molecular basis of cell recognition during neuronal development.

In order to answer some of these questions, one needs to isolate the molecules involved or the genes that code for them. The use of monoclonal antibodies has already revealed a specificity that is not apparent on morphological inspection. In the grasshopper, the monoclonal antibody *Mes*-2 recognizes an antigen that is present on only four neurons out of 1000 in a hemisegment. *Mes*-3 and *Mes*-4, on the other hand, distinguish the MP1-dMP2 nerve bundle from the other 25 longitudinal fascicles that occur at this developmental stage.

These findings mean that we are on the threshold of unravelling simple circuits in terms of cell-surface molecules. What is still unknown, is whether these phenomena are controlled by single genes or multigene families and whether there are many different proteins, or whether the proteins are few but their modification, by glycosylation, leads to a diverse array of cell-surface types.

Since the wiring diagram of the central nervous systems of both vertebrates and invertebrates is so poorly described, it is difficult to take the alternative genetic approach of mutagenizing the genome and looking for specific connectivity dislocations in anything but giant fibres or anatomically distinct cell types. This approach, however, has also begun in *Drosophila*.

In the vertebrate brain the molecular situation is almost palaeoneuro-biological. Even so, the surprises are already with us. One will recall that, relative to other tissues, the polyA$^+$ RNA complexity of the vertebrate brain is enormous, near 120 million nucleotides, whereas that of liver and kidney, for example, is of the order of 30 million nucleotides. Milner and Sutcliffe (1983) and Sutcliffe et al. (1983) have isolated polyA$^+$ RNA from adult rat brain, converted it to cDNA and have found that brain specific RNAs are larger and rarer than most metabolic transcripts. They estimate that there are at least 30,000 polyA$^+$ RNAs in the brain, most of which are brain specific. This is a larger number of transcripts than all other somatic tissues combined. Furthermore, brain-specific messages contain an 82 bp core sequence, termed an identifier (ID). There are about 100,000 of these sequences per genome and they are generally found in the introns of brain genes where they have been hypothesized to be important in brain-specific gene control.

Sutcliffe et al. (1983) also determined the nucleotide sequence of one of their clones (IB236), identified the sense strand and its long open reading frame and translated it into the amino acid sequence. They then chemically synthesized short peptides, raised antisera against such molecules, and then probed thin brain sections in immunocytochemical reactions. These experiments revealed a neuronal specificity which was not apparent from neuronal morphology alone. Clone IB236 is synthesized in certain cell bodies and transported via certain axons to particular areas of the cerebellum, hippocampus and cortex.

These experiments, although only in the pioneering stage, lift the lid off the problems, complexities and evolutionary hurdles that need to be faced in the central nervous systems of vertebrates and invertebrates. Some of the structural components will clearly be held in common, as we saw from the *Drosophila* monoclonal cross-hybridization to human brain, but others will clearly be distinctive. Furthermore, if brain-specific transcripts, or the protein products, really are larger than conventional metabolic transcripts, then the developmental pathways so involved may have a greater complexity. It is already known that several neuropeptide precursors such as pro-piomelanocortin are processed to yield several proteins.

All these results, both anatomical and molecular, are still too patchy to give us a firm idea of the underlying principles involved in setting up the architecture of the brain. However, the pioneering approaches we have illustrated show that the gap between genes and phenotype is rapidly narrowing.

Genes, Phenotype and Evolutionary Studies

When one examines the controversies that still rage in population genetics, namely selectionists versus neutralists, gradualists versus punctualists, and regulatory versus structural genes as prime movers of evolutionary novelty, one is struck by how far removed they are from the real problems of evolution. In the case of the structural versus regulatory gene argument, it has quite correctly been pointed out by MacIntyre (1982) that no single compelling observation was responsible for this spotlighting of the putatively significant evolutionary role of regulatory genes. In fact '... in the absence of detailed molecular and developmental studies, any interpretation of interspecific differences involving gene regulation are strictly hypothetical'.

We turn finally from the reductionist molecular approach, which is now pouring out information on genomic architecture, and ask how much of it impinges on evolutionary studies. Here, much of the data are being used to examine rates of change in nucleotide and protein sequences, and to refine phylogenies by the estimations of evolutionary distances. Very little of it addresses itself to the underlying mechanisms of phenotypic change that have given rise to morphological novelties. Molecular evolution is currently still in a DNA data collecting phase, and does not address itself to what many consider to be the main tasks of population genetics, namely '... the description and prediction of changes in the genetic and phenotypic composition of populations responding to the forces of mutation, recombination, selection, migration and systems of mating'.

Multigene Families and Molecular Drive

In general, the mechanisms by which new genes and new phenotypes arise is largely based on speculation. Multigene families were, until recently, not an integral part of modern theories, and as yet they still sit uneasily with Neo-Darwinists.

The concept of molecular drive as developed by Dover (1982), and its implication for evolutionary change as well as phenotypic novelty has exposed the fragility of some aspects of Neo-Darwinism. This concept is based on a feature of many multigene and non-coding families (ribosomal DNA, globins, immunoglobulins, histones and repetitious 'satellite' sequences) in which a continuous homogenization process produces a greater within-species homogeneity than between-species homogeneity.

In utilizing turnover in multigene families, drive focuses on the spread and establishment of novelties within populations by internal processes of flux such as unequal exchange, transposition and gene conversion. This literally leads to the transformation of a population under its own steam without the involvement of any conventional selective forces. The example of the bat-eared fox used by Dover (1984), illustrates the situation, though it could be applied to any presumed adaptation.

Bat-eared foxes have abnormally large bat-like ears and the conventional 'explanation' for such extremities is that they have evolved under the aegis of natural selection for increased efficiency in tracking prey. They would conventionally be considered as evidence of an adaptation that increases the survival of the animals possessing them, as most adaptations are so interpreted. There are alternative explanations, however, particularly if multigene families are involved, and we consider them below.

What Dover's theory encapsulates is that once the nexus is broken between chromosome behaviour and gene behaviour, as occurs by internal turnover mechanisms in the genome, then there can be unexpected transformations of a population with the changes occurring in a cohesive manner. Different individuals in a population undergoing molecular drive will be expected to show little variance between members of the particular multigene family in question. Individual members of a multigene family are not constrained by a chromosome, since chromosomes no longer provide refugia from the genomic turnover processes, which can be both inter- and intrachromosomal. Thus when the rate of spread of a variant between chromosomes differs from the rate of gametic mixing, a multigene family can be homogenized in a synchronous fashion in all individuals. This means that a *population* is slowly changing from one DNA composition to another *without* a large variance between individuals. Thus if the bat-like ears of the fox are largely determined by one or a number of multigene families (and this is as yet unknown), then the change to large ears may well be a consequence of genomic turnover processes. This increase in ear size affects all members of the population because it is a *cohesive* molecular change. The change in phenotype can be gradual or sudden, and there are no biological incompatibilities between individuals.

This last point circumvents the problem of how a *single* mutant individual arising in a population can lead to subsequent fixation of a new phenotype. Under molecular drive, the DNA of all individuals can be converted slowly and synchronously. Thus the change of ear size in foxes could simply be a fortuitcus by-product of changes in multigene families that impose no real problems of differential survival. Nor do such changes need to be adaptive.

We see from examples such as these that much of the fabric of population genetics theory is under pressure to accommodate the molecular data bases. Furthermore, it is particularly striking that the field not only shows strong resistance to any challenge of orthodox beliefs, but is itself preoccupied with what Levin (1984) has referred to as the 'two camp syndrome':

Some hypotheses are more often championed than tested. Positions (almost two for any given issue) are fiercely defended even when they are not mutually exclusive. New ideas are liable to attack for no reasons other than their real, or even apparent, violation of orthodoxy.

It is clear to us that this preoccupation with Darwinian orthodoxy enshrouds the field to little benefit. Dover's concept of molecular drive is a bold new incursion into the orthodox realm. It is based on known mechanisms of genomic turnover processes, whereas selection has always been, and still is, more a logical construct than a well-documented process. Neo-Darwinism as an 'explanation' for evolutionary change is in fact in need of a serious new outlook. It needs to focus back to the developmental gene circuits that are critical for phenotypic change and hence for evolutionary change.

9

Gene Variability

Evolutionary changes rely on two fundamental aspects of genetics or biological inheritance: *constancy* and *variability*. Constancy resides in the genetic observation that like produces like and derives from the ability of nucleic acid macromolecules to replicate. The evolutionary significance of constancy is that all living processes, from biochemistry to behaviour, depend on the transmission of constant or reliable information from previous generations. In contrast, variability resides in the genetic observation that like also produces unlike—the replication of biological information is not always constant or exact—thus producing inherited biological changes (*mutations*). The obvious evolutionary significance of genetic variability is that it fuels evolution by allowing organisms to differ from their ancestors. This chapter briefly reviews some fundamental concepts of genetic constancy and variability.

Cell Division

In modern organisms the transmission of biological information coordinates with cellular division so that both parental and daughter cells carry copies of the same information. Such cell division processes must have originated early in the history of life as a solution to the problem of how membrane—enclosed organisms could grow and expand without enlarging themselves to the point of endangering their existence. Since biological information is coded as long-chain nucleic acid molecules, it is basic to cell division that the structures that bear these molecules, the *chromosomes*, must replicate and divide.

In prokaryotes that have a single chromosome, the two products of chromosomal replication attach to different points on the cell membrane. As the cell elongates to form two daughter cells, the two chromosome products separate, each becoming enclosed in a separate daughter cell. In eukaryotes, more than one chromosome is usually present, and more complex cell division processes occur, divided into *mitosis* (cell division of somatic or body tissues) and *meiosis* (cell division of gamete-producing tissues).

In mitosis, the parental and daughter cells have exactly the same numbers and kinds of chromosomes, and this form of cell division provides all the body cells of an organism with the same chromosome constitution, or *karyotype*. Mitosis first becomes obvious during the *prophase* stage, in which the gradually condensing chromosomes have replicated so that each chromosome now consists of two *chromatids* connected at their *centromeres*. These chromosomes next arrange themselves on a *metaphase* plate in which the two members of a pair of chromatids connect to spindle fibers that go to opposite poles. Thus, when chromatids separate at the *anaphase* stage, each daughter cell receives one chromatid replicate (which is now an individual chromosome) derived from each of the original chromosomes in the parental cell. For example, if a parental cell has four chromosomes, A^1, A^2, B^1, and B^2, mitosis divides these into chromatids A^1-A^1, A^2-A^2, B^1-B^1, and B^2-B^2, to produce two daughter cells, each of the chromosome constitution (karyotype) A^1, A^2, B^1, and B^2.

In meiosis, the cell division process performed by eukaryotes engaged in sexual reproduction, gametes form that contain only one representative of each kind of chromosome. For example, a gamete containing two kinds of chromosome, A^1 and B^1, fertilizes a gamete containing the same two kinds of chromosome, A^2 and B^2, to form a zygote that now has one pair of each, A^1A^2 B^1B^2. If these events occur in an organism whose life cycle is primarily *diploid* (i.e., cells possessing two sets of chromosomes), mitosis replicates the A^1A^2 B^1B^2 karyotype in every somatic cell, until meiosis forms gametes in the male and female gonadal tissues. Each such gamete now contains one *haploid* karyotype (i.e., only one set of chromosomes), either A^1B^1, A^1B^2, A^2B^1, or A^2B^2. If the zygote forms in an organism whose life cycle is primarily haploid, the meiotic process produces non-gametic cells carrying one of these four haploid karyotypes.

In both cases, whether haploid or diploid, meiosis generally follows the stages that is primarily based on close *homologous pairing* between similar kinds of chromosomes called *homologues*, such as between $A^1 \leftrightarrow A^2$ and between $B^1 \leftrightarrow B^2$. Pairing between such homologues lets them separate (disjoin) from each other at the end of the first meiotic

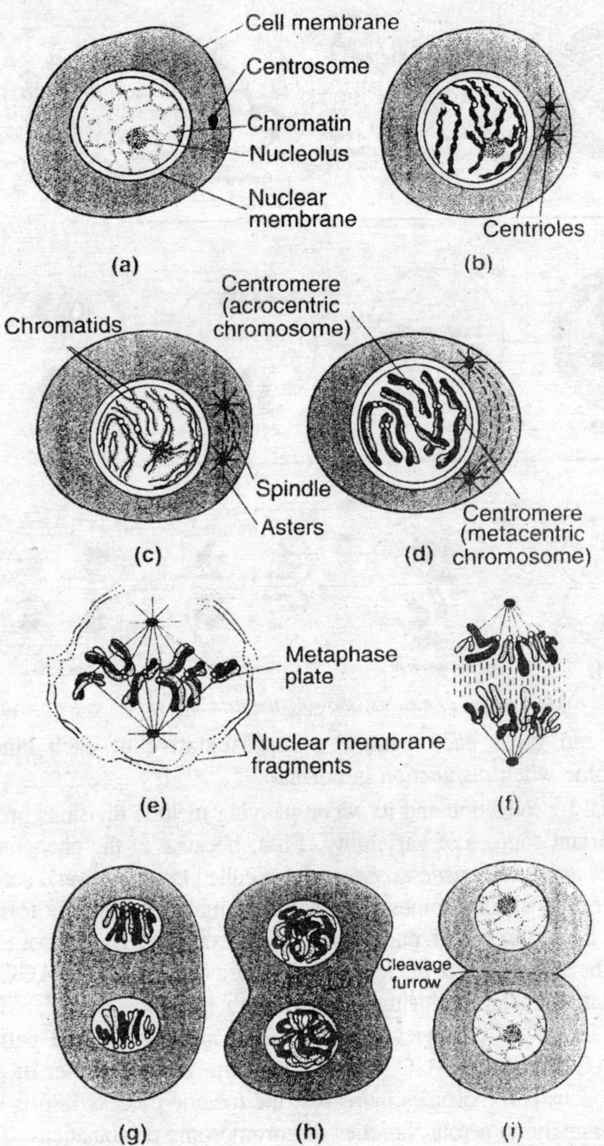

Fig. 9.1. Diagrammatic presentation of various stages of mitosis in a somatic cell. (a) Interphase; (b, c, and d) early, medium, and late prophase stages, showing increased thickening and condensation of chromosomes; (e) metaphase; (f) anaphase. During the telophase stage (g, h, and i), a nuclear membrane re-forms around each polar group of daughter chromosomes, and these chromosomes then revert to the more extended interphase state.

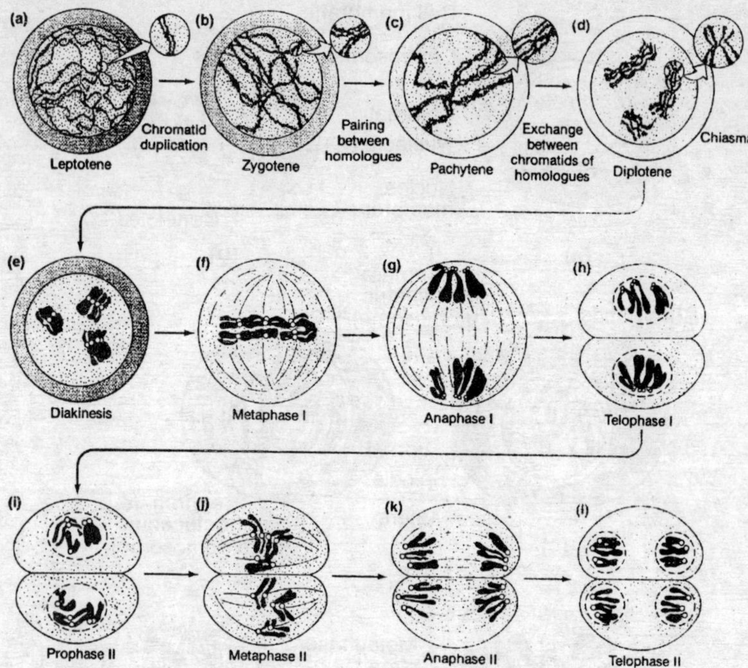

Fig. 9.2. Principal stages of meiosis, showing the first and second meiotic diviions.

division and gives each gamete a representative of each kind of chromosome when *disjunction* is normal.

Sexual reproduction and its accompanying meiotic divisions provide two important sources of variability. First, because of the phenomenon of *recombination* or *genetic exchange* (also called *crossing over*), sections of homologous chromosomes can exchange material, thereby forming different linear arrays of nucleotides. For example, chromosome A^1, bearing the hypothetical nucleotide sequence AGC · AGC · AGC ..., can exchange material during meiosis with chromosome A^2, AGC · TCG · TCG . . . , to yield products that are part A^1 and part A^2, such as AGC · TCG · AGC. . . . In addition, as the number of pairs of homologous chromosomes increases, the meiotic process assorts them into increasingly numerous varieties of chromosome combinations. Thus, as previously described, two homologous pairs of chromosomes (A^1 A^2, B^1 B^2) segregating in a cross can yield four different kinds of gametes (A^1B^1, A^1B^2, A^2B^1, A^2B^2), whereas segregation of three pairs of homologous chromosomes (A^1 A^2, B^1 B^2, C^1 C^2) can yield eight different gametic combinations ($A^1B^1C^1$, $A^1B^1C^2$, $A^1B^2C^1$, $A^1B^2C^2$, $A^2B^1C^1$,

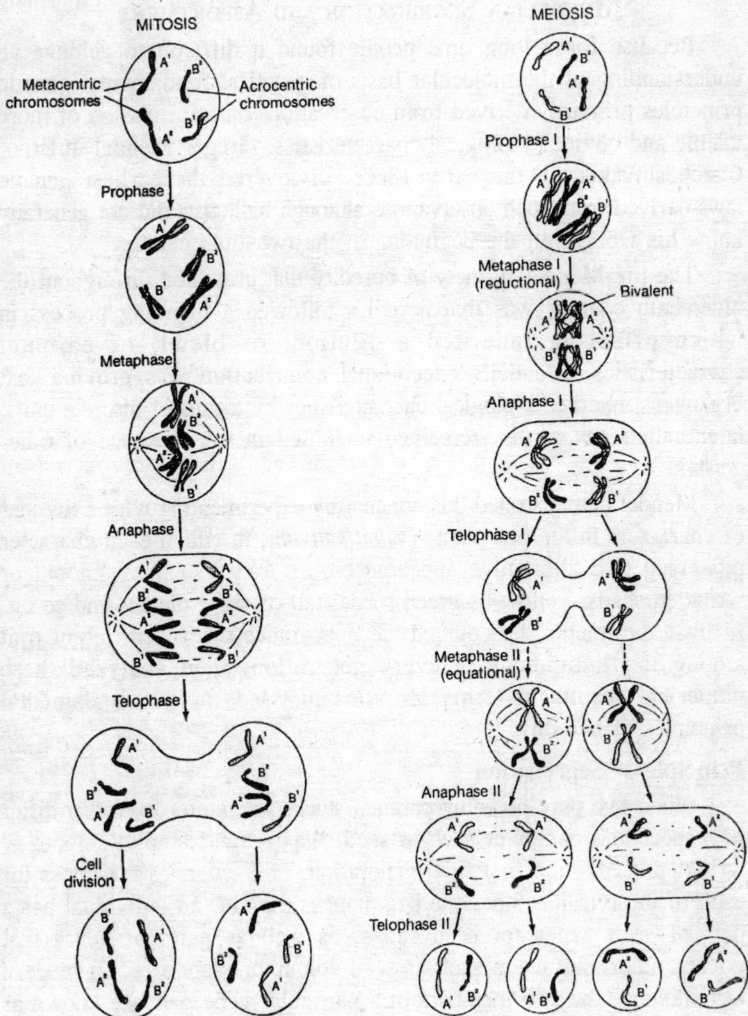

Fig. 9.3. Diagrammatic comparison of mitosis and meiosis, using an organism that has two pairs of homologous chromosomes, A^1 and A^2, and B^1 and B^2.

$A^2B^1C^2$, $A^2B^2C^1$, $A^2B^2C^2$). If we extend this to four pairs of chromosomes, 16 different gametic combinations are possible, the rule being that the number of possible different kinds of gametes equals 2^n, where n represents the number of pairs of chromosomes undergoing meiosis. In humans with normal karyotypes of 23 chromosome pairs, the different kinds of gametes that can form in this simple chromosome assortment process alone is 2^{23}, or more than 8 million varieties.

Mendelian Segregation and Assortment

Because for a long time people found it difficult to achieve an understanding of the molecular basis of genetics, fundamental genetic principles primarily derived from observations on transmission of more visible and obvious biological characteristics. Gregor Mendel in Brno, Czechoslovakia, in the early 1860s, discovered the earliest genetic laws derived from such observations although biologists did not generally know his work until the beginning of the twentieth century.

The pre-Mendelian view of heredity that prevailed throughout the nineteenth century was that heredity followed a blending process in which offspring inherited a dilution, or blend, of parental characteristics. Mendel's exceptional contribution was proving that organisms inherited biological characteristics by means of discrete units, later called *genes*, that remained undiluted in the presence of other genes.

Mendel demonstrated this concept by experimenting with a number of characters in the pea plant, *Pisum sativum*, in which each character possessed two alternative appearances, or traits, such as smooth or wrinkled seeds, yellow or green seeds, tall or short plants, and so on. In his experiments, he counted the appearance of each different trait among the individuals in every generation, then analyzed these numerical results in terms of ratios that led to two fundamental principles of heredity.

Principle of Segregation

When two pure-breeding parental stocks are intercrossed that differ in respect to a character such as seed shape—for example, smooth × wrinkled seeds—the first filial generation, or F_1, carries the genes for each of these traits. Since the pea plant is diploid, an individual has a pair of such genes for seed shape, as well as pairs of genes that govern other characters such as seed colour or plant size. In modern terminology, the two members of a particular gene pair are known as *alleles*. For simplicity, we can designate alleles by italic alphabetical letters; for example, the smooth and wrinkled alleles are, respectively, *S* and *s*. These alleles then segregate in the gametes of the F_1 hybrid, which then unite to form a second filial generation (F_2) in predictable proportions. These F_2 proportions arise because the allele for one trait (e.g., *S*) has *dominant* effects over the other *recessive* allele (*s*) when both are present together in *Ss* individuals. That is, *Ss* peas appear smooth (as are *SS* peas), and the only wrinkled peas are *ss*. Individual organisms that carry two different alleles for any particular character

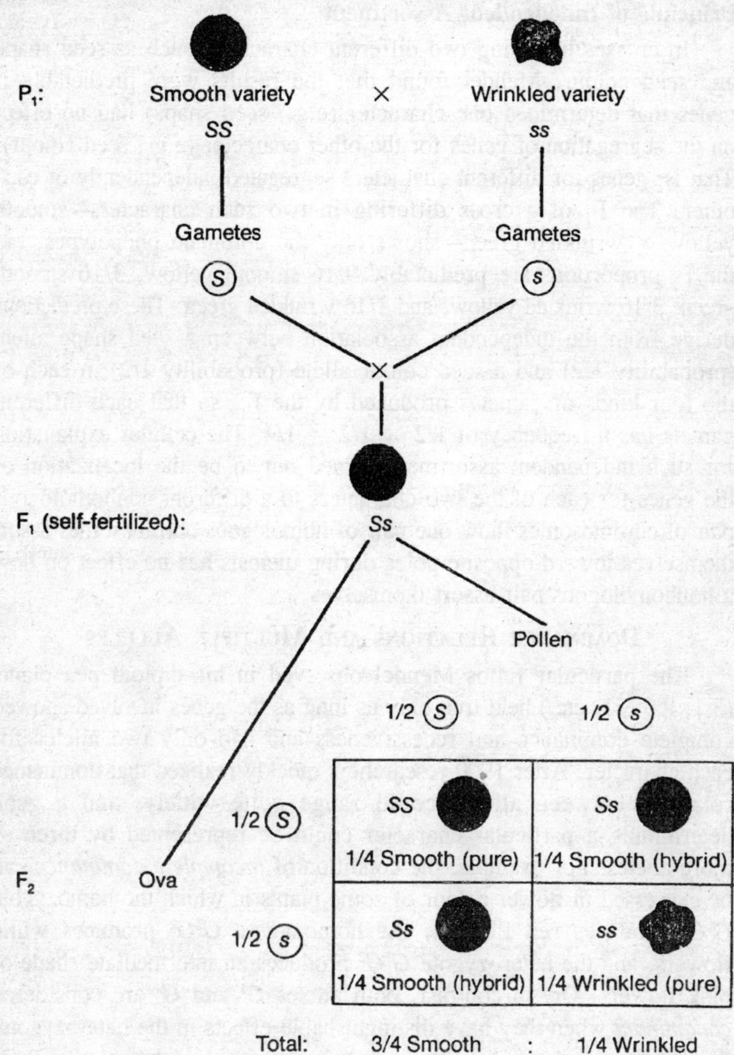

Fig. 9.4. An explanation of Mendel's observed F_1 and F_2 results for the inheritance of the seed shape character in garden peas, based on the segregation of alleles S (dominant) and s (recessive).

(e.g., *Ss*) are *heterozygotes*, and those which carry two identical alleles (e.g., *SS*, *ss*) are *homozygotes*. Mendel's experiments showed that these alleles had obviously not been changed or blended in the heterozygote but segregated from each other to be transmitted as discrete and constant particles between generations.

Principle of Independent Assortment

In crosses involving two different characters, such as seed shape and seed colour, Mendel found that the results were predictable if genes that determined one character (e.g., seed shape) had no effect on the segregation of genes for the other character (e.g., seed colour). That is, genes for different characters segregated independently of each other. The F_1 of a cross differing in two such characters—smooth yellow × wrinkled green—shows only the dominant phenotypes, but the F_2 proportions are predictably 9/16 smooth yellow, 3/16 smooth green, 3/16 wrinkled yellow, and 1/16 wrinkled green. These predictions derive from the independent association between a seed shape allele (probability 1/2) and a seed colour allele (probability 1/2) in each of the four kinds of gametes produced by the F_1, so that each different gamete has a frequency of $1/2 \times 1/2 = 1/4$. The cellular explanation for such independent assortment turned out to be the localization of the genes for each of the two characters to a different nonhomologous pair of chromosomes; how one pair of homologous chromosomes assort themselves toward opposite poles during meiosis has no effect on how a nonhomologous pair assort themselves.

Dominance Relations and Multiple Alleles

The particular ratios Mendel observed in his diploid pea plants (3:1, 9:3:3:1, etc.) held true only as long as the genes involved showed complete dominance and recessiveness and had only two alleles for each character. After 1900 researchers quickly realized that dominance relations between alleles could range quite widely, and a gene determining a particular character could be represented by three or more alleles. For example, the condition of *incomplete dominance* can be expressed in flower colour of some plants in which the homozygote G^1G^1 produces red flowers, the homozygote G^2G^2 produces white flowers, and the heterozygote G^1G^2 produces an intermediate shade of pink flowers. Or, in contrast, both alleles G^1 and G^2 are considered *codominant* when they have distinguishable effects in the heterozygote; that is, they each produce a uniquely recognizable substance, such as a special sugar compound, so that the heterozygote G^1G^2 has both compounds.

Any gene may also have more than two alleles, G^1, G^2, G^3, . . . producing, for example, a different colour or different compound in each different kind of homozygote, G^1G^1, G^2G^2, G^3G^3, ...The source for such *multiple allelic* systems arises from the fact that a gene consists of a linear array of hundreds or thousands of nucleotides, and

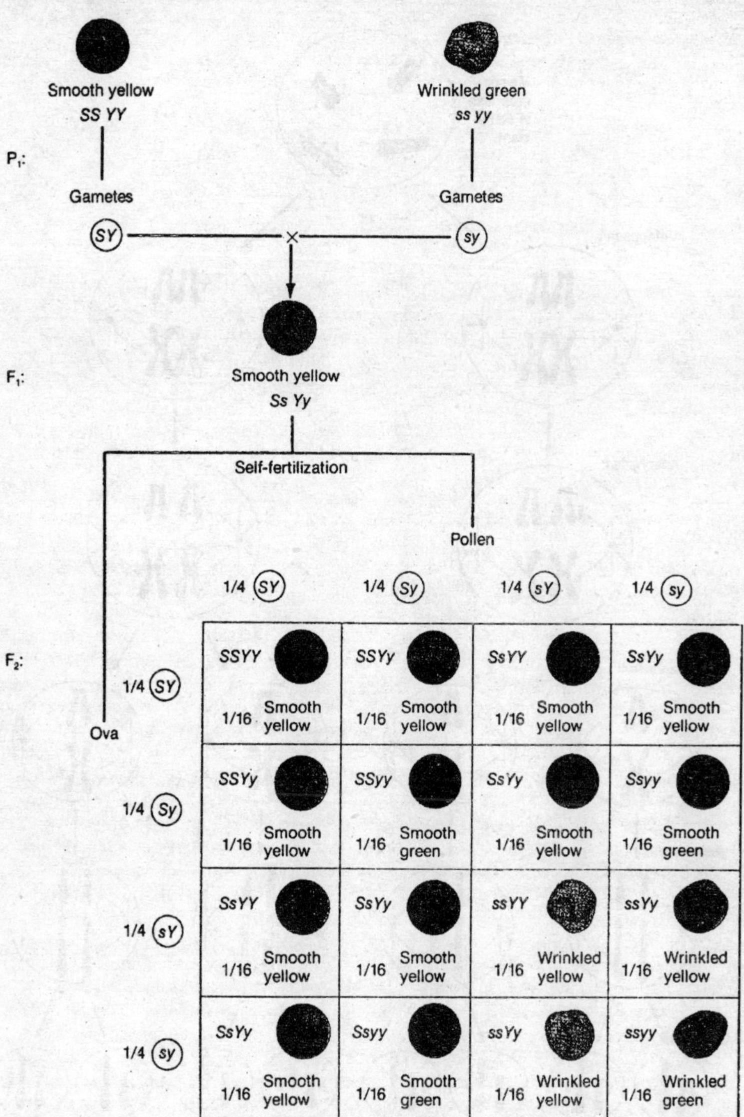

Fig. 9.5. Explanation of Mendel's results for the segregation and assortment of alleles at two pairs of genes; S and s for seed shape, and Y and y for seed colour.

an allelic difference may be caused by only a single nucleotide change at one of many positions along its length. Moreover, there may be different dominance relationships between the alleles in such a system so that G^1, for example, produces a codominant effect with G^2, but

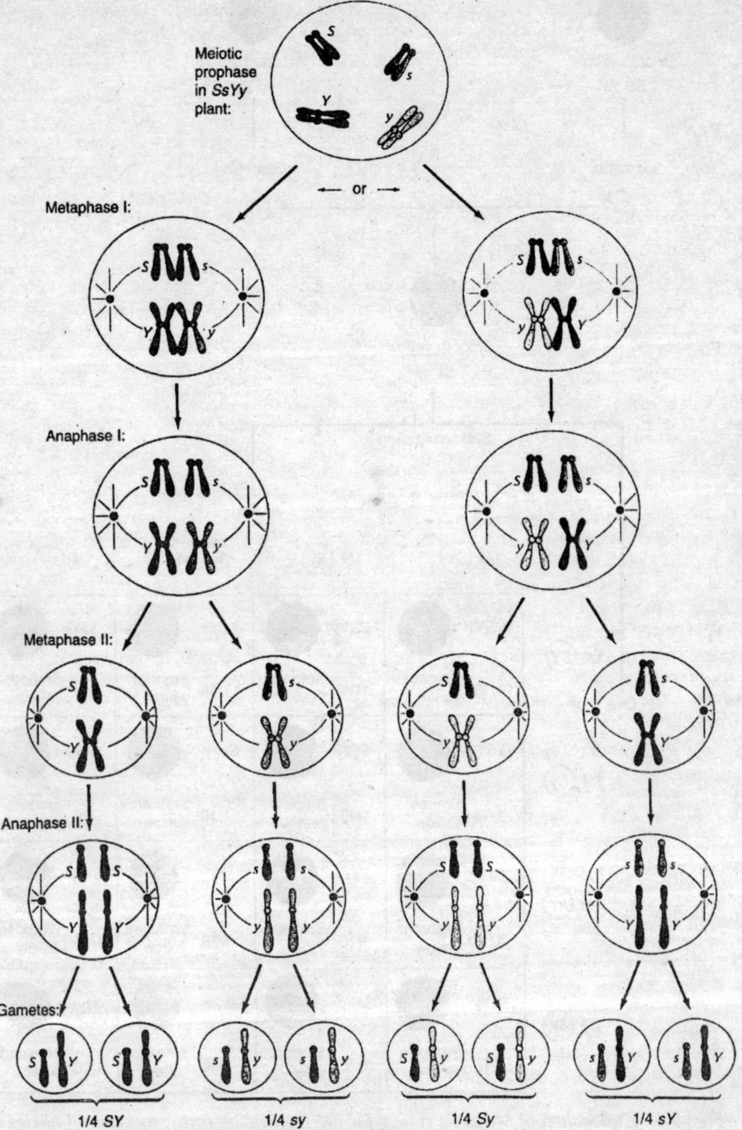

Fig. 9.6. Explanation for the segregation and independent assortment of seed shape and seed colour in Mendel's experiments in terms of factors or genes (S, s, and Y, y) localized on different chromosomes.

alleles G^1 and G^2 act as though they are each completely dominant over G^3. In addition, interactions between genes in different allelic systems may occur so that the expression of allele G^1, for example, changes because certain alleles are present at a different gene pair, such as H^1 or H^2. Among such interactions are those which can modify the dominance relations of a particular allele so that the effect of G^1, for example, is dominant over G^2 in certain genotypes (e.g., $H^1I^1J^1K^3$) but is recessive when the background genotype changes (e.g., $H^2I^1J^3K^2$). Complications of this kind have been given the name *epistatic interactions*, to describe when one or more gene pairs change the effect caused by other gene pairs. Another term, *modifiers*, is used for quantitatively measurable effects of genes on other genes.

In general, most common, or *wild type*, alleles have been evolutionarily selected to be dominant in diploid organisms because they produce advantageous products in the presence of other alleles whose products may not be as advantageous or may even be deleterious. On the molecular level, such wild-type allelic products are mostly functional proteins such as enzymes, whereas the products of mutant alleles often appear nonfunctional, only partly functional, or even absent (*null alleles*).

Because so many allelic differences, as well as different dominance and interaction effects, are possible, the variability thus generated is far beyond that described previously for the independent assortment of chromosomes during meiosis. To take a simple example, a gene with four alleles (G^1, G^2, G^3, G^4) can produce ten different possible diploid genotypes (G^1G^1, G^1G^2, G^1G^3, G^2G^1, G^2G^2 etc.), and 100 such genes can produce 10^{100} possible genotypic combinations—a figure larger than the estimated number of protons and neutrons in the universe. Since any cellular organism carries more than 100 genes, each with many possible alleles, its potential genetic variability is probably numerically greater than any conceivable aspect of reality.

Exceptions to Mendelism

Although Mendel was not aware of the cause, his principles were based, as we can see, on the meiotic process of gamete formation. When scientists discovered his contributions in 1900, they soon recognized the association between Mendelian factors and meiotic chromosome distribution, which provided the foundation for the new science of genetics. Within a short period, many other studies echoed Mendelian ratios and their variations and—because eukaryotes shared meiotic division processes—Mendelism seemed generally applicable to most, if not all, eukaryotic organisms. However, two important deviations from Mendelism became apparent in the decades that followed; both circumventing conventional meiotic expectations and both therefore having genetic and evolutionary significance.

The first unusual finding was that some traits did not follow a nuclear inheritance pattern but transmitted through egg cytoplasm. Such *extranuclear inheritance* (also called *cytoplasmic* or *maternal inheritance*) arose because cytoplasmic organelles such as mitochondria and chloroplasts had their own DNA genetic material. During fertilization of the zygote, practically all the egg cytoplasm including these organelles was maternally derived, while the sperm contributed little else beside nuclear genetic material. It was, in fact, through cytoplasmic inheritance that genes were first identified in mitochondria and chloroplasts, supporting proposals that these organelles had an independent evolutionary origin.

The second departure from Mendelism was the finding that some nuclear genes did not segregate as expected in heterozygotes. As mentioned previously, the four haploid products produced by the meiotic process in heterozygotes should lead to four gametes, of which two carry one allele and two the other allele (e.g., $Rr \rightarrow 2R{:}2r$). In a few cases, however, segregation is not normal, and the heterozygote for certain alleles produces more of one allele than the other (e.g., $Uu \rightarrow 4U{:}1u$). Researchers called such biased transmission *segregation distortion* (also *meiotic drive*), and found the cause to be associated with a few rare genes, most notably *Segregation Distorter in Drosophila* and the *tailless* genes in mice. Although relatively uncommon, meiotic drive can obviously lead to the persistence of a segregation distorter gene even though the gene may have deleterious effects.

Perhaps another deviation from Mendelism—really an elaboration rather than a departure—was the finding that genes do not necessarily assort independently of each other if they are linked together on the same chromosome. These investigations began with inferences that some genes were localized to the X (sex) chromosome, sex-linked genes.

Sex Linkage

Early in the twentieth century researchers discovered that various genes could be localized to a particular chromosome associated with sex determination. A prominent example of this is the "bleeder" disease, hemophilia, localized to the X chromosome of humans and other mammals. In these organisms males and females differ in respect to a pair of sex chromosomes so that males are XY and females are XX. Thus males (the *heterogametic sex*) produce sperm that contain either X or Y chromosomes, whereas females (the *homogametic sex*) produce

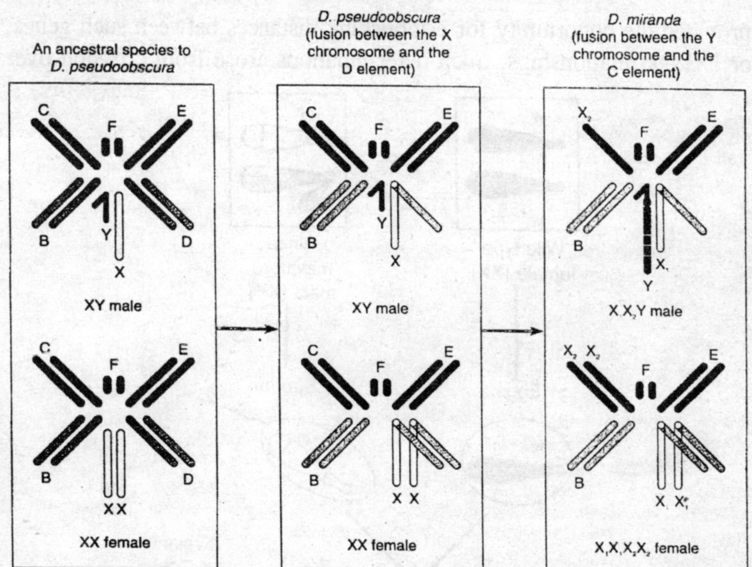

Fig. 9.7. Diagrammatic view of sex chromosome changes in a lineage leading to Drosophila miranda.

only X-bearing eggs. If the proportions of X- and Y-bearing sperm are equal, fertilization restores the two sexes in equal frequency: 1/2 X-bearing sperm × X-bearing eggs → 1/2 XX females; 1/2 Y-bearing sperm × X-bearing eggs → 1/2 XY males. However, since males carry only a single X chromosome, and the Y chromosome is mostly inactive, alleles present on the X chromosome express their effects in males, although such alleles may be recessive when in females. Thus a single hemophilia-producing allele on the X chromosome of a *hemizygous* male causes the classic hemophilia disease, whereas in XX females two such alleles are necessary to cause hemophilia.

Interestingly, of the more than 100 such sex-linked genes geneticists have now identified in humans, many are also sex linked in other mammals. The conservation of the same genes on different mammalian X chromosomes indicates that a large part of this chromosome has persisted throughout mammalian evolution, at least for 90 million years. Ohno, in fact, has proposed that we can expect any sex-linked gene in one mammalian species to be sex linked in other mammals as well.

Linkage and Recombination

The localization of different genes to the X chromosome, which geneticists first demonstrated in the fruit fly, *Drosophila melanogaster*, provided the opportunity for establishing distances between such genes, or *linkage* relationships. Such determinations arose from crossing over

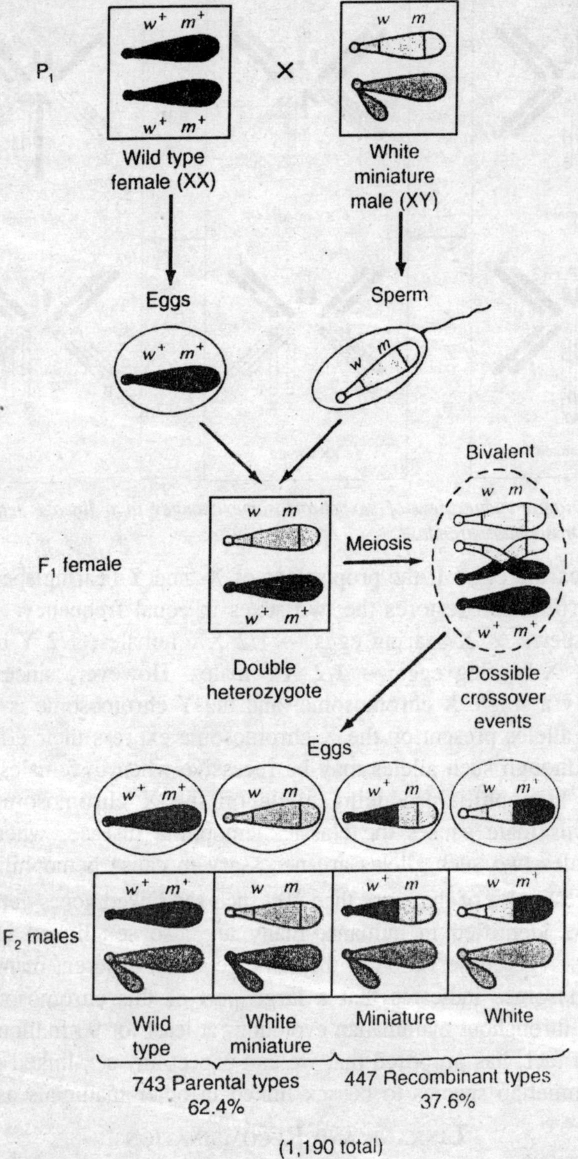

Fig. 9.8. Recombination between the sex-linked genes white and minature in D. melanogaster as evidenced by recombinant classes among F_2 males.

or recombinational events in which exchanges occurred between the chromatids of paired homologous chromosomes. For example, two *Drosophila* sex-linked genes, those involved in *white* eyes (w^+ and w) and *miniature* wings (m^+ and m), recombined with a frequency of about 38 percent to produce new chromosomal combinations.

This frequency of recombination provided a measure of the *linkage distance* between genes on the same chromosome: the greater the recombination frequency, the greater the distance. Thus, since the recombination frequency between the genes for *white* eyes and *cut* wings was about half the frequency of that between *white* and *miniature*, geneticists could assume the linkage distance between *white* and *miniature* to be about twice that between *white* and *cut*. Such experiments with both sex-linked and nonsex-linked (autosomal) genes provided linkage maps, indicating that chromosomes consist of linear arrays of genes, or *loci*, whose relative positions are additive (if three genes link in the order *H-G-I*, then the *H-I* linkage distance is the sum of *H-G* + *G-I* distances).

In humans and other mammals, geneticists have used techniques involving somatic cell fusions rather than mating between sex cells to obtain linkage relationships. These somatic hybridization techniques have been extraordinarily successful, and a large number of genes are now localized to many of the 23 human chromosomes. Moreover, as with similar sex-linked genes in mammals, genes localized to particular human chromosomes (identified by special banding patterns) can be localized to similar chromosomes in other mammals (e.g., sheep, cattle, cats, rats, mice)—a phenomenon known as *synteny*. In addition, linkage relationships among genes with similar functions in these mammals also often persist.

In bacteria and viruses, recombinational techniques are considerably more advanced than in eukaryotes because of the very small size of these organisms, their rapid generation times, and the ease with which we can identify biochemical mutant genes. These advantages have led to elaborately detailed linkage maps. In addition, new methods that permit exact determinations of nucleotide sequences in both prokaryotes and eukaryotes are now providing information beyond anything conceived even a decade ago. All these studies indicate the immense variability that can generate through recombination: any chromosome may come to differ from its homologues by carrying a distinctive combination of alleles, for example, … $G^1H^2I^1J^3K^2$ … versus … $G^2H^1I^3J^2K^1$ … versus … $G^1H^1I^1J^3K^1$ …, and so on.

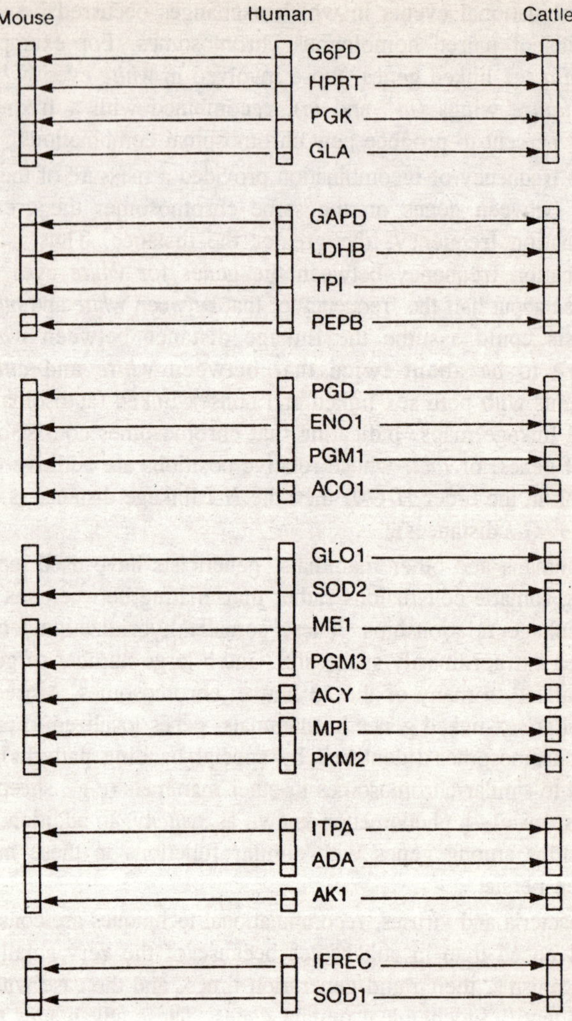

Fig. 9.9. Linkage relationships between 24 human gene loci compared to those of similar genes in mice and cattle.

Chromosomal Variations in Number

Forms of variation that are often microscopically observable are changes in the number or structure of chromosomes. The numerical chromosome variations are of two major kinds: changes in the number of entire sets of chromosomes, *euploid variations*; and changes in the number of single chromosomes within a set, *aneuploid variations*. Since most sexually reproducing eukaryotes are diploids with two sets of chromosomes (2n), euploid variations may extend from the haploid or monoploid condition (1n) to various levels of *polyploidy* (3n, 4n, ...) .

Such events may be caused by more than one sperm fertilizing an egg, or by cell division failures in which, for example, a diploid rather than haploid gamete results.

The appearance of extra sets of chromosomes within a species itself, *autopolyploidy*, seems to be a common mode of evolution in many plant groups such as mosses, apples, pears, bananas, tomatoes, and corn. Polyploids that originate from crossing between different species, *allopolyploids*, also appear in some plants such as wheat. For example, a gamete from species *A* may fertilize species *B* to produce the diploid hybrid *AB*, which can then undergo polyploidy to form the allotetraploid *AABB*, or other variations. Researchers have produced both auto- and allopolyploids with chemicals such as colchicine that break down spindle fiber microtubules and thereby interfere with chromosome segregation during cell division.

Interestingly, the first laboratory-created species was the product of a cross between two tobacco plants, *Nicotiana tabacum* (a diploid

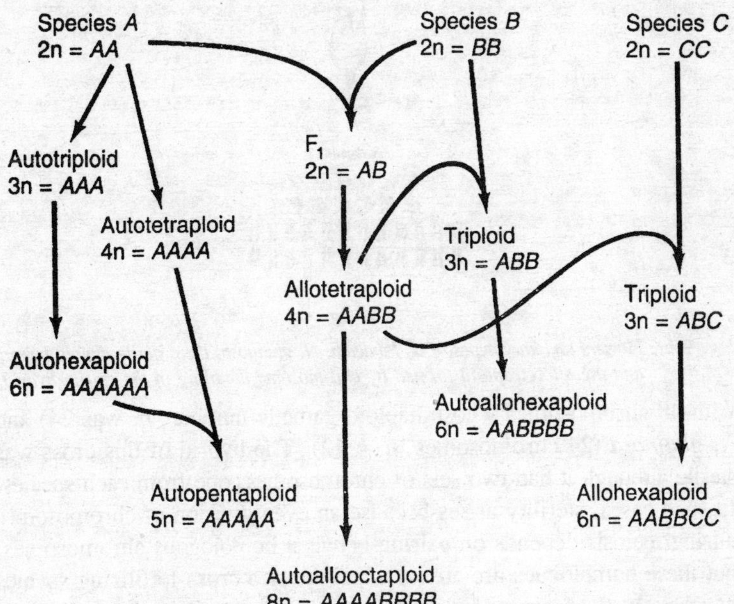

Fig. 9.10. Terminologies used for different polyploids, and some of the pathways by which they originate.

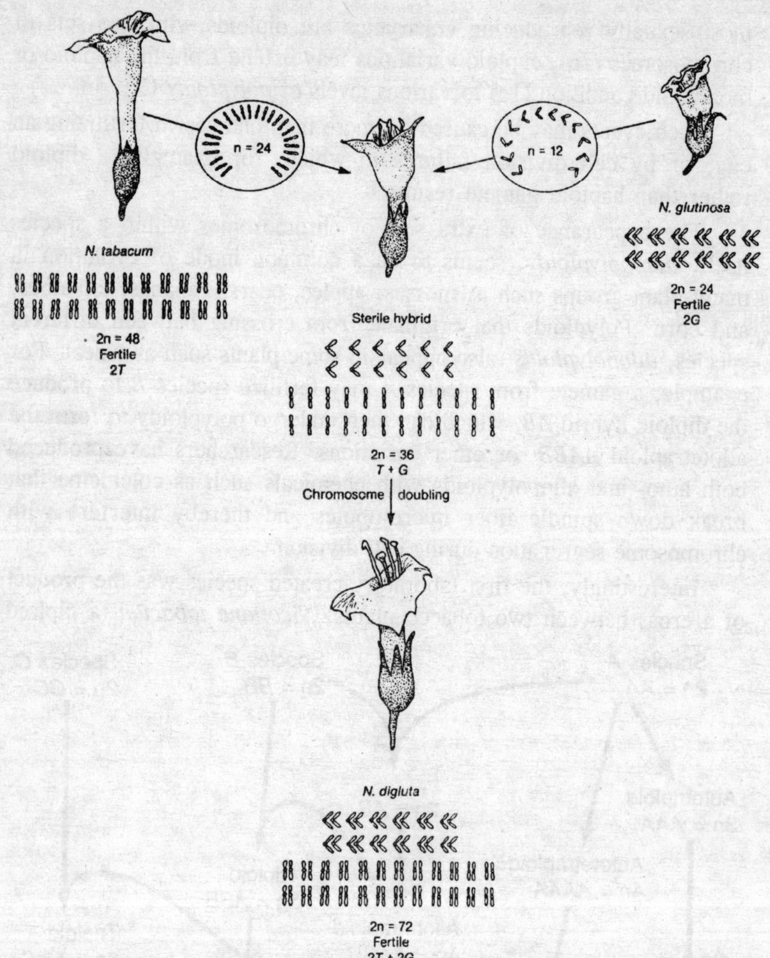

Fig. 9.11. Flowers and karyotypes of N. tabacum, N. glutinosa, their sterile diploid hybrid, and the allotetraploid formed by chromosome doubling in the sterile hybrid.

with 48 chromosomes whose haploid gametic number, n, was 24) and *N. glutinosa* (24 chromosomes, $n = 12$). The hybrid of this cross was sterile although it had two sets of chromosomes, one from each species. In such cases, sterility arises because an even division of chromosomes during meiosis depends on pairing between homologous chromosomes, but these homologues are absent when the two cross-fertilizing species have evolved chromosomal differences between them. Nevertheless, although sterile, the hybrid plant could continue to grow and increase by vegetative cuttings.

Eventually, a chromosome-doubling event produced an allopolyploid (also called *amphidiploid*) that was fully fertile because each chromosome now had a pairing mate, and normal meiosis could take place. Thus, at one stroke, a new species with 72 chromosomes, *N. digluta*, was created that was only fertile when crossed with itself (all its chromosomes could undergo homologous pairing) but could not produce fertile offspring when crossed to either of the parental species (only some of its chromosomes could pair homologously with those of either parent).

In animals, polyploidy is much rarer than in plants because most animals show much greater developmental sensitivity to even a small change in chromosome number. This sensitivity extends also to chromosomal sex-determining mechanisms, so that animal polyploids face difficulties in maintaining the same proportions of X and Y chromosomes present in normal diploids. For example, polyploidy in a species in which males are XY and females XX could lead to establishment of XXYY males and XXXX females. The subsequent combinations of gametes produced by these individuals (XY sperm + XX eggs) might well produce XXXY individuals that are not completely male or female. Therefore, when animal polyploid species do occur, they usually have some form of *asexual reproduction* such as *parthenogenesis* (embryonic development of eggs without fertilization).

Nevertheless, polyploidy probably has some advantages. In both plants and animals, the extra polyploid chromosomes may act as multiple buffers in various organismic processes, improving vigor and enabling such individuals to face new and drastic conditions. Also, the additional chromosomes may provide the chance to evolve new functions for their extra sets of genes.

In aneuploids, a wide range of variations may result when chromosomes are either added or subtracted from a normal set. Such events may occur because meiotic disjunction between homologous chromosomes is abnormal (*non-disjunction*). Again, plants seem to be more tolerant of such chromosomal variations than animals, but aneuploids are found in both groups.

Chromosomal Variations in Structure

Deletions or Deficiencies

The terms *deletions* and *deficiencies* describe losses of chromosomal material. In general, the severity of a deletion depends on how extensive it is and on what nucleotides or genes are missing. If functional genes are involved, deletions can be quite harmful in both diploids and haploids but not necessarily harmful in polyploids or aneuploids, where such genes may be present in extra chromosomes. Researchers have detected various deletions in populations, and many of these produce "buckles" meiotically observed when the deleted and nondeleted chromosomes pair up closely in heterozygotes.

Duplications

Duplications are either short or long segments of extra chromosome material originating from duplicated sequences within a genome. Since numerous *gene families* of similar or identical genes occur in many species, duplications show evidence of being common during evolution. For example, various eukaryotes produce the RNA components of ribosomes in clusters of long tandem arrays of duplicated genes. Other instances, such as the genes involved in producing different hemoglobin-type proteins show that duplicated genes have evolved in different pathways, enabling them to perform different functions. As in deletions, researchers can detect many chromosome duplications in heterozygotes by the buckles formed during meiotic pairing.

Inversions

Inversions are reversals in chromosomal gene order that researchers can sometimes observe in the formation of loops during meiotic pairing in heterozygotes. Inversions generally lower the recombination frequency within the inverted sequence, because crossing over within such sequences in heterozygotes can lead to chromosomal abnormalities. As a result, the genes included within an inversion tend to remain together as a nonrecombinant block, called a *supergene* by some workers.

Documented in various eukaryotes are two major kinds of inversions, *paracentric* that do not include the centromere, and *pericentric* that include the centromere. Among the possible effects of pericentric inversions is a shift in the relative position of the centromere. In some cases, this shift may be drastic, moving the centromere from the chromosomal center (metacentric position) to one end (acrocentric). In the deer mouse genus, *Peromyscus*, there are species whose 48 chromosomes are entirely metacentric (e.g., *P. collatus*) and species whose 48 chromosomes are almost entirely acrocentric (e.g., *P. boylei*).

Translocation

The term *translocations* primarily refers to the transfer of material from one chromosome to a nonhomologous chromosome. When the exchange of such material is mutual, it can result in the kind of *reciprocal translocation*. Such translocations are recognizable by a cross-shaped configuration between translocated and non-translocated

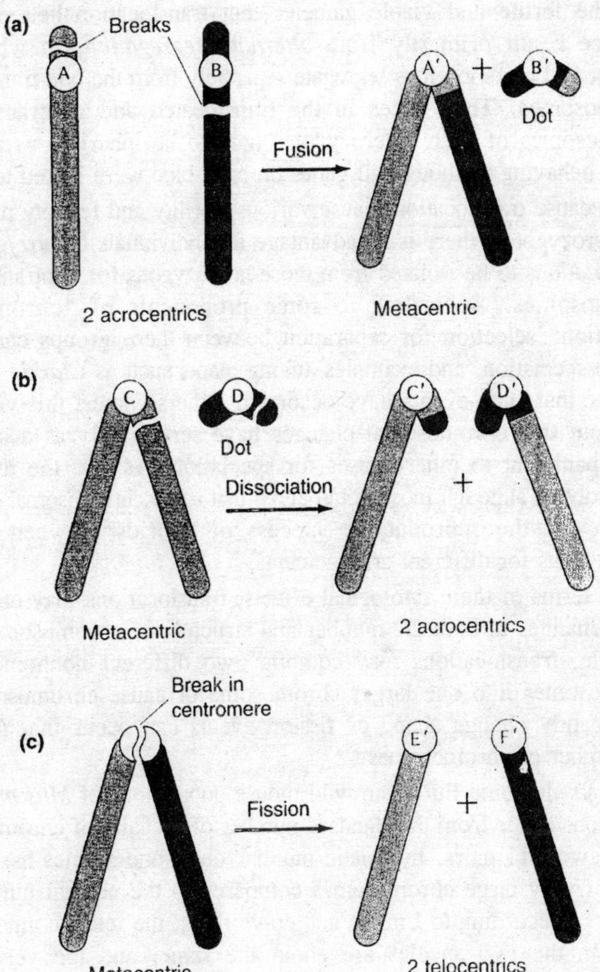

Fig. 9.12. (a) Translocations that lead to fusion between the arms of two acrocentric nonhomologous chromosomes (A and B) to form one metacentric chromosome (A') and a "dot" chromosome (B') that carries only a small amount of chromosomal material. (b) The reverse process of dissociation involves reciprocl translocations between the metacentric (C) and dot (D) chromosomes leading to C' and D' acrocentrics. (c) The fission mechanism proposed to explain the origin of presumed single-armed telocentric chromosomes (E', F') from two-armed metacentrics or acrocentrics.

chromosomes during meiotic pairing in the heterozygote. Since such meioses can produce gametes containing duplications and deficiencies, translocation heterozygotes often show sterility.

The fertile and viable gametes that translocation heterozygotes produce result primarily from *alternate segregation*, in which the translocated chromosomes segregate separately from the non-translocated chromosomes. Thus genes in the translocated and nontranslocated chromosomes of such heterozygotes tend to be inherited as separate blocs, behaving as though all genes on each bloc were linked together.

Because translocations cause various sterility and fertility problems in heterozygotes, there is an advantage for individuals homozygous for translocations to be isolated from those homozygous for nontranslocated chromosomes. According to some proponents of "chromosome speciation" selection for separation between these groups can easily lead lo speciation, and examples among plants such as *Clarkia* (Lewis) indicate that such events have occurred. Others contest this view and point out that chromosomal changes have served only as a sporadic accompaniment to other causes for speciation. As yet, the matter is not resolved, although most would agree that some chromosomal changes may lower the reproductive success of hybrids between groups homozygous for different arrangements.

In terms of their cytological effects, translocations may obviously cause changes in both the number and structure of chromosomes. For example, translocations may combine two different nonhomologous chromosomes into one larger chromosome or cause chromosomes to significantly change shape or fission events can occur that increase the number of chromosomes.

Thus, in some European wild mouse populations of *Mus musculus* reductions occur from the standard number of 20 pairs of chromosomes to as few as 11 pairs. In Asiatic muntjac deer, one species has only 3 pairs of very large chromosomes compared to the normal number of 25 pairs. According to Liming and co-workers, the relative amounts of DNA in the two species are about the same, and the very large muntjac chromosomes are derived from successive translocations that combined the smaller muntjac chromosomes. Chromosomal conservatism, in contrast, may be a strong feature in other mammalian groups. For example, present species of the 30-million-year-old lineage that includes Asian (Bactrian) and African (Dromedary) camels as well as the South American guanaco, vicuna, llama, and alpaca, all have the exact same number of morphologically similar chromosomes, 74. Qumsiyeh proposes that the number of chromosome pairs reflects a species' adaptive capability: larger numbers allow increased genetic recombination, thereby producing more variability and environmental flexibility; whereas smaller numbers allow genetic combinations to persist, which lead to greater specializations for specific habitats.

CHROMOSOMAL EVOLUTION IN *DROSOPHILA* AND PRIMATES

In those instances where researchers can identify linear sections of chromosomes through distinctive bandings, we can chart chromosomal

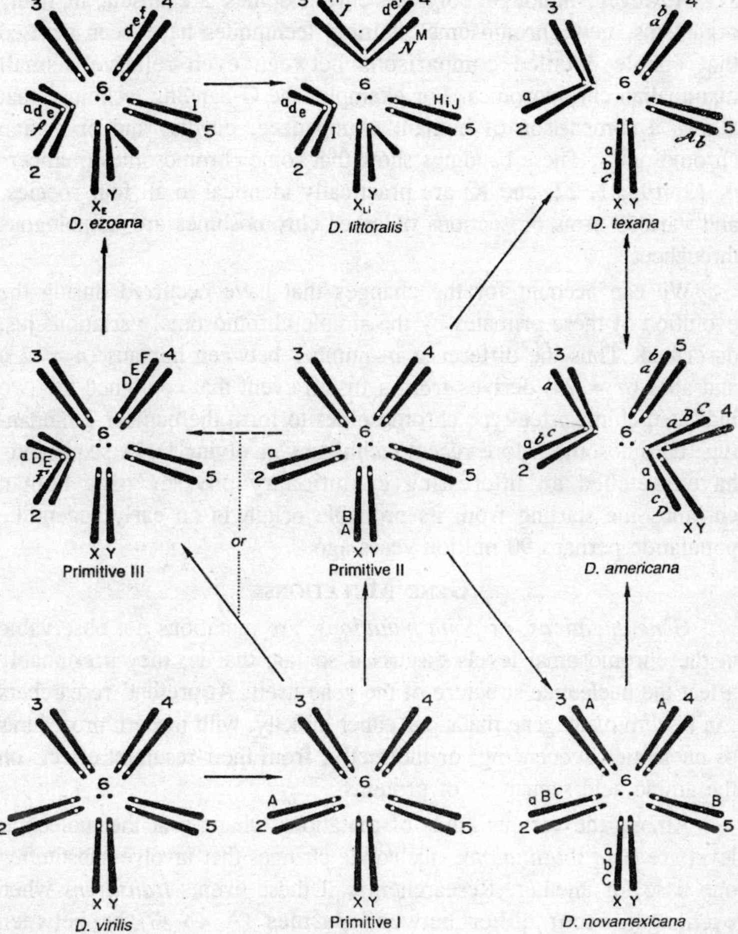

Fig. 9.13. Paths of chromosomal evolution in some species of the virilis group of Drosophila.

evolution in great detail. In *Drosophila* species and other dipteran insects, the chromosomes of salivary gland cells and other tissues have replicated many times over, yet each replicate still remains closely apposed to other such replicates within the same nucleus. As a result, such *polytene chromosomes* are tremendously enlarged and show highly detailed banding configurations along their lengths. These distinctive banding arrangements enable even minor chromosomal changes to be traced. Geneticists have described practically all the chromosomal changes in the evolution of hundreds of these fly species.

However, although polytene chromosomes are absent in many organisms, new chromosomal staining techniques have been devised that enable detailed comparisons between even relatively small mammalian chromosomes. For example, the G-banding technique that allows a comparison of human, chimpanzee, gorilla, and orangutan chromosomes. These bandings show that some chromosomes (numbers 6, 13, 19, 21, 22, and X) are practically identical in all four species, and various arms or sections of other chromosomes are homologous throughout.

We can account for the changes that have occurred during the evolution of these primates by the simple chromosomal variations just described. Thus the difference in number between humans ($n = 23$) and apes ($n = 24$) derives from a fusion event that combined the two indicated chimpanzee-type chromosomes to form the number 2 human-type chromosome. More recent techniques involving DNA sequencing have sketched an interesting evolutionary odyssey of a human chromosome starting from its probable origin in an early rodentlike population perhaps 90 million years ago.

Gene Mutations

Gene mutations, or *point mutations*, are mutations not observable at the chromosomal levels discussed so far; that is, they presumably affect the nucleotide structure of the gene itself. At present, researchers can discern many gene mutations either directly, with modern procedures of nucleotide sequencing, or indirectly, from their resultant effects on the amino acid sequences of proteins.

Among the various kinds of mutational changes at the molecular level are *base substitutions*, nucleotide changes that involve substituting one base for another. Researchers call these events *transitions* when exchanges occur either between purines (A $\leftrightarrow$ G) or between pyrimidines (T $\leftrightarrow$ C) and *transversions* when purines exchange for pyrimidines or vice versa (A, G $\leftrightarrow$ T, C).

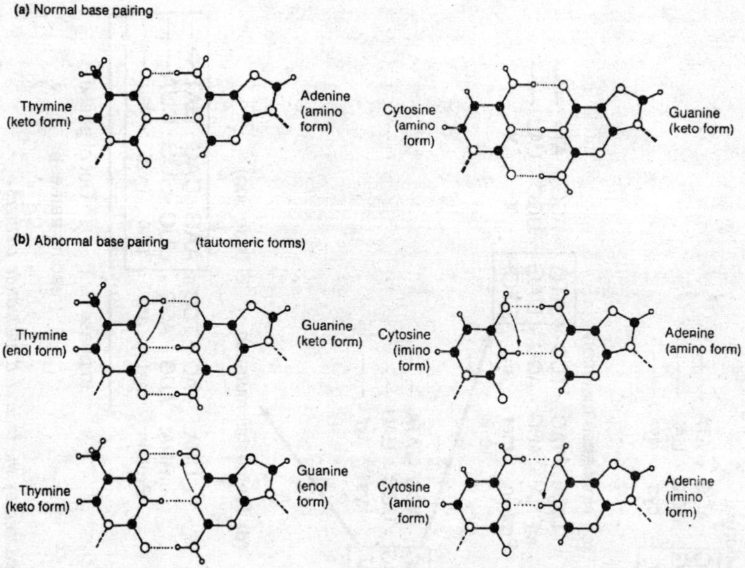

(a) Normal base pairing

Thymine (keto form) — Adenine (amino form)

Cytosine (amino form) — Guanine (keto form)

(b) Abnormal base pairing (tautomeric forms)

Thymine (enol form) — Guanine (keto form)

Cytosine (imino form) — Adenine (amino form)

Thymine (keto form) — Guanine (enol form)

Cytosine (amino form) — Adenine (imino form)

Fig. 9.14. (a) Normal complementary pairing between nucleotide bases during DNA replication. (b) Modified base-pairing relationships that result from tautomeric molecular changes.

Some such substitutions may occur spontaneously through copying errors caused, for example, by rare *tautomeric* nucleotide base changes that enable complementary pairing between adenine and cytosine or between guanine and thymine. Other base substitutions may arise from the action of mutagenic agents such as hydroxylamine, nitrous acid, and nitrogen mustards. Also, in some instances specific nucleotide sequences cause increased mutations among adjacent nucleotides. Such "hot spots of mutation" may act by coiling the DNA molecule in ways that influence DNA polymerase enzymes to produce replication errors, the replication accuracies of polymerase enzymes also differ; some strains carry enzymes that are apparently more prone to produce mutational errors than others.

Other mutational events at the molecular level include nucleotide deletions, duplications, and insertions. These can arise spontaneously within the cell or be evoked by externally applied mutagenic agents such as acridine dyes. Rearrangements of nucleotides also occur, such as inversions (reversals in nucleotide order) or transpositions (movement of nucleotide sequences to new positions).

In general, mutational effects caused by these mechanisms may express at two levels of gene activity: (1) changes within the gene

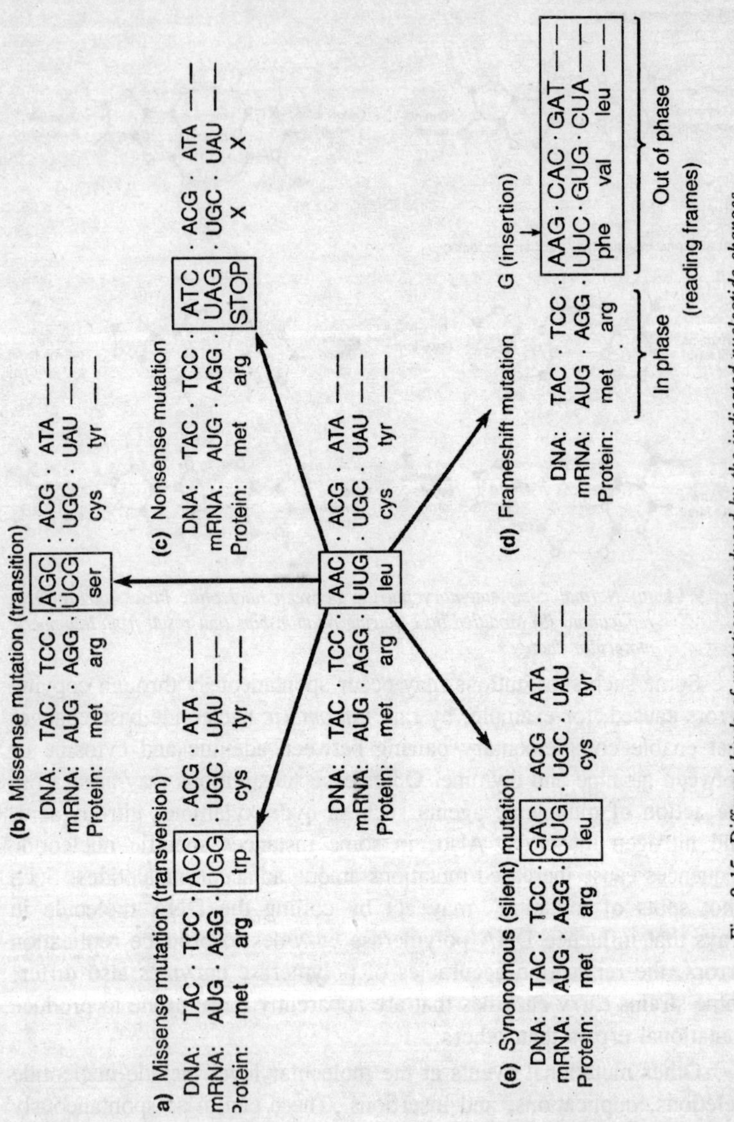

Fig. 9.15. Different kinds of mutations produced by the indicated nucleotide changes.

product itself—for example, in the amino acid constitution of a particular protein: and (2) changes in the regulation of a gene product—meaning that the gene product is not itself affected, but the timing of its appearance is different from normal.

Mutational effects that result in a changed gene product may arise because of nucleotide changes that cause

- A substitution for one or more of the amino acids in a protein (missense mutations).
- Changes that insert protein termination ("stop") codons in the middle of a gene sequence, thus causing premature termination of polypeptide chain synthesis (nonsense mutations).
- Nucleotide insertions or deletions that modify the messenger RNA (mRNA) protein translation reading frame so that a new and different sequence of codons appears (frameshift mutations).
- In addition, since practically all amino acids are coded by more than one kind of codon, synonymous mutations can change an amino acid codon without producing an amino acid substitution.

Depending on its position, a nucleotide mutation may have important consequences for the organism although it causes only a single amino acid substitution in a long-chain protein. A prominent example of such an effect is that caused by the *sickle cell* mutation in humans, a gene that, in the United States, is almost entirely confined to blacks. Normally, the adult hemoglobin molecule in human blood cells consists of four polypeptide globin chains, two α's and two β's, each about 140 amino acids long with its own specific sequence. However, in homozygotes for the sickle cell gene (Hb^s/Hb^s) all β-globin chains differ from normal β's at the number 6 position because of a transversion that changed the glutamic acid codon GAA to the sickle cell valine codon GUA.

The effects of this single genetic mutation are profound, causing a variety of phenotypic changes that often lead to inviability. (*Pleiotropy* is the name given to multiple phenotypic effects of a single gene.) Sickle cell disease is known to kill more than 10 percent of American black homozygotes before the age of 20, and probably has even more lethal effects in Africa, where medical facilities are limited. The high frequency of this gene in black populations is related to the selective advantage of sickle cell heterozygotes in malarial regions.

Regulatory Mutations

Regulatory mutations are those that affect the rates at which gene products are produced, although the products themselves may be unaffected. Among such examples are the *thalassemias*, genetic diseases in which the production of either α or β chains is absent or diminished. Although such mutations, like the sickle cell allele, often cause lethality in homozygotes, they apparently also offer protection against the dread malarial parasites. Presumably this special advantage of some thalassemia heterozygotes explains the frequency of these genes in human populations.

Viruses and prokaryotes have, so far, provided much more information on regulatory mechanisms than have eukaryotes. The prokaryotic system in *Escherichia coil* bacteria governs the production of enzymes involved in lactose sugar metabolism. Since *E. coli* bacteria do not commonly encounter lactose, a *repressor protein* that occupies a specific regulatory *operator site* normally prevents the genes used in *lac* enzyme synthesis (*structural genes*) from being transcribed into mRNA. Molecular binding between the repressor and operator DNA prevents the RNA polymerase enzyme from attaching to its *promoter site*, near which transcription normally begins. As a result, transcription of *lac* enzyme genes into mRNA is prevented.

However, when bacteria encounter lactose sugar in the medium, some lactose molecules convert to a form called *allolactose*, which acts as an *inducer* that binds with the repressor. This combination releases the repressor from the operator site, thus allowing the RNA polymerase to transcribe the genes necessary to metabolize lactose.

Because of the complexity of regulatory systems, various kinds of mutations can obviously affect the quantity and timing of gene productivity. For example, some mutations in the *I* regulator gene that produces the *lac* repressor can prevent it from binding to the operator, thereby causing *lac* enzyme synthesis to occur even in the absence of inducer. Conversely, other *I* mutations produce repressor proteins that cannot bind to inducer molecules, thus persistently preventing *lac* enzyme synthesis by keeping the repressor attached to the operator site even in the presence of inducer. In addition, various mutations of DNA at the promoter site may either increase or decrease the rate of transcription by preferentially enhancing or diminishing the attachment of RNA polymerase enzymes.

In eukaryotes, regulatory systems involve special sites on DNA sequences, called CAAT and TATA boxes, to which special proteins attach that allow transcription. Some investigators have also demonstrated that the DNA double helix at some eukaryotic regulatory sites changes from a right-hand to a left-hand form, accompanied by a zigzag placement of phosphate groups.

Since regulation plays an essential role in the timing and placement of all metabolic reactions, regulatory mutations can easily affect both the morphology and function of any organism. For example, the *bithorax* locus in *Drosophila melanogaster*, which governs the placement of

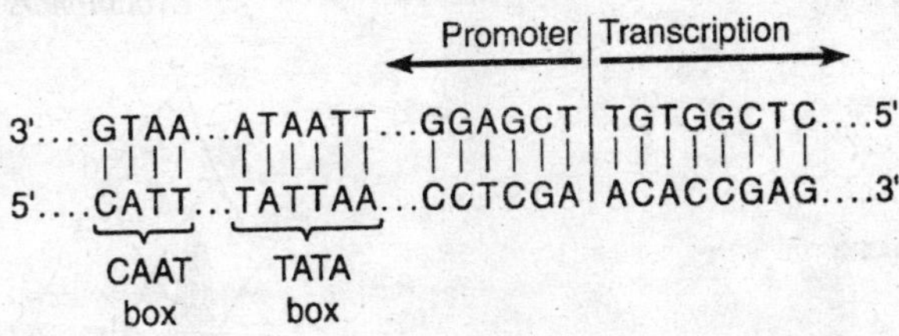

Fig. 9.16. A eukaryotic nucleotide sequence regulating the gene that produce the thymidine kinase enzyme.

structures in various segments, may have mutations that produce an extra set of wings. Researchers have long suggested that simple regulatory changes that affect developmental growth coordinates can produce changes in the shapes of various related fishes, as well as in many other species groups.

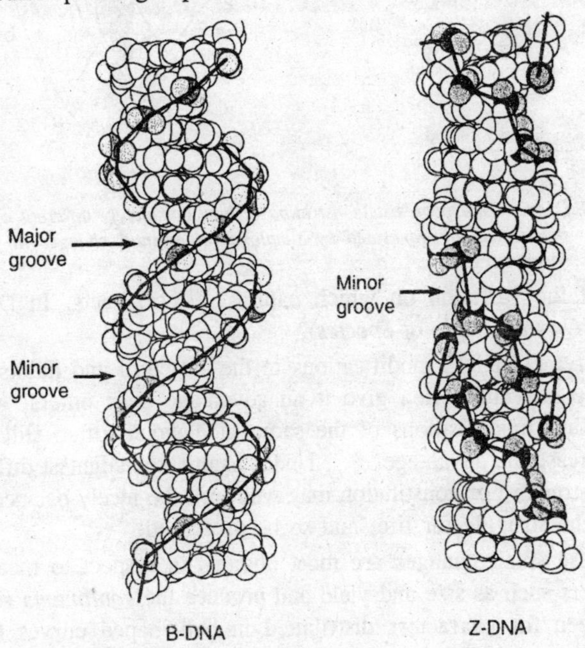

Fig. 9.17. Space-filling models of B-DNA and Z-DNA double-helix molecules.

Quantitative Variation

Although large regulatory changes can explain some major differences between groups, the extent to which they account for most other evolutionary events is still unclear. Beginning with Darwin himself, many evolutionists suggested that rather small heritable changes provide

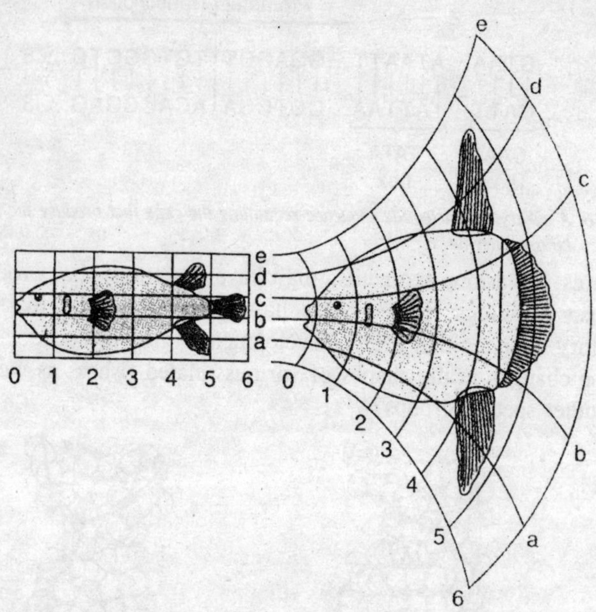

Fig. 9.18. D'Arcy Thompson's demonstration of how completely different organismic shapes can be generated by simple developmental changes in geometric coordinates.

most of the variation on which natural selection acts. In Darwin's words (*On the Origin of Species*).

> Extremely slight modifications in the structure and habits of one species would often give it an advantage over others; and still further modifications of the same kind would often still further increase the advantage. . . . Under nature, the slightest differences of structure or constitution may well turn the nicely-balanced scale in the struggle for life, and so be preserved.

These small changes are most obvious in respect to measurable characters such as size and yield and produce the *continuous variation* often seen for characters distributed in bell-shaped curves (normal distributions), such as human heights.

Various studies showing that small differences can accumulate through selection to give large differences have supported the importance of continuous variation in evolution. For example, selecting for the presence or absence of white spotting in Dutch rabbits can lead to completely coloured or completely white strains, the genetic causes for these changes are genes with small phenotypic effect, called *multiple factors* or *polygenes*, that can produce the familiar normal distributions when they assort independently. Thus, a mating between heterozygotes for three pairs of genes, each with two alleles—one coloured, one white—produces the phenotypes, ranging from all coloured to all white. Experiments and analyses of this kind therefore demonstrated that we could explain selection for quantitative characters on the basis of the segregation and assortment of simple Mendelian genes whose individual small effects may add up to large phenotypic differences.

Mutation Rates

Considering everything discussed so far, we see that the opportunity for all kinds of mutations obviously derives from various sources and may have various phenotypic effects. For newly arisen mutations, these effects will most likely be harmful because prevailing genotypes are generally well adapted for their particular environments, and most changes are unlikely to improve them further. Detecting new mutations in most organisms therefore usually goes along with observing newly inherited harmful effects.

The observed mutation rates are generally low, on the order of about one mutation per 100,000 copies of a gene. In organisms such as humans, carrying an estimated 100,000 genes per haploid genome, this means that each sperm and egg may well carry one newly arisen mutation, or an avenge of two such mutations in a diploid fertilized zygote. Multiplied by the large number of individuals in most populations, it is clear that almost any species will carry a large number of newly arisen mutations, many or most of which are harmful.

However, mutation rates are not necessarily constant. Among the causes that can modify mutation rates are genes for polymerase enzymes that replicate DNA. Some alleles of these genes act as *mutator genes* that can increase mutation rates manyfold, whereas other alleles act as *antimutators* to decrease mutation rates. Mutation rates, like other essential traits, seem therefore adjusted to optimum values in most, if not all, organisms. Nevertheless, external causes may considerably affect mutation rates, including, surprisingly, infectious elements that can be transmitted from other individuals. For example, some viruses, such as herpes simplex, rubella (German measles), and chicken pox, can cause breaks and deletions in chromosomes because they release nuclease enzymes that attack host DNA.

Transposons, Repeated Sequences, and Selfish DNA

Other sources of mutational change in both prokaryotes and eukaryotes are *transposons*: nucleotide sequences that can promote their own transposition among different genetic loci. The transposon produces special transposase enzymes that let copies of the transposon insert into various target sites. For example, the IS1 transposon makes staggered cuts at each side of a nine-nucleotide base pair sequence, and a copy of IS1 inserts within the gap these cuts produce. Depending on where transposons insert, mutations of all kinds may arise, marked by target site repeats, in which similar sequences of nucleotide bases appear at each end of the insertion but in inverted order.

Researchers have therefore used inverted repeats to detect the presence of transposable elements in various species, and such repeats indicate that a transposon can pick up DNA sequences and transfer them to other DNA locations by recombination. Antibiotic resistance genes, for example, can be transmitted between bacterial strains by small, circular DNA particles called *plasmids* that have received transposon insertions. Such instances indicate that some hereditary traits that transposons carry may have passed "horizontally" between individuals of the same generation, rather than through normal "vertical" (i.e., gametic) transmission between generations. Once established in their new hosts, horizontally transmitted traits can undergo further mutation, leading to entirely new characteristics. Thus, as an example of rapid adaptive evolution, various pathogenic bacteria have widened their spectrum of resistance to many different antibiotic drugs in a decade or less.

Moreover, such horizontal transmission may occur even between different species and between different phyla. Thus, the *P* transposon, originally a transposable element common to *Drosophila willistoni*, first spread to *Drosophila melanogaster* about 50 years ago and is now present in all wild populations of that species. The likelihood that the original *willistoni-melanogaster* event involved horizontal transfer is supported by the fact that the two species differ significantly in their gene nucleotide sequences, indicating a separation of about 20 million years, yet their *P* transposons differ by only a single nucleotide. Although discriminating between horizontal and vertical transfer can be difficult, other transposon elements (such as *Hobo*) also show evidence of horizontal transmission, and the phenomenon may be more frequent than most geneticists anticipated. As Kidwell and others point out, one important advantage transposons gain by horizontal transmission is to circumvent the barriers of reproductive isolation between species, and thus escape their inevitable extinction in vertical lineages when a species dies out.

Also of interest is the presence of many such sequences within species. In primates, for example, a 300 base pair sequence with transposon-like features called *Alu* is represented by perhaps more than one million copies per human diploid cell. Smaller repetitive sequences of the type are also widely prevalent in various eukaryotes. For example, only 3 percent of human DNA codes for proteins, whereas the remainder are noncoding sequences comprising many kinds of repetitive elements. What explains the widespread distribution and persistence of what appears to be extraneous DNA?

According to some writers many transposable elements and other forms of *repeated sequences* contribute little, if any, function to their host cells. Since the DNA replication process cannot discriminate between functional and nonfunctional sequences, it replicates any introduced DNA sequence. Transposon DNA and repeated sequences may therefore perpetuate parasitically as either "junk" or "selfish" DNA. Other explanations propose that some such sequences may function as essential elements in regulating gene activity, or help maintain the structure and integrity of the chromosome, or serve as origins of DNA replication, or act as mutator genes that occasionally provide new adaptive mutations. No agreement exists on how to weigh the selfishness or unselfishness of such sequences, and possibly some repeated sequences fulfill both roles.

Randomness of Mutation

Until the 1950s many bacteriologists felt that bacteria had a unique "plastic heredity" in which appropriate mutations arise as an immediate response to the needs of the environment. This concept seemed supported by observations in which bacteria exposed to some virus or antibiotic would quickly develop a resistant form. The explanation offered was that mutations do not originate on a random basis before exposure to some selective agent (*preadaptive* mutations), but rather that appropriate mutations are stimulated to arise only after bacteria have encountered a selective agent (*postadaptive* mutations). As with Lamarck, the postadaptive mutation concept had to rely on some unknown, perhaps mystical, agency that allowed the environment to directly cause the appearance of new adaptive hereditary factors instead of employing Darwin's natural selection to choose among preadaptive hereditary factors already present.

J. and E. Lederberg performed an impressive test that evaluated these pre- and postadaptive models, using a novel technique called *replica plating*. They transferred samples from a "lawn" of bacteria growing on a petri dish (master plate) to other petri dishes (replica plates), which contained selective media such as viruses (bacteriophages) or antibiotics (streptomycin). They made these transfers in a manner enabling bacteria derived from specific clones on the master plate to be localized to the same positions on the replica plates. The finding that resistant bacteria occupied identical positions on various replica plates indicated that they arose from the same clone on the master plate, a clone that must have been present before exposure to the selective medium on the replica plates. In other words, these mutants shared a common preadaptive origin on the master plate and did not arise postadaptively because of a Lamarckian response to a stimulus by the selective medium.

Genetic Polymorphism: Widespread Nature of Variability

New mutations that have an immediate beneficial effect on the organism seem generally to be quite rare, although some others are either neutral in their effect or harmful only when they occur in relatively rare homozygotes. Such neutral and deleterious but recessive mutations can therefore often accumulate in a population without any immediately serious unfavourable effects, thereby furnishing a reservoir of genetic variability.

Genetic variability, or *polymorphisms*, may also include the maintenance of different kinds of chromosomal anomalies such as inversions, translocations, and extra chromosomes. In *Drosophila pseudoobscura*, for example, populations in different localities in the western United States are polymorphic for a wide variety of third chromosome gene arrangements, and the frequencies of such arrangements may change seasonally. This indicates not only that chromosomal polymorphism is generally adaptive in this species, but also that certain polymorphic variations are preferentially adaptive in helping populations adjust to their specific environment at specific times.

Variation in chromosome number between very similar species can also associate with adaptive features. According to Nevo and co-workers, stressful ecological conditions such as periodic aridity and other unpredictable hardships appear to correlate with chromosome numbers in Israeli and Turkish populations of the mole rat *Spalax*. They suggest that the chromosomal increase caused by dissociation or fission leads to increased genetic diversity by increasing the numbers of different possible chromosome combinations, and thus allows species in such localities to specialize for a rise in ecological variation.

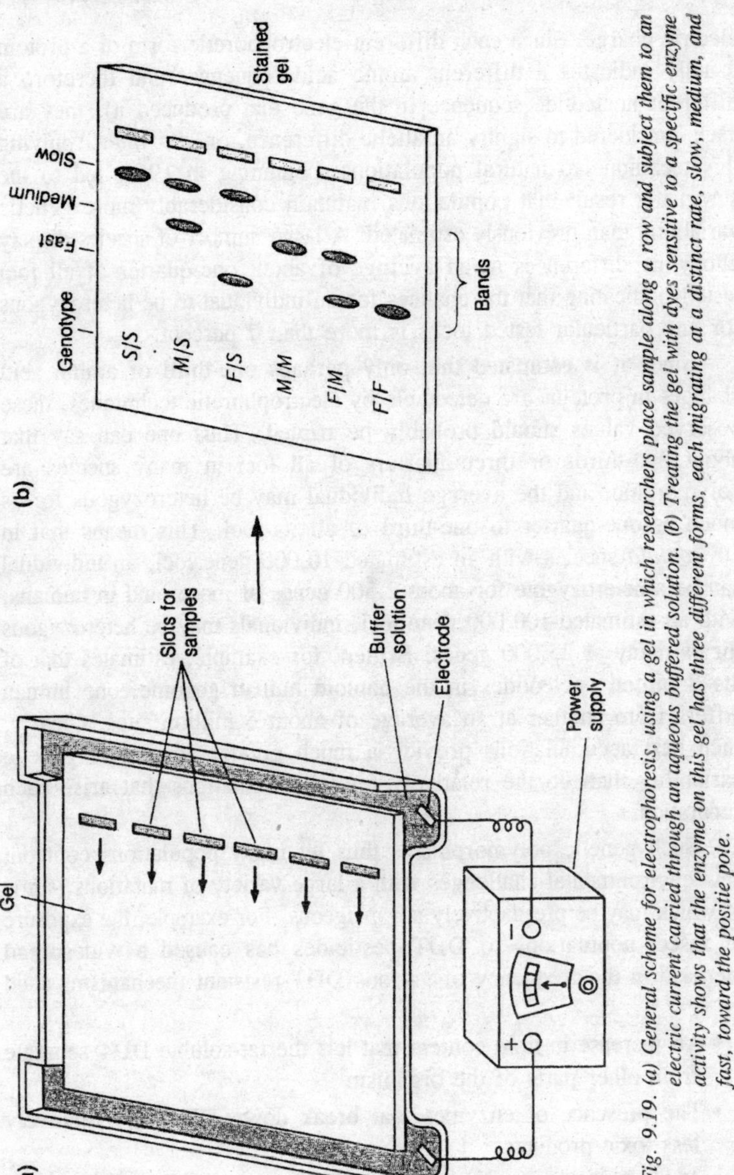

Fig. 9.19. (a) General scheme for electrophoresis, using a gel in which researchers place sample along a row and subject them to an electric current carried through an aqueous buffered solution. (b) Treating the gel with dyes sensitive to a specific enzyme activity shows that the enzyme on this gel has three different forms, each migrating at a distinct rate, slow, medium, and fast, toward the positie pole.

On the gene level itself, researchers can discern the magnitude of polymorphism by techniques that make allelic differences visible. Among such studies are electrophoretic methods that measure the mobility of a protein in an electric field, distinguishing even slight variations in electric charge. Since each different electrophoretic form of a protein usually indicates a different amino acid sequence (and therefore a different nucleotide sequence in the gene that produced it), they are each considered to signify an allelic difference, or *allozyme*. Applying this technique to natural populations, beginning in 1966, led to the surprising result that populations maintain considerably more genetic variability than previously estimated. A large number of species display allozymic differences at an average of about one-quarter of all loci tested, indicating that the chances for an individual to be heterozygous for any particular tested locus is more than 7 percent.

Since it is estimated that only perhaps one-third of amino acid changes in proteins are detectable by electrophoretic techniques, these observed values should probably be tripled. Thus one can say that about two-thirds or three-quarters of all loci in many species are polymorphic, and the average individual may be heterozygous for as much as one-quarter to one-third of all its loci. This means that in *Drosophila* species with an estimated 10,000 gene loci, an individual can be a heterozygote for about 2,500 genes or more; and in humans, with an estimated 100,000 gene loci, individuals may be heterozygous for as many as 25,000 genes! Britten, for example, estimates that of the 3 billion nucleotides in the haploid human genome, one human differs from another at an average of about 5 million sites. Clearly, such past accumulations provide a much greater amount of genetic variability than do the relatively few new mutations that arise each generation.

Such genetic polymorphisms thus let many populations confront new environmental challenges with a large variety of mutations, some of which may be preadaptively advantageous. For example, the exposure of insect populations to DDT pesticides has caused a widespread increase in the frequency of various DDT-resistant mechanisms such as

- An increase in lipid content that lets the fat-soluble DDT separate from other parts of the organism
- The presence of enzymes that break down DDT into relatively less toxic products
- A reduced toxic response of the nervous system to DDT
- Changes in the permeability of the insect cuticle to DDT absorption
- A behavioural response that reduces contact with DDT

It is therefore not surprising that insecticide resistance is associated with numerous genes. For example, the genes responsible for DDT resistance in *Drosophila* are located on all major chromosomes, each gene acting as a polygene with a small incremental effect.

On the whole, it seems clear that most populations do not await the lucky incidence of new favourable mutations to provide for their evolutionary needs. Instead, populations tend to use their reservoir of genetic variability, consisting of many historically accumulated mutations. We can therefore consider evolutionary potential and genetic variability as two sides of the same evolutionary coin.

10

HOMEOSTATIC EVOLUTION

The tendency to increase in size is, in a sense, the most fundamental feature of living things. We are all familiar with it, and take it for granted as a 'natural' process. Yet even slight study will show that growth is very far from simple and that we do not properly understand it. Moreover, it is very 'unnatural', since it is unusual among physical phenomena. It involves incorporation of less-ordered material into the highly ordered living systems, contrary to the general tendency for increase of entropy. It is indeed worth while to study growth closely, and we may expect to have surprises when we do so. There are many ambiguities even in attempting to define growth. It has been called 'the net balance of mass produced and retained over mass destroyed and otherwise lost'. But is all increase of mass to be called growth? Is addition of storage fat or carbohydrate as much 'growth' as addition of protein? And what about the addition of water? How are we to define the mass of any organism? It fluctuates from moment to moment.

Features other than mass may serve to define growth. Height would be regarded as a good criterion by many children. Increase in cell number is obviously linked in some way with the concept of growth, but individual cells may grow, so the increase of number of cells alone cannot define the process of growing. From this point of view we might distinguish between (a) the addition of material to cells, (b) increase in their number, and (c) increase of intercellular material.

It would be satisfactory to be able to put forward some concept that would unify these various aspects of growth. The differences between them are semantically puzzling but trivial. A true measure of growth might be related to the rate of increase of order that is implied by the incorporation of non-living into living matter. The increase of DNA in a population might serve as the basis for an assessment of the degree of order, which would evade these difficulties about fat, water, etc. This statistic (amount of DNA) is indeed very useful for some purposes, but we may also have to be content to measure growth by whichever of the paprameters discussed above is most readily available to us. If a general phrase is wanted we can perhaps say that growth is 'the addition of material to that which is already organized into a living pattern'.

We shall be concerned here not so much with definitions or even with finding quantitative expressions for growth (though this will follow). It is more important to try to see the relation of growth to the processes of self maintenance. In fact we begin by asking why should there be growth in order to ensure homeostasis? This may seem a very large question. And indeed so it is, for it touches the very nature of living organization. A system of steady-state interchange is able to continue only because it is provided with 'instructions' that ensure correct responses in the situations that are likely to arise. To put it in another way, an organism an continue to live only because it has been provided with the necessary organs. The instructions thus constitute a system for forecasting the situations and risks that the creature is likely to meet. But by the very nature of the randomness of events, something will happen sooner or later that destroys part of, and eventually the whole, organism. Continued survival in a steady state by any one given set of materials is therefore impossible. Survival of the organization can, however, be ensured by a suitably arranged system of replacement. This is, therefore, the 'function' of growth. It operates at a whole series of levels, from the continuous replacement by turnover within cells, to continuous replacement of cells, to replacement of whole individuals by others, in the processes we ordinarily call reproduction, and in the replacement of one species by another in evolution. All these are a single series of phenomena by which the power of replication and of incorporation into the living system serves to ensure the maintenance of life.

Exponential Growth

One of the most unexpected findings for the layman is that growth, if it is not restricted, continues without limit. What grows is itself capable of growing. This is well illustrated by bacteria. When bacterial cells are isolated in a culture medium there is first a lag period and

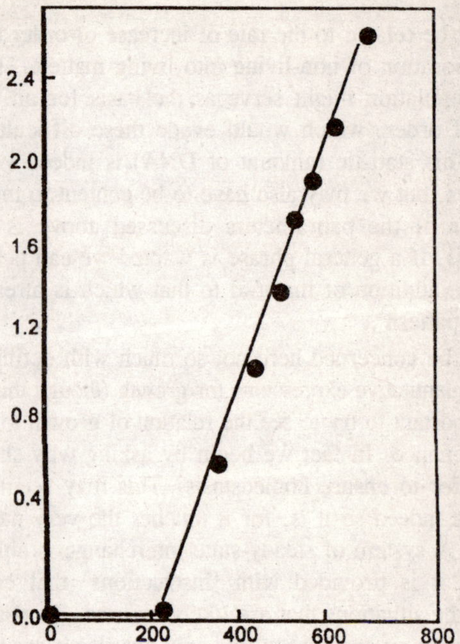

Fig. 10.1. Growth of bacterial cells in culture afer explanation.

then exponential growth begins. Each bacterium divides about once every 30 to 40 minutes and the products then 'grow' until they reach the size of the original. This process continues until the supply of nutrients is exhausted, after which special conditions set in, which do not interest us here. If the culture medium is renewed, growth continues indefinitely. Use is made of this for continuous cultivation methods. If all the growing organisms were to be supplied with the required nutrients there would have to be an exponential increase of the size of the vessel and the amount of inflow. This is obviously not possible, but if part of the volume of multiplied micro-organisms is withdrawn at the same rate as nutrient medium is supplied, then a steady state can be maintained in which all the materials are utilized. More penetrating observations have shown that we must think more carefully about continuous growth. Does each individual grow at an increasing rate per unit of mass, as would be required if what grows is capable of growing?

Mitchison studied the growth of single cells of yeast, using interference microscopy to determine the 'dry weight' of single cells continuously. He kept the yeast in a sterile medium so that there was no question of interference by the growth processes of other organisms. Each yeast cell grew at a constant rate independent of mass, until late in the process of division, which occurred every four hours, and then the rate doubled and remained the same throughout the next generation. The rate of synthesis therefore depends upon a group of components, perhaps the ribosomes, which may double their numbers before cell division. The whole process of division must obviously be related to the replication of DNA and this, in many organisms, occurs at about the middle of the period between divisions.

The determination of the DNA cycle probably depends upon a complex relation between the replicating material and the associated polymerase enzymes, but the system is not fully understood. The whole question of the basic cycles that regualte cell division is still obscure. The cell does not divide because it has reached a certain size. By various means the growth can be stopped prematurely, but division will take place, though delayed. Addition of material is therefore one factor influencing division, but not the only one. Probably replication of the DNA is a second, and perhaps cycles of the RNA and ribosomes a third. A fourth may be a rhythmic change in the energy-yielding mechanisms and other enzymatic process of the cell, perhaps related to the cycle of reducing sulphur-containing compounds (e.g. glutathione).

In examining the fundamental process of control of increase we find, therefore, many facts that are interesting and sometimes paradoxical, but difficult to reduce to any single system of control. Life may perhaps once have been simple, but today even the lower organisms maintain their balance through a web of interacting control mechanisms. We have, however, the important conclusion that the general condition is to continue growth indefinitely, at least in these simple organisms. The situation with which we as humans are familiar, namely, decline and cessation of growth, followed later by death, thus appears as a limitation of the general condition. Our thesis will revolve around the concept that our life cycle indeed includes special sets of limitations upon continuous growth, serving to ensure that the population as a whole can continue to survive by means of a particular way of life.

Differentiation and Growth Control

Perhaps it should not surprise us to find that continuous growth is the general condition. Since continuous replacement is the absolutely necessary condition of self-maintenance we should expect to find it everywhere in living processes. The growth must of course be regulated, especially in the higher organisms, where survival depends upon the presence of a most complicated system of tissues, differentiated for particular functions. Differentiation is in this sense the inverse of growth; indeed, it often follows growth. We shall expect to find, then, that the study of growth consists largely in the study of the control and limitation of growth, especially in the higher organisms.

Intermittent Growth

When we look more closely, it thus appears that even in bacteria growth is not continuous. Addition to the mass of a whole culture masks the cycles in the growth of each individual. Replication of the DNA occurs at set intervals, and something is known of the means for regulating this. In the absence of limitation by lack of raw materials, it seems that the single strand of DNA in each bacterium is replicated by the attendant polymerases at intervals that are not closely related to the total increase in mass. The factor that regulates this cycle thus serves to control the whole growth potential of the species. Presumably, however, this cycle is itself genetically controlled. This question of the power of increase of a set of cells comes very close to the heart of all biological problems.

Unfortunately, we have only a glimpse of it in bacteria and viruses and known almost nothing of it in higher animals and plants. It is difficult to conceive of the process of growth as a wholly continuous phenomenon, for the simple reason that organisms themselves are essentially heterogeneous. We can think about the continuous synthesis, say, of a given protein. But the more interesting proteins do not simply accumulate, they associate with other substances to form membranes, organelles, fibres, and so on. When any one of these has been formed the continuity is in a sense broken as the building of another organelle is begun. The point is not trivial, for organisms are assemblies of such units. Higher organism are assemblies of assemblies of them (cells, tissues, colonies, populations). Since the components of such systems do not increase together and continuously we have to think of them separately and individually. The origin and fate of each mitochondrion or man is therefore inevitably a large part of the study of biology. Investigation of the cycles of their multiplication and growth may reveal some general principles, but is also likely to show that there is a *great diversity of rules of growth*.

Turnover and Growth

A further complication is that addition to living systems is not a one-way traffic. We must also include the subtraction that is involved in breakdown, and then compute the growth as the net gain. It seems so obvious that growth is the resultant of anabolism and catabolism that we do not pause to ask 'Why should there be breakdown at all?' More practically, we may seek an answer by looking at lower organisms to see whether turnover is always present. Perhaps it is necessary feature for the maintenance of a steady state. Viruses cannot be said to show turnover, but they are not capable of independent self-maintenance.

In bacteria there is turnover, btu at a simple level than can hardly be distinguished from reproduction. In the more familiar situation of multicellular organisms, it has been clear since the time of Bizzozero (1894) that some parts of the body become changed much faster than others. Whether we are considering cells alone or their components (fibres, organelles, etc.) it is convenient to think of three types.

1. Labile populations, multiplying throughout life (vegetative intermitotic cells of Cowdry (1952)).
2. Stable populations, usually multiplying only during development, but capable of doing so in the adult after damage (reversible postmitotics).
3. Permanent populations, never multiplying once formed (fixed postmitotics).

As applied to cell populations in mammals this would produce the following classification:

Labile (renewing) populations

Epidermis and its derivatives

Endodermal epithelium

Transitional epithelium of ducts

Blood-forming tissues (in part)

Endometrium of uterus

Germ cells of the testis

Stable populations (able to expand when necessary)

Liver

Kidney

Exocrine glands

Lens

Connective tissues

Skeletal tissues (in part)

Permanent populations

Striated muscle
Neurons
Germs cells of the ovary

Turnover and the Anticipation of Risks

In spite of our considerable understanding of the functions of the various tissues in the economy of the body it is not always clear why some tissues should remain intact and others be regularly replaced. Moreover, little is known about the important question of what determines the rate of production of the labile tissues or stops the stable ones from further expansion. The products of the epidermal line are ultimately keratinizeid. Thus removed from metabolic sustenance, they are subject to the cruder forms of wear and must be replaced. This simple situation may serve as an example of the principles that we may search for. The body as a self-maintaining system inevitably suffers the physical hazards imposed by a more or less tempestuous environment. Some of these risks are of simple physical nature (like the wearing away of the skin), others are chemical changes or fluctuations of temperature. No portion of the body can be regarded as permanently immune from risks. However strong the bones, in some individuals one or another of them will ultimately be broken. To make them still stronger would involve not only collecting much material but, worse, carrying it around.

The tortoise provides himself with a house of inert material that is hard to break. Within it he lives longer than, perhaps, any other creature, but tortoises are not among the most widespread animals and have not been able to colonize may habitats, for example the trees or the air. Such massive over-insurance has not proved to be the ideal policy for life. In general, the best protection seems to be provided, not by excessively strong or inert materials, but by a system of turnover that allows for replacement.

Cell replacement may proceed continuously, as in the skin, or be called into play when needed, as in the repair processes of the connective tissues or skeleton. In this sense the continuous change of the renewing populations is an aspect of homeostasis. It provides a permanent insurance against the inevitable damage to the components that is imposed by the environment. That this is likely to be an efficient policy is clear, for example, in the skin, whose material is directly subject to wear. It is also easy to understand for the lining of the alimentary canal, which must deal with all the chemical influences that are introduced with the food, including the products of the bacteria that thrive there.

In other tissues it is less clear why continuous cell-replacement is needed. There is no obvious need for rapid renewal of the red corpuscles of the blood. Possibly the limiting factor here is the energy needed to pump them round the body. It is cheaper to pump sacks of almost pure haemoglobin then to destroy them after a while, and replace them from a factory, than to pump also the instructions, machinery, and materials for their renewal. Incidentally, birds find the latter course better. Their red cells, though they are still end cells, have nuclei, and this enables them to manage with a small factory and so to lighten their bones with air sacks, while haemopoiesis on a much reduced scale proceeds in the liver, spleen, and elsewhere.

Cell replacement is obviously not the only possible means of renewal. In some tissues replacement is probably undesirable for functional reasons. Thus the activities of the nervous system depend upon an elaborate system of connectivity between the cells. To renew this would need complicated instructions and in fact regeneration can be achieved only in simpler animals and simpler parts of the nervous system. No effective regeneration takes place in the brains of mammals (or of cephalopods). Where the brains contain memories, capable of holding records of the experiences of the individual, it may obviously be difficult to provide also instructions to replicate these records if cells are destroyed. Correspondingly the neurons lose all power of cell division. They do not, however, lose their powers of synthesis. On the contrary, protein turnover is especially active in the nervous system.

Nerve cells are characterized by a large amount of ribonucleic acid in their cytoplasm—the basophilic Nissl bodies. Probably there is a continuous flow of material down the nerve fibres from the cell bodies. What happens to it at the peripheral end is uncertain. It is clear, therefore, that these static post-mitotic' cells are not really different from the rest. Indeed it may be that they are provided with especially active mechanisms for self-renewal. For, unlike the other populations, they cannot be replaced if damaged, and must therefore continuously repair themselves. This presumably applies to striated muscle as well as to nerve.

The very large number of nuclei in each syncytial muscle fibre suggest the operation of continuous processes of replacement (though myosin itself turns over little it at all). It becomes clear, therefore, that the phenomenon of growth, that is to say, synthesis of new material, is much more than the production of a new individual, as we commonly regard it. Growth is one of the fundamental features of homeostasis; growth provides the means of replacement of what will inevitably sooner or later be damaged—a guarantee, so far as possible ,against the risks of destruction.

The growth with which we rare familiar, that of the child, after the act of reproduction, we shall find to be the special from of this guarantee, since it allows survival *in face of almost all risks,* a promise for the future that would seem impossible to give, but which has so far served to ensure life for 3000 million years or more. For the purpose of a general study of growth, then, our conclusion is that addition to the existing mass of living material is a fundamental property, presumably existing since life began. The numerous special forms of life that exist today have been arrived at partly by limiting this general power of growth. Thus all the higher organisms are colonies of different types of cells, serving different functions in the homeostatic system. the number and position of each type is regulated to produce a smoothly working whole. The study of growth thus becomes, in effect, the study of what limits growth in each tissue, during development and in the adult.

Repair of the Individual

Maintenance, Repair, Replacement, and Regeneration

These are a series of similar processes, linked both in their significance for continued maintenance and through the methods by which they operate. All are processes by which the individual life is enabled to persist in spite of actual wear or damage. As we have seen in the last chapter, in most tissues there is continual turnover and in some the cells are regularly replaced. These maintenance operations anticipate the damage or destruction of the parts in question, and they are found particularly in such vulnerable areas as the outer skin and lining of the gut. Other parts of the body, though liable to damage, may suffer it only 'accidentally', that is, occasionally or not at all. For these the genetic system provides processes of healing, repair and replacement that are set into action only when called for as a result of traum.

'Regeneration' is usually kept as the name of a similar process but on a larger scale, leading to the production of a whole new part such as a limb. These powers are therefore, as it were, a second line of defence when the forecasting mechanisms that avoid damage have failed. For example, the mechanisms of heredity and development provide a mammal with bones of strength adequate to meet most of the stresses likely to fall upon them. If greater stresses are met and a bone breaks, then a specific mechanism of regeneration is available to mend it. All these processes involve growth, and are linked with the daily turnover within the tissues. They also involve cell migration and differentiation, and in this are linked with normal development from the egg.

All of these repair operations that we usually separate conceptually for convenience under the headings given above form a series running from those more closely related to turnover to those that partake more nearly of reproduction and development. Each uses specific mechanisms to contribute to the homeostasis of the system. In this sense all are 'normal' processes. They differ in the regularity and timing of the wear or accidental events that they are designed to correct. Processes of 'maintenance' are designed to meet the damage that occurs in certain tissues regularly every day, for example in the skin of man, or the incisor teeth of a rodent. 'Repair', 'replacement,' and 'regeneration' are processes that come into operation only intermitently, when the organism has suffered one of the various hazards that are likely to befall it. In this sense they approach nearer to the mechanism of reproduction, which is the act that initiates the replacement of the whole creature before eventually something happens by which it is destroyed. The earlier categories, maintenance and repair, are thus more directly concerned with minor, recurrent, and temporary damages; the last two, especially regeneration, constitute major acts of differentiation, approaching reproduction.

Indeed, in some organisms it is hard to distinguish between regeneration and reproduction by budding (e.g. sea squirts (tunicates) or tapeworms). Since the processes of repair, replacement, and regeneration are characteristically brought into operation after accidents or unusual events in the life cycle, they are obviously of particular concern for the medical man. He is, for many people, the specialist who is called in to help with the results of such untoward stresses, rather than the guardian of longer-term homeostasis (though he may well play a great part in this to, and will perhaps do so even more in the future).

In order to be able to assist the repair processes of the body effectively, the doctor must obviously know about their means of operation. It is to some extent the function of pathology to provide this understanding. Here, however, we meet with the complication that, as a prelude to the study of repair, the pathologist also investigates the processes of damage themselves, especially as inflicted by invasive agents such as bacteria and viruses. We shall not attempt to study pathological conditions here but shall try to investigate the nature of the healing processes as exemplified in a few examples in which they have been set into action by controlled surgical experiment. Elementary principles can be illustrated in this way, but while the reactions to particular pathogens are alike in the early stages, later the picture may change specially for each disease condition.

Maintenance

Of the four categories that we are considering this approaches most closely to the normal turnover processes. The concept of maintenance, as borrowed from engineering and household management, implies the repair of wear and tear that is normal in the sense perhaps of 'expected'. We have to paint our houses and renew electric light bulbs or motor-car tyres or brake-linings, because the designers of these items were not able to make them more permanent. A similar distinction can be made for the body, though it is not perhaps altogether appropriate. The body has no permanent components (unless it be the instruction in the DNA), but it has some parts that turn over very little (e.g. the collagen of tendons, actomyosin of muscles, or antibodies), while others turn over so efficiently that we do not ordinarily think of them as suffering wear (e.g. the continual replacement of red cells or intestinal epithelia). Perhaps, however, this is indeed a wrong attitude and we should recognize that it is exactly because of the inevitable molecular wear and degradation that the turnover of these tissues is so active.

In addition, of course, it allows the continual adjustment of the level of their functioning by adaptation. The conceptions of maintenance and repair should be extended for living organisms to the molecular level. Some enzyme molecules persist only for a few minutes. It is not clear in what sense that is due to wear, but in muscle there is an increased turnover after activity. The replacement of the surface of the skin or nails from deeper layers is more obviously an act of maintenance. We find it easy to conceptualize this process because we can literally *see* it at work on our own bodies.

There is no reason to doubt that with sufficient insight we should equally recognize that the replacement of our red corpuscles and all the other replacing tissues shows the operation of specific mechanisms under the control of the DNA, ensuring maintenance of the integrity of the system in spite of wear. As a steady-state open system, the organism is continually subject to stresses resulting from the tendency to merge with its surroundings. These knocks provide the stimuli for the operation of the synthetic growth processes, which are made possible by the replicative powers of the instructional molecules within. Each tissue is thus maintained at the appropriate level of functioning by a *double dependence*, on stresses from without and synthetic forces from within. The problem for the student of growth is to learn to identify and measure these forces so that he can estimate the probability of successful continuance and show how to increase it by appropriate action.

To take a simple example, in managing the healing of bone the physician needs to know how the effect of mechanical stress will influence the process at each stage (specifically, when and how much to let the patient with a broken leg walk.). On the other hand, he needs to know whether the bone-growth potential needs the assistance of particular dietary factors; say, vitamin C (for collagen), vitamin D, or extra calcium. It would also, of course, be of great importance to know of any specific agents that stimulate or retard healing, a question to which we shall return.

Nerve Repair as an Example of Healing

Healing, repair, replacement, and regeneration are so closely related that there is no sense in trying to distinguish sharply between them. All are distinguished from maintenance in that they are processes initiated by the occurrence of some incident that is likely to happen but does not recur regularly. The instructions of the organism nevertheless contain provisions to meet such accidents, the occurrence of which is in this sense predicted and expected, although arhythmic. We may use the name as headings under which to describe the repair of damage of increasing severity, by operations ranging from simple clotting of the blood to replacement of a whole limb. Healing has many meanings, but refers especially to such processes as the closure of wounds in the skin or of other organs, where the damge is due not so much to the loss of substance, which may be minimal, as to the interruption of the continuity of the tissues. The effect of severing a nerve provides an interesting example.

The physical break in the nerve trunk damages simply when it interrupts, not because of loss of tissue, but because the contained nerve fibres are dependent for their trophic maintenance on continuity with the nerve-cell body. After severance, therefore, only the central part remains alive. The peripheral axons undergo degeneration and if the two stumps are close together the old axons are then replaced by outgrowth from the central end. We can therefore regard the activities that reunite the stumps as an example of healing and the function of the new fibres as replacement or regeneration. After any wound (in a mammal) the repair can be considered to proceed in three phases: (1) inflammatory, the first-aid operations of closure and removal of debris, (2) proliferative, the advent of fibroblasts, able to lay down new structural materials, and perhaps of specific types of cell; (3) remodelling, such as the maturation of scar tissue or, in nerve repair ,the growth to maturity of some nerve fibres and the atrophy and disappearance of others. The first aid is provided by clotting of the blood. This is produced by a special set of enzymes and other proteins provided for the purpose.

The inactive enzyme prothrombin of the blood is activated by an enzyme thrombokinase, liberated by the damaged tissue, which, in the presence of calcium, converts prothrombin to the active enzyme thrombin. The latter then turns the protein fibrinogen, circulating in the plasma, into threads of fibrin. These help to close the openings of the vessels, stopping the loss of blood. Contraction of the vessels themselves assist this. The fibrin clot serves also immediately to meet the first need of repair, which is to fill up the space left by the wound. But the clot is not itself strong enough to resist any great compression or tension and must be replaced by stronger tissues.

Almost at once, therefore, it becomes infiltrated by white blood corpuscles. The first are small microphages (polymorphs), able to remove debrisi and infective agents; later, larger macrophages move in. together these cells remove the fibrin clot, probably by fibrinolytic enzymes, made available by breakdown of the neutrophil white blood corpuscles and release of their lysosomes. These enzymes work differentially on the fibrin molecules, attacking first those that are not under tension. The clot thus comes to present orientated strands. Along these migrate the fibroblasts, which then proceed to produce collagen to unite the severed surfaces more firmly. When a nerve trunk has been cleanly severed, without other damage, the process of healing varies according to whether the two severe ends remain close together. Let us assume that they do so, or that, having been separated, they have been sewn together by a surgeon. Such clean interruptions may occur when a nerve is cut by glass, say in the wrist when a hand is pushed through a window.

There may often be complications from damage to tendons, and nerves may be equally or more seriously damaged by being pulled, as when a humerus is broken and the radial nerve is bruised by contusion. However, if the severed ends of the nerve are together, a blood clot forms between them. After three or four days the fibrin will be dissolving away and fibroblasts will move in from both the central and distal stumps. Peripheral nerve normally contains much connective tissue, and the fibroblasts presumably come from this, though what activates them will have to be discussed below. The fibroblasts, moving along the strands of fibrin, soon begin to lay down their characteristic protein, collagen, and mucopolysaccharides. If the two joined ends of the nerve are under light tension, then the new collagen fibres will run longitudinally between the ends, in the direction that is most calculated to join them firmly. It is probably desirable, therefore, hat nerve stumps should be left under slight tension, rather than pressed together, which would produce a scar of collagen fibres not all orientated along the line of the nerve.

A similar argument applies even more strongly to the union of the severed ends of a tendon. The strands of fibrin and new strands of collagen serve a further function besides this resistance to stretch. They also guide the course of the outgrowing nerve fibres, or rather of the satellite Schwann cells, which precede the latter. Vertebrate nerve fibres are surrounded by these Schwann cells, which in some make an insulating myelin sheath that greatly increases the speed of nervous conduction. Even those peripheral nerve fibres that have no myelin all have some from of Schwann cell. In regeneration these satellites have a special function. They cease to be ensheathing cells and move out into the site of the wound by active amoeboid movement. They creep along the threads of fibrin and collagen, froming a web of nucleated tissue between the two stumps. They emerge mainly (perhaps wholly) from the lower (distal) stump, where they are released from their enshething function by the gradual breakdown and dissolution of the axons, which have been separated from their cell bodies and are therefore degenerating. A piece of protoplasm separated from its nucleus sooner or later dies. Nerve fibres, being very elongated, are especially vulnerable to this hazard.

The nuclei of the nerve fibres that run to the toes are in the spinal cord and must therefore sustain the protoplasm of threads about a metre long, but only to μm in diameter. It is not known what influence passes along to do this, but there is evidence of an actual transport of material down the fibre. Therefore, 2-3 days after a nerve fibre in a mammal has been interrupted, the isolated part breaks up, first at a few widely separated points and then into a string of avoids and ultimately spheres. The process is not complete for many days; indeed, some pieces of axon remain for several weeks (or longer in cold-blooded animal). Meanwhile the Schwann cells of the peripheral stump divide mitotically, forming strands within the tubes of the supporting tissues of the distal stump. The strands emerge from the cut end and grow up to join the central stump. Here they may make contact with the severed central ends of the nerve fibres. These have formed swollen amoeboid tips, putting out 'feelers' into the tissues between the stumps. When one such feeler touches a strand of Schwann cells it makes some specific from of union and the tip moves away down into the distal stump, spinning a fine thread, the new axon, behind it. We shall discuss its further course later on as an example of regeneration.

Nerve-Growth Factor

Some information is available about chemical factors that stimulate the growth of nerve fibres in the embryo, and these may be related to those of the adult. If tumour cells of mice are implanted in place of the limb bud of a 3-day-old chick the nerve fibres of the latter wander around among the cancer cells. The sensory or sympathetic cells from which the fibres originate are stimulated to increase in number several times so that their nerve fibres invade all the tissues of the host embryo. They even enter the blood-vessels in such numbers as to interrupt the circulation (Levi-Montalcini 1964, Levi-Montalcini and Angeletti 1968). In the search for the chemical nature of the 'nerve growth factor' (NGF) responsible for the increase it was found that snake venom has a great power to produce the phenomenon, and that an even more potent source of NGF is the salivary glands of a male mouse. The substances are both proteins but no identical. That from snake venom has a molecular weight of about 40,000.

The effect of the factor isolated from the male mouse salivary glands and injected into young mice is to produce a specific hypertrophy only of the sympathetic system. All the rest of the nervous system remains unaffected, even the parasympathetic ganglia. A further development was the preparation of an antiserum that *suppressed* growth of the sympathetic cells. The purified NGF protein was injected into other foot-pad of a rabbit. Serum from the rabbit injected into new-born mice then caused the latter to develop with as little as 3 per cent of their normal quota of sympathetic cells, the growth of the rest being suppressed by the antibodies. We cannot yet relate these findings to the processes of normal growth and development of nerve, but they provide further valuable evidence of the existence of specific chemical factors able to stimulate or suppress the growth of particular types of nervous tissue. With further knowledge of such factors it should eventually be possible to understand how the nervous system becomes provided with the many types of channel that constitute the essential feature of its system of coding and signalling.

Processes of Healing

The union of two nerve stumps provides a typical example of the basic processes of wound healing. We might list the following as the components.

1. Quickly filling the defect, achieved by the fibrin clot—as it were, a provisional emergency repair.
2. Next, filling the defect with cells, first polymorphs, then macrophages, fibroblasts, and Schwann cells, and finally the characteristic cells of the tissue, nerve fibres.
3. Giving physical strength to the new tissue by collagen, making good the mechanical damage.
4. Repairing also the vascular defect, by ingrowth of new vessels.
5. Finally, repairing the actual functional defect, in this case by maturation of the nerve fibres to make effective lines of nerve conduction.

It is obvious that many specific growth processes are concerned. A few of them are simple enough for us to be able to see the reactions in some detail. Thus, the clotting of the fibrinogen and formation of the new collagen are the results of change or synthesis of relatively simple proteins. It is much harder to understand how the orderly sequence of so many processes is achieved. In particular, we should like to know what activates the cells to divide and to migrate after wounding and what stops them when their aim has been achieved.

Theories of Wound Healing

The various theories of the stimulus to wound healing offer different levels of explanation.

1. The simplest is that there are wound hormones produced by the damage to the cells. Thee are two versions of this, (a) that the substances stimulate the cells to divide and migrate, (b) that they remove an inhibition on these activities.

2. A second group of theories suggests that as the repair process serves to fill up the defect left by the injury, it is this deficiency or lack that somehow sets in motion the actions of repair.
3. The third theory is that it is the functional lack that is sensed by a suitable system and thus initiates repair.

The second and third approaches seem at first sight much vaguer than the first, but may have more value than appears. In particular, we have to study not only what initiates repair but also what ends it at the correct point. The fact that the defect has been filled or the function restored may provide the relevant information that stops the tissue growing.

It is unfortunately true that at present we do not know what initiates or controls repair mechanisms. This is indeed a great weakness, but biological and medical science are faced here with a fundamental difficulty of method. One cannot distinguish between theories of the three types outlined above because one can not produce a loss of tissue without producing traumatized cells. Conversely, one can not do damage without causing some from of deficit. In plants it has been established that injured cells release a substance, traumatic acid ($HOOC(CH_2)_8CH{=}COOH$), which stimulates cell division near by. No similar relatively simple wound hormones have been identified in animals.

Indeed, the whole question of growth-promoting substances is highly complicated. From the earliest days of tissue culture it was found that growth is stimulated by extracts, especially by extracts of embryos, but also of other tissues. But such extracts are very complex mixtures and it is still not clear to what extent their effects are due to the nutrients they supply or to specific growth stimulants. There is, however, a considerable body of scattered evidence that the growth of each tissue can be stimulated or inhibited by specific substances produced within that tissue. Examples of these substances are the epidermal growth factor in the skin, nerve growth factor, and erythropoietin, which stimulates the bone marrow. Extracts of various adult and embryonic tissues have been shown to stimulate growth in tissue culture under carefully controlled conditions. The stimulating power is associated with protein.

Wound-Healing in the Skin

Repair of damage to the surface has perhaps a special human interest since we all observe it so often in ourselfes. Moreover, its superficial position makes the skin especially easy to study. After damage to the epidermis and underlying dermis there is first the formation of a fibrin clot and then its dissolution after a few days. Cells of the neighbouring epidermis are stimulated to division. They are normally, of course, in a cycle of continuous division, and presumably the mechanism for ensuring the increase is related to this. After 12 to 24 hours the cells move in such a way as to cover the damaged area. Meanwhile the cells of the underlying dermis also proliferate, and polymorphs, macrophages, and fibroblasts proceed to operate as in the nerve regeneration already mentioned. The response of the epidermis is especially interesting. In a small wound its cells continue to proliferate and to migrate until the form a complete covering.

The migration is apparently triggered by the absence of contact between the cells and stops when the space is filled by contact inhibition which stops the fibroblasts. The cells then proceed to contract, drawing the separated edges of the wound together. This process of wound contraction is distinguished from the 'wound contracture' by which collagen fibres may draw together the edges, at a much later stage, in a large, poorly healing wound leaving a bad scar. The

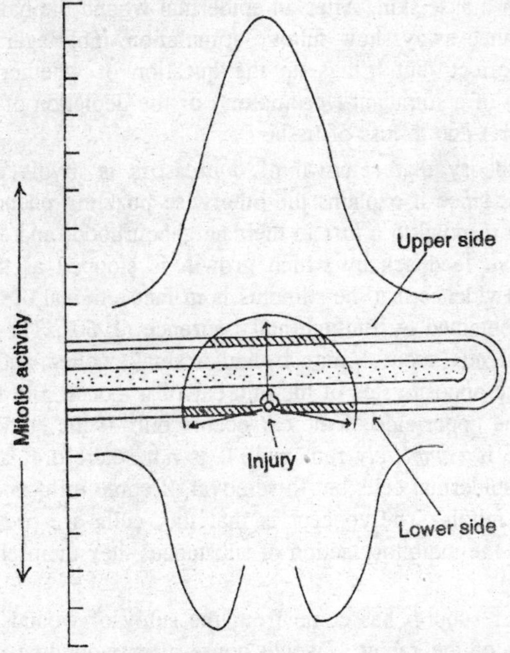

Fig. 10.2. The relation of epidermal hyperplasia to a wound.

mechanism of skin-healing thus shows all the factors we have seen before: filling the space, adding new cells, and restoration of function. Here, however, there is some information about the stimulating factors. The stimulus to cell division is specific to the cells that are damaged. If a flap of mouse skin is cut beneath the dermis, then the overlying epidermis shows no increase in mitosis. Conversely, damage to the epidermis causes mitosis only there and not in the dermis. There is some evidence that the stimulation is produced by an extractable *epidermal growth factor*.

Extrats of epidermis have been reported to increase the rate of epidermal wound healing when applied either locally or by in traperitoneal injection. Particularly convincing evidence has been obtained in the bug *Rhodnius*. A small skin wound is surrounded by a zone of mitotic cells and this area is increased by the application of an autolysed epidermal cell extract. However, the effect is not specific and is produced by a variety of degarded protein extracts. It may be due simply to damage of the cells of the recipient by the extract. So we have again the same problem that the healing response appears only when tissues have been removed. A somewhat similar phenomenon is found in mouse skin. After an epidermal wound the hair follicles at some distance away show mitotic stimulation. This again suggests a chemical effect, but brings up the question of whether this is the production of a stimulant ('promotor') or the depletion of an inhibitor ('depressor') due to loss of tissue.

The theory that removal of depressors is involved has many attractions, since it explains the otherwise puzzling phenomenon that the tissues respond to a *loss* in their neighbourhood, and also suggests the negative feedback by which growth is stopped as the defect is repaired. Evidence that the stimulus is in fact removal of an inhibition has been obtained by Bullugh and Laurence (1960). They found that mitosis extends about 1 mm around a small wound and will occur even on the opposite side of the thin ear of a mouse. But the response of, say, the upper side of the ear occurs only if the epidermis of the under side is *completely* removed. It is not, therefore, a reaction to damaged epidermal cells but to removal of some influence previously inhibiting mitois. The concept is that the; cells are prevented from mitosis by the inhibitory action of substances they themselves produce ('chalones').

Further support has come from the study of wounds of the lens epithelium of the rabbit. Twelve hours after wounding, cells in the neighbourhood of the injury begin to synthesize DNA (as shown with tritiated thymidine). Shortly afterwards they divide. The region of DNA activation and division then spreads peripherally away from the wound. The explanation seems to be that after injury cells migrate into the wound. This depletes the neighbouring zone, reducing the concentration of inhibitor and setting in motion the procedure of DNA replication. Meanwhile cells further out have moved in, leading to a wave of depletion and hence to a wave of mitosis that is propagated outwards at about 17 μm/h. Twenty-four hours later the cells near the wound again incorporate DNA and a second wave passes outwards. This is indeed a possible interpretation, but there is no direct evidence of a chemical inhibitor.

Another possibility is that the cells begin to synthesize DNA only when they have lost contact with their neighbours, as happens in some tissue cultures. We thus have the outlines of a hypothesis about the control of repair in epidermal structures. Once again, the replication of DNA is seen to be the central phenomenon. The process of repair appears to be triggered by the most obvious stimulus, namely the absence of a component previously present in the neighbourhood. If this is due to the absence of a substance normally produced by the cells themselves, then we have a mechanism by which the repair is terminated when the appropriate concentration has been restored. It is unfortunate, however, that the specific substances concerned have not been isolated or identified. But there is enough evidence to make it well worth while to continue the search for suitable sources of them. We now have to inquire whether similar principles apply to replacement of larger quantities of material and to regeneration of whole organs.

Replacement and Regeneration of Parts after Loss

Regeneration of the Liver and Other Glands

The replacement of a single mass of a tissue such as liver is easy to study experimentally, although perhaps a not very common event in nature, except through disease. In tissue that renew by cell multiplication (such as the epidermis) the increased production needed for regeneration is presumably started off by acceleration of whatever system controls daily replacement. In the so-called 'expanding' tissues, however, such as the liver, where mitotic activity in the adult is normally very low, special mechanisms must be brought into play to produce replacement after loss of a large mass. These mechanisms are indeed exceedingly powerful. After reduction of the liver of a mouse to as little as 30 per cent of its mass it will return to normal within 6-8 days. A kidney will enlarge if its fellow is removed, and a host of similar compensatory responses can occur throughout the body. We may perhaps distinguish situations in which the loss affects a systemic function, and produces hypertrophy elsewhere inthe body, from purely local losses.

Thus, after removal of the salivary glands of a rat on one side those of the other side hypertrophy. Such examples are numerous and suggest that the response to removal is basically governed by functional consideratins. This, however, does not give us an insight into the mechanisms concerned, except to suggest that they are specific to each tissue and have been highly selected. Indeed, action of these compensatory processes. The liver of rats has been especially well investigated. After partial hepatectomy the cells and nuclei enlarge and at about 18 hours after removal mitosis begins. It quickly reaches a maximum of nearly 3 per cent of the cells (normal 0.01 per cent) and then after about 30 hours begins to decline towards normal. The original weight is recovered withina week.

The functions of the liver are so varied that it has been impossible to decide which of them it is whose deficiency triggers the cell multiplication. It is probably partly the lowering of the plasma protein and the induction of specific protein synthesis. Other factors may be the failure of the digestive functions, or simply a result of the distortion of the; pattern of vascular flow. Similarly, after kidney removal there are many physiological repercussions, any of which may trigger the hypertrophy of the remaining kidney. This is not due simply to the need for the single kidney to produce more urine. One ureter can be diverted to enter the peritoneal cavity or the gut. Its secretion is absorbed and makes more work for the other kidney, but the latter does nto hypertrophy.

Simlar unexplained discrepancies are abundant inthe literature of replacement. Thus one effective way of producing hypertrophy of the salivary gland of the rat is repeated amputation of the lower incisor and the effect is unilateral on the salivary gland only of the same side. Here some very special mechanisms (presumably nervous) must be involved to relate the secretion to the wear on the tooth. The wear would usually be symmetrical and it is an acident that the mechanism has remained unilateral.

Compensatory growth of the endocrine organs is very marked and is normally produced by stimulation from other endocrines. Thus the pituitary stimulates the thyroid, adrenal cortex, sex glands, and others.

In each case any deficiency produced by removal s timulates the stimulator, but the details of how it does so are obscure. Increased activty alone does not necessarily produce increase, as is shown by stimulating an exocrine gland repeatedly, say wtih a drug. Neither the pancreas nor the salivary glands hypertrophy after repeated stimulation by pilocarpine. Curiously enough, another drug that stimulates salivary secretion, isoproterenol, does produce hypertrophy, so that the glands may become five times larger than normal. These examples really serve only to emphasize our ignorance of the mechanism that sets off the hypertrophy.

Functional Regeneration in the Nervous System

Special processes are available for repair of some parts of the nervous system in all animals and of major parts in simpler animals. This is the extreme example of repair of a tissue with fixed cell-number. In mammals, defects of the peripheral nervous system are made good by outgrowth from the existing neurons, without any production of new ones. It should be possible, therefore, to see the factors that are at work initiating and controlling the 'hypertrophy', uncomplicated by cell division. It is indeed true that we can discern the principles of control but there is no detailed knowledge of their chemical basis. Although nervous tissue is post-mitotic it probably undergoes continual replacement even in the absence of injury. As in other tissues, therefore, the regenerative processes are probably an extension of those normally at work.

The axons of the central stump swell to make irregular masses of material, probably of amoeboid nature, from which filaments are spun along the; bands of Schwann cells to reach the peripheral stump. Once within the tubes of the stump they advance at a rate of 5 mm/day (in mammals) and are robably led by the tube all the way to a peripheral organ. However, the question of how correct connections are made is still highly controversial. The nervous system differs from all others in the extent of its division into specifically different units, which must be correctly connected to function properly. Effective regeneration therefore depends upon the outgrowing fibres making connection with the right peripheral tubes. Ths is assuming that there is no other means of finding the right peripheral organ or alternatively of remodelling the central connections. The latter is improbable, at least in adults of higher organisms. On the other hand, there is strong evidence of special mechanisms for ensuring correct connections at the end of the axon.

The optic nerve can regenerate inmany vertebrates, from fishes to birds. The new nerve fibres may become much interwoven in the nerve trunk, but each of them ultimately comes to re-make connections with its original partners inthe optic lobe of the mid-brain. Unfortunately there is really no satisfactory view as to how this astonishing and very important result is produced, and it would nto be profitable to discuss the interesting evidence here. It emphasizes that very precise mechanisms must be at work. A little is known about the types of factor that operate for the control of the development of new fibres in a regenerating peripheral nerve stump. In a normal nerve the fibres with different functions differ in diameter. Thus the fibres that carry signals to the muscles to make them contract are large, and conduct fast.

Afferent fibres that carry signals about the tension on the muscle are also large. Fibres signalling touch are rather smaller. Fibres controlling blood flow (sympathetic) are very small and conduct slowly, presumably because they change the state of the blood-vesels only slowly and maintain them in a given s tate for relatively long periods. When new nerve fibres are first laid downtheyare all very thin, whether in the embryo or in a regenerating nerve. Some remain small, others grow larage and acquire sheaths, the process of 'maturation'. There must be specific factors at work to control the growth of each to the correct size. In fact each type of nerve fibre is highly specified from some early embryonic stage, probably by virtue of the connections made at the periphery, though we know little indetail abotu this.

During regeneration the process is at least to some extent repeated. For example, motor firbes that do not reach to their normal periphery do not return to their original size. Some influence must pass *up* the nerve from the muscle or other end organ to control the synthesis of new axoplasm. Unfortunately, once again we must admit to complete ignorance of the nature of this influence. New axoplasm is probably produced mainly in the nerve-cell body. After the axon has been interrupted there are quite explosive changes in the RNA and synthetic mechanism in the cell, as indicated by the process known as retrograde degeneration or chromatolysis. This again shows that some reverse signalling takes place, by which the damaged lower end of the neuron evokes the production of extra supplies of axoplasm.

Presumably a further set of signals switches off the synthetic proceses when sufficient axoplasm has been produced to return the fibre to its appropriate diameter, large or small. The end result of the nerve regeneration is thus to fill the space left as a result of the injury. However, the effectivensss of the repair lies in the capacity for useful signalling that it confers. Whether as pecific mechanism will be present to repair any part of the nervous system therefore depends on whether repair is likely to increase the survival value of the individual and of the species. Regeneration of the peripheral nervous system is no doubt party associated with a normal wear and tear process, for example at the tips of the nerves in the skin of the fingers.

Interruption of a few small peripheral nerve fibres will not be sufficiently in capacitating to make life impossible, and repair processes to make the part useful again are well worth while. On the other hand, damage to the central nervous system is probably such a serious handicap that it is not worth while to attempt to make it good. A mammal or bird with the spinal cord severed is so severely handicapped that in nature it could never stay alive long enough for regeneration to take place. Indeed, it is surprising that it is worthwhile for many fishes to have the capacity to regenerate the CNS when the spinal cord is severed. Perhaps it is because the swim-bladder and the waterprovide support that is lackig on land. It is interesting that frogs also lack power to regenerate the spinal cord whereas this is present in newts and tadpoles. Similarly, the higher nervous centres are so important to the life of a bird or mammal that survival is impossible without them and there is no mechanism for their repair. That the fore-brain and rectum can be regenerated in some fishes and newts thus really shows how little use the animal makes of them, since it can survive for a while without them! It may be asked what form of damage to the brain is likely to leave a fish in a condition in which repair is worth while. Presumably not after mechanical injury, for if the jaws of a redator damaged the brain, they would hopelessly disorganize the skull. It may be that the power of regeneration in these animals has developed to repair the damage of parasites rather than predators.

The brains of fishes are often found to be packed with helminth (worm) parasites. The brains of higher vertebrates apparently have methods that prevent the entry of such invaders and it may be for this reason that they have lost the power of regeneration. But of course with increasing complexity and detail of connections the morphogenetic processes needed for regeneration must become exceedingly complex. The capacity to regenerate the optic nerve in many vertebrates, up to and including birds, shows presumably that one eye can compensate for the missing one for a time, but also that to have both significantly increases the probability of survival. To understand this situation properly we should need to know more of how the animals use their eyes in the normal environment.

Regeneration of Limbs

The same principle applies to regeneration of limbs and of internal organs. Mechanisms able to produce repair have been evolved for organs that are useful but not indispensable for the life of the individuals. This raises interesting questions about the selective value of carrying whatever load is imposed by processes that are used only intermittently. Of course the rate of loss varies greatly. Ninety-three per cent of individuals of some species of crustaceans have been found to show loss of major parts. Siince crabs and lobsters have many liimbs they can afford to be without one while it is growing. There is little dubt, from the fact that the power of regeneration is used so often, that is has survival value. It is sometimes suggested that this capacity is merely a 'casually persisting pristine power of morphogenesis.' It is much mor elikely that the basic powers of metabolic turnover have been specially selected and improved upon to produce regeneration of the more vulnerabel members. Proof that these are specially selected powers comes from the animals such as lizards that have evolved autotomy mechanisms, regions at which a limb or tail is cast off when it is seized and remains wriggling to distract the attacker. There are special mechanisms to close the wound, and regeneration takes place faster following autotomy than after injury elsewhere. Autotomy may be useful even when not followed by regeneration. The crane-flies ('daddy-longlegs', tipulids) often cast off a leg when captured, but cannot replace it.

Regeneration and Evolutionary Advance

It is widely held that lower organisms have greater regenerative powers than higher ones. The attempt to produce quantitative scales either for evolutionary status or regenerative power is hazardous but has been bravely made by Needham (1964). He concludes that regenerative power is inversely related to evolutionary grade and that the reason is probably that morphological, and particularly histological, differentiation are directly related to the grade, so that if de-differentiation is necessary as a prelude to regeneration it becomes increasingly difficult or uneconomic. He recognizes, however, that the matter is very complicated and many factors may influence the presence or absence of regenerative power. Thus the alertness and nimbleness of the higher animals as potential prey saves them from serious wounds short of capture and total destruction, for which the appropriate biological defence is not regeneration but reproduction. On the other side there are indications that the regenerative power is not *necessarily* lost in all higher animals.

The capacity of stags to regenerate huge antlers each year is a conspicuous example. It seems unreasonable that the skin of a small area should have the power to produce the luxuriant velvet of new hair follicles, much more perfect tan the skin that is regenerated over wounds. Yet this beautiful new covering is then 'rubbed off in tatters soon after antler growth is complete'. It is clear that we are still a long way from understanding the biological significance of many phenomena of growth and regeneration. We think that we can appreciate repair processes that are of value to ourselves, such as healing of wounds or of bones. But we know so little of the significant features of the life system of other species that some of their major repairs seem to us to be simply a waste. It is doubtful if this is a sound judgement.

11

GENE LOCI IN POPULATION

We begin with an example of a character that is controlled by a multi-locus genotype. The set of genes that an individual inherits from one of its parents form a "haplotype," and the theory of population genetics for multi-locus systems traces haplotype frequencies through generations. This chapter has two main purposes. First, it introduces the population genetic theory for multi-locus systems, and the distinctive concepts that apply to it but not to one-locus models; we examine the multiple-locus concepts of linkage disequilibrium (together with its causes), recombination, and multiple-peaked fitness surfaces. Second, it describes the conditions under which single-locus models of evolution are inadequate and multi-locus models are necessary—the main condition is the existence of linkage disequilibrium.

The characters with which we dealt in earlier chapters were controlled by single genetic loci. Enzymes, such as alcohol dehydrogenase, are encoded by a single gene, and it is not much of a simplification to treat the polymorphism in the peppered moth as a set of genotypes at one locus. We will now move on to consider evolutionary changes at more than one locus.

The first example concerns a multi-locus polymorphism, and we can lead into it via a similar polymorphism that is controlled by a single locus. Both examples come from the same attractive group of butterflies called swallowtails. The swallowtails have a global distribution, and *Papilio* is the largest genus of them; their most striking characteristic is a "tail" on the hindwing. They come in many colours—gorgeous greens, subtle shades of reds and orange, and marbled patterns in white and gray—but the most common type has stripes of black and yellow. The North American tiger swallowtail *Papilio glaucus* is easy to recognize by its tiger stripes, as it flutters through woodland lanes or humid valleys. More accurately, *most* tiger swallowtails are easy to recognize in this way. In part of the species' range (roughly, to the southeast of a line from Massachusetts to south Minnesota and from east Colorado to the Gulf Coast), the standard form of the tiger swallowtail lives alongside another form of the same species. This second form is black, with red spots on its hindwings, and is called *nigra*, it is only found in females. The *nigra* form is not poisonous, but mimics another species, the pipevine swallowtail *Battus philenor*, which is poisonous. The *nigra's* geographic distribution fits that of the pipevine swallowtail, and *nigra* is well protected there from predatory birds that have learned by stomach-churning experience not to eat butterflies that resemble pipevine swallowtails. The tiger swallowtail, therefore, has a *mimetic polymorphism*. It has both the standard non-mimetic tiger morph of yellow and black stripes, and a black mimetic morph.

The polymorphism of the tiger swallowtail *P. glaucus* looks almost simple when compared with the amazing array of female forms in the species *Papilio memnon*. *P. memnon* lives in the Malay Archipelago and Indonesia. Its male is non-mimetic, although its colour is deep blue rather than yellow and black stripes. However, instead of one mimetic female form, *P. memnon* females come in a seemingly infinite variety. Their forewings show different geometric patterns of black and white; their hindwings, as well as varying in shape, can be coloured in yellow, orange, or blood-red, and may or may not have a bright white spot; some have tails, others do not; the abdomen varies in colour; and a spot at the butterfly's "shoulder" (i.e., at the base of the forewing near the head) called the epaulette, may be present in various shades of red.

Clarke and Sheppard, who have studied the species, suggest that each female form mimics a different model. Their evidence is not strong, as it comes only from the geographic ranges of mimic and model, and from superficial similarity of appearance (which is not exact in all cases). Good evidence for mimicry requires experimentally showing that birds that have learned to avoid the model will also then avoid the mimic; this research has been performed for the *nigra* form of *P. glaucus*, but not for *P. memnon*. However, we can accept as a working hypothesis that the apparently mimetic morphs of *P. memnon* are, indeed, mimetic. (*P. memnon* has other, non-mimetic forms as well, but they are not essential to the discussion here.)

Clarke and Sheppard were interested in the genetic control of this complex mimetic polymorphism. Crosses between the various morphs initially suggested the existence of a single genetic locus with many alleles. When two forms are crossed, the offspring usually either all resemble one of the parents, or contain a mixture of the two parental types, as will happen with one locus and simple dominance relations among alleles. For instance, if one morph has genotype A_1A_1, another A_1A_2, and A_2 is dominant to A_1, then an $A_1A_1 \times A_1A_2$ cross produces the same two classes of offspring (A_1A_1 and A_1A_2) as were present in the parents.

Soon the genetic story grew even more complicated. In addition to the mimetic and non-mimetic morphs of *Papilio memnon*, all of which exist in reasonable frequencies in nature, some much rarer morphs have been found. An example, in Java, is the rare morph called *anura*. A specimen found in Borneo was sent to Clarke and Sheppard in Liverpool. When it was crossed with a known *P. memnon* morph, it behaved like another allelic form of the mimicry "locus." A closer look at *anura* suggests a different interpretation. *Anura's* morphology mixes patterns from two of the common morphs—it has the wing colour pattern of the morph *achates*, but it lacks *achates*' tail.

Clarke and Sheppard's interpretation is that *anura* is not actually an allelic variant, but a recombinant; they also decided that the mimetic patterns of *P. memnon* are not controlled by one locus but by a whole set of loci. If *anura* is a recombinant, then there must be at least one locus (call it T) controlling the presence (allele T_+) or absence (T_-) of a tail and at least one other locus (C) controlling the colour patterns (C_1 for *achates*, and C_2, C_1, . . . alleles for other colour morphs). In this case, *achates* would have a genotype consisting of one or two sets (depending on whether the alleles are dominant) of the two-locus genotype T_+C_1, and *anura* would have T_- C_1, after recombination between a tailless morph and *achates*. The loci in question are so tightly linked that these recombinants practically never arise in the lab—which is why the different multi-locus genotypes appear, when crossed, to segregate like single-locus genotypes. We can predict that if more than one locus really is involved, a sufficiently large number of crosses should be able to break one of the "alleles" (such as the *anura* "allele") into several real combinations of alleles at several loci. Darlington and Mather defined a set of genes that are so tightly linked that they behave like a single locus in a breeding experiment as a *supergene*.

From *anura* alone, it seemed that there must be at least two loci controlling the mimetic polymorphism of *P. memnon*, but other rare types have also been found. Some, for example, combine the forewing colour of one morph and the hindwing pattern of another, suggesting that separate loci control the colour of forewings and hindwings. When all the inferred recombinants are considered together, there appear to be at least five loci in the mimicry supergene: *T*, *W*, *F*, *E*, and *B*. These loci control, respectively, presence or absence of tail, hindwing patterns, forewing patterns, epaulette colour, and body colour. Thus, *anura* is a recombinant between the *T* locus and the other four loci. The common morphs; which mimic natural models, should each consist of a particular set of alleles at the five loci. The morph mimicking model species number 1, for example, might have genotype $T_+ W_1F_1E_1B_1/T_+ W_1F_1E_1B_1$, and another morph (mimicking a second model) might have $T_- W_2F_2E_2B_2/T_- W_2F_2E_2B_2$ or $T_- W_2F_2E_2B_2/T_+ W_1F_1E_1B_1$. The recombinant genotypes such as $T_+ W_1F_1E_2B_2$ do not exist naturally, except as very rare forms like *anura*.

The point to remember is that each of the morphs of *Papilio memnon* is thought to be controlled by a multi-locus genotype. This mechanism does not operate like the camouflage polymorphism in the peppered moth, in which the different morphs are controlled by genotypes at one locus. A whole set of one-locus genotypes is needed to produce each of the swallowtail butterfly morphs. The genetics are not definitively confirmed, but we can use the idea for purposes of discussion.

Coadapted Genotypes

How will natural selection act on a rare recombinant morph of *Papilio memnon*, such as *anura* in lava? Successful mimicry requires as complete a resemblance as possible between a mimic and its model. A potential mimic that mixes the patterns needed to resemble two species will mimic neither as successfully as an organism that resembles one model in all respects. In fact, it will probably be selected against. The *anura* butterfly has the colour pattern of *achates*, but will not mimic the model species of *achates* because it lacks a tail on its hindwings. The models of tailless morphs, in turn, have different colour patterns, and *anura* will not mimic them either.

In general, natural selection will act against any recombinants between the mimetic five-locus genotypes. A five-locus genotype that mimics one model species in all five respects will be favoured, but a swallowtail collage, mimicking one model in three aspects and another in two different aspects, will look like neither and be selected against.

The genes at the five loci in this situation are said to be *coadapted*, or to show *coadaptation*. Coadaptation means that a gene (or genotype), like T_+ (or T_+/T_+) is favoured by selection if it is in the same body as a particular gene (or genotype), such as W_1 (or W_1/W_1), at another locus but selected against when combined with other genes (or genotypes), such as W_2 (or W_2/W_2) at that locus. Selection favours $T_+ W_1F_1E_1B_1/T_+ W_1F_1E_1B_1$ and $T_- W_2F_2E_2B_2/T_- W_2F_2E_2B_2$ individuals, but (if the alleles with the 2-subscript are dominant) works against $T_+ W_2F_2E_2B_2/T_+ W_1F_1E_1B_1$ individuals. It has not been empirically confirmed that selection works against the recombinant forms of *Papilio memnon*, but the argument is quite convincing.

Note that the terms "supergene" and "coadaptation" are related. Coadaptation refers to the fitness interactions among genes, and supergene to their linkage relations. Extreme coadaptation is likely to lead, over evolutionary time, to the evolution of a supergene, but physical linkage and fitness relations are conceptually distinct.

GENIC CONTROL OF MIMICRY

The butterflies of the genus *Heliconius* make an interesting comparison with *Papilio memnon*. Two species of *Heliconius*, *H. melpomene* and *H. erato*, are found in South America, both of which have multiple mimetic forms. The colour patterns are again controlled by many loci: 15 loci in *H. erato* and 12 in *H. melpomene*. However, in both species the loci are scattered at random among the chromosomes rather than being tightly linked in a supergene. When two morphs of a *Heliconius* species are crossed, the offspring contain a kaleidoscopic variety of non-mimetic recombinant forms that resemble neither parent nor any known morph of the species.

Why do the genetics differ in *Heliconius* and *Papilio memnon*? The reason is probably geographic. At any one site in the range of *P. memnon*, several morphs often live side by side. Crosses between them will occur with high frequency naturally. With *Heliconius*, however, only one morph usually appears at any one place; the different morphs are geographically separated and will not interbreed in nature. The non-mimetic recombinant *Heliconius* are generated only when morphs from different places are brought together in the lab. In *Heliconius* it does not matter if the mimicry genes are scattered around the chromosomes, because the non-mimetic progeny are not usually produced. On the other hand, it does matter in *P. memnon*. If the mimicry genes were not linked in a supergene, the recombinants would be produced—and be killed by predators.

Haplotype Frequencies

The theory of population genetics for a single locus is concerned with gene frequencies; the analogous variable in two-locus population genetics is *haplotype* frequency. (The term haplotype has two meanings. Here it refers to a combination of alleles at more than one locus. It is also used, in DNA sequencing, to refer to the base sequence of one of an individual's two sets of DNA.) For two loci with two alleles each (A_1 and A_2, B_1 and B_2), four haplotypes are possible: A_1B_1, A_1B_2, A_2B_1, A_2B_2. A diploid individual's genotype will be something like A_1B_1/A_1B_2: it has two haplotypes, one inherited from each parent, just as a one-locus genotype contains two genes from the two parents. If the *A* and *B* loci are on the same chromosome, each haplotype represents a gene combination on a chromosome. Haplotypes can also be specified for loci on different chromosomes. The frequency of a haplotype in a population can be counted as the number of gametes bearing a particular combination of genes.

A haplotype can be specified for any number of loci. We will mainly discuss two-locus haplotypes, but the haplotypes in the *Papilio memnon* example had five loci, and the mimetic patterns of *Heliconius* are controlled by 12 or 15 loci. As we will see, to understand the evolution of haplotype frequencies, we need some concepts that do not exist for gene frequencies. Two-locus population genetics is, therefore, not simply a doubled-up version of single-locus population genetics.

Role of Haplotype Frequencies

We can begin by asking a question like the one that led to the Hardy-Weinberg theorem for one locus. In the absence of selection) and in an infinite population with random mating, what will be the equilibrium frequencies of haplotypes? The question for multiple loci will lead us to another important concept, called *linkage equilibrium*.

The simplest case is for two loci with two alleles each. The crucial trick is to write the haplotype frequencies in terms of the gene frequencies at each locus, plus or minus a correction factor, called *D*. Let the gene frequency in the population of $A_1 = p_1$, $A_2 = p_2$, $B_1 = q_1$, and $B_2 = q_2$. Then

Haplotype	Frequency in Population
A_1B_1	$a = p_1q_1 + D$
A_1B_2	$b = p_1q_2 - D$
A_2B_1	$c = p_2q_1 - D$
A_2B_2	$d = p_2q_2 + D$

The total frequencies add up to one (that is, $p_1q_1 + p_1q_2 + p_2q_1 + p_2q_2 = 1$), and the sum of the two $+D$ and two $-D$ factors is zero. The important term to understand is D, which measures "linkage disequilibrium." Linkage equilibrium occurs when $D = 0$ and means that the alleles at the two loci are combined independently. The two B alleles would then be found with any one A allele (such as A_1) in the same frequencies as they are found in the whole population. If we examine all of the A_1 genes, q_1 of them are linked with B_1 genes and q_2 with B_2 genes; likewise, q_1 of the A_1 genes are linked with B_1 genes and q_2 with B_2 genes. At linkage equilibrium, the frequency of the A_1B_1 haplotype is p_1q_1. D measures the deviation from linkage equilibrium. If $D > 0$, an excess of A_1B_1 (and A_2B_2) haplotypes have arisen: A_1 is more often found with B_1 (and less often with B_2) than would be expected if they combined at random.

D is conventionally defined by adding the frequencies of A_1B_1 and A_2B_2 and subtracting that total from the sum of the frequencies of A_1B_2 and A_2B_1. It could also have been defined the other way around. Although other measures of the non-random combination of genes are possible, the points of the linkage equilibrium principle can be made with D alone.

Papilio memnon is an example of high linkage disequilibrium. If Clarke and Sheppard are correct, the allele T_+ is almost always combined with the other alleles W_1, F_1, E_1, and B_1 rather than with W_2, W_3, or W_4 (and equivalent alleles at the other loci). There is a large excess of the haplotypes $T_+ \ W_1F_1E_1B_1$, $T_- \ W_2F_2E_2B_2$, $T_- \ W_3F_3E_3B_3$, and so forth, while haplotypes such as $T_+ \ W_2F_2E_2B_2$, $T_+ \ W_1F_2E_2B_2$, or $T_+ \ W_1F_1E_2B_2$ are almost absent. The linkage disequilibrium in *P. memnon*, as we have seen, is caused by selection. In this section, however, we are asking how a set of haplotype frequencies should change through time in the absence of selection.

Let us return again to the haplotype A_1B_1. Its frequency has been defined as a in one generation. What will its frequency be in the next generation? (We can again use the notation a' as the frequency of A_1B_1 one generation on.) In the absence of selection, the frequencies of each gene will be constant, but the frequencies of the haplotypes can be altered by recombination. The frequency of A_1B_1 cannot be altered by recombination in double or single homozygotes. The number of A_1B_1 haplotypes coming out of an A_1B_1/A_1B_1 individual or an A_1B_1/A_1B_2 individual must equal the number going in, whether or not recombination occurs. The frequency can be altered only by recombination in the double heterozygotes A_1B_1/A_2B_2 and A_1B_2/A_2B_1. When recombination takes place in an A_1B_1/A_2B_2 individual, the number of A_1B_1 haplotypes is decreased; when it takes place in an A_1B_2/A_2B_1 individual, the number of A_1B_1 haplotypes is increased. To be exact, half of the genes of an A_1B_1/A_2B_2 double heterozygote are A_1B_1. When recombination hits between the loci the frequency of A_1B_1 decreases by an amount $-^1/_2$. Similarly, recombination in an A_1B_2/A_2B_1 individual increases the frequency of A_1B_1 by an amount $+^1/_2$.

The frequency of A_1B_1/A_2B_2 heterozygotes in the population is $2ad$ and of A_1B_2/A_2B_1 is $2bc$. The frequency of recombination between the two loci is defined as r. (Theoretically, r can have any value up to a maximum of 0.5, if the loci are on different chromosomes; its value lies between 0 and 0.5 for loci on the same chromosome depending on how tightly linked they are.) Thus

$$a' = a - \frac{1}{2}r2ad + \frac{1}{2}r2bc$$

$$a' = a - r(ad - bc)$$

The expression $(ad - bc)$ is simply equal to the linkage disequilibrium, D. (We can confirm this equivalence by multiplying out $ad - bc$ from the definitions of a, b, c, and d.) If $D = 0$ (i.e., if the genes are randomly associated), the haplotype frequencies are constant, and $a' = a$. If there is an excess of A_1B_1 haplotypes, the excess decreases by an amount rD per generation. The same relation holds true for any successive pair of generations. We can see what is happening graphically if we substitute for a in the equation:

$$a' = p_1q_1 + D - rD$$

$$a' - p_1q_1 = (1 - r)D$$

The difference between a and p_1q_1 is the amount of "excess" of the A_1B_1 haplotype (i.e., the amount by which the frequency exceeds the random frequency). It is also equal to the linkage disequilibrium ($D = a - p_1q_1$). Therefore

$$D' = (1 - r)D$$

In the absence of selection and in an infinite random mating population, the amount of linkage disequilibrium undergoes exponential decay at a rate equal to the recombination rate between the two loci. In other words, the difference between the actual frequency of a haplotype such as A_1B_1 (a) and the random proportion (p_1q_1) decreases each generation by a factor equal to the recombination rate between the loci.

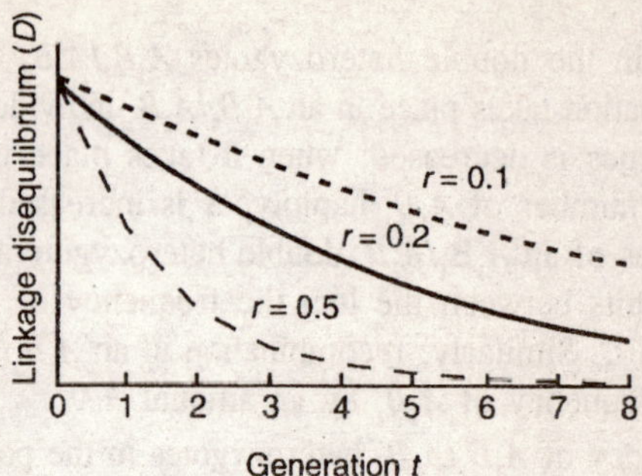

Fig. 11.1. Non-random associations between genes at different loci are measured by the degree of linkage disequilibrium (D).

Over time, any non-random genie associations will disappear, as recombination will destroy the association. The higher the rate of recombination, the more rapid the destruction. The highest possible value of r is $^1/_2$, which is true when the two loci are found on different chromosomes. Genie associations persist longer for tightly linked loci on the same chromosome, as we should intuitively expect.

The equilibrial haplotype proportions have $D = 0$. At equilibrium, the following frequencies apply:

Haplotype	Equilibrial Frequency
A_1B_1	$a = p_1q_1$
A_1B_2	$b = p_1q_2$
A_2B_1	$c = p_2q_1$
A_2B_2	$d = p_2q_2$

These haplotype frequencies, which we have encountered before, are called linkage equilibrium. It is called an "equilibrium" because, in the absence of selection, the action of recombination will drive the haplotypes to these frequencies and then keep them there.

Recombination randomizes genie associations over time. If an excess of one haplotype, such as A_1B_1 exists, recombination will tend to break it down, and A_1 will end up with B_1 and B_2 in their population proportions (q_1 and q_2) and B_1 will be found with A_1 and A_2 in their population proportions (p_1 and p_2). At linkage equilibrium, each of the two alleles at the A locus, A_1 and A_2, are then associated with B_1 in the same proportion.

In a way, linkage equilibrium is, for a two-locus system, an analogy to the Hardy—Weinberg equilibrium for the one-locus system. It describes the equilibrium that is reached in the absence of selection, and in an infinite, randomly mating population. Linkage equilibrium, however, is a property of haplotypes, not genotypes. A diploid individual has two haplotypes, and at the equilibrium the genotypes at each locus will be in Hardy-Weinberg proportions while the haplotypes are at linkage equilibrium. Notice also that whereas the Hardy-Weinberg equilibrium for one locus is reached instantly in one generation, it takes several generations for linkage equilibrium to be reached.

Linkage equilibrium has three important aspects for our purposes. The Hardy-Weinberg theorem for one locus was the simplest model in single-locus population genetics; it illustrated how to construct a model with recurrence relations for gene frequencies. The model of linkage equilibrium is, likewise, the simplest model for two loci and shows us how to construct a recurrence relation for haplotype frequencies.

Also like the Hardy-Weinberg theorem, linkage equilibrium provides a theoretical baseline telling us whether anything interesting is occurring in a population. Deviations from Hardy-Weinberg proportions in a natural population suggest that selection, non-random mating, or sampling effects may be operating. Likewise, if two loci are in linkage disequilibrium, we may also suspect that one or more of these variables are at work. If the first thing we had discovered about *Papilio memnon* had been its high linkage disequilibrium, we would have been led on to study how selection was operating on the loci, and perhaps ended up discovering the mimetic polymorphism. In fact, the direction of research in *P. memnon* took the opposite tack, but the general point—that linkage disequilibrium indicates something interesting—holds true.

Linkage equilibrium can also tell us whether the more complex two-locus theory is needed in a real case. To a rough approximation, the theory of population genetics for a single locus is satisfactory for populations in linkage equilibrium. It is when genes become non-randomly associated that a two-locus model is needed. The case we have been discussing can show why. At linkage equilibrium, A_1 and A_2 are equally associated with B_1. To understand evolution at the A locus we can then ignore the relative fitnesses of B_1 and B_2, because if B_1 is fitter than B_2, the association with B_1 will benefit A_1 and A_2 equally (and B_2 will equally detract from them). If A_1, for instance, is more associated with B_1 than is A_2 (i.e., linkage disequilibrium exists), then any advantage of B_1 over B_2 will passively give rise to an advantage to A_1. To understand frequency changes of A_1 we then need to know the relative fitnesses of B_1 and B_2, and the degree of association A_1 has with them. We need a two-locus model.

Human HLA Genes

The HLA system in humans consists of a set of linked genes on human chromosome 6 that control "histocompatibility" reactions. When an organ is transplanted from one individual to another, it is immunologically rejected by the recipient in a matter of days—skin grafts last only 2-15 days, for instance. The rejection implies that the immune system can distinguish between "self" and "foreign" cells, and the distinction is generally believed to be achieved by the products of the HLA genes. The best evidence for the HLA genes' role comes from the time course of kidney transplant rejection among siblings that either have or have not been matched for their HLA genes. For kidney transplants between HLA-matched siblings, more than 90% of transplants will survive after 48 months; among HLA-unmatched siblings, 90% survive for 4 months and only about 40% for 48 months.

The HLA system contains a number of genes; we will concentrate on two of them, called HLA-*A* and HLA-*B*. Each HLA locus, in a human population, is highly polymorphic. The *B* locus alone may contain 16 alleles with frequencies of 1-10% and many more rare alleles. For example, a sample of 874 people in France contained 31 different alleles at the *B* locus and another 17 alleles at the *A* locus. These degrees of variability are exceptionally high. More typical loci (outside the HLA) might have 1-5 alleles, many less than the number found in the HLA. The reason for the high variability is uncertain, but it would allow the HLA genotype of an individual even in a large population to be unique, which is presumably important in the distinction of self from foreign cell types.

Particular HLA alleles are associated with specific diseases. The strongest association found so far is between ankylosing spondylitis and the allele *B27*; 90% of people with the disease have the *B27* allele, compared with only 7% in the population at large.

The diversity of HLA types may reflect a history of coevolution between humans and disease agents. Over evolutionary time, disease agents may have tried to fool the immune system into treating the agent as part of the body, with the human population then responding by evolving new HLA alleles as new, reliable indicators of "self." This type of evolution would provide a further advantage to variability in the HLA loci. A heterozygous individual with two HLA proteins can compare itself with a possible invader in two ways. The invader must match a homozygote in only one respect, but a heterozygote must be matched in two independent respects. The HLA loci, therefore,

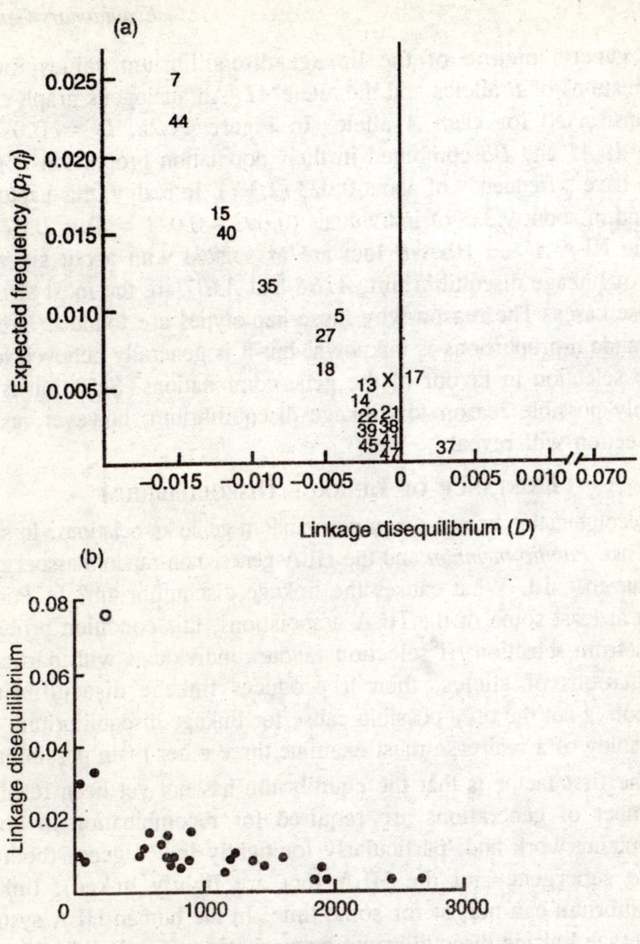

Fig. 11.2. (a) One example of the pattern of degrees of linkage disequilibrium (D) in the HLA: the linkage disequilibrium between 21 B alleles and the A1 allele. (b) The linkage disequilibrium between 8 HLA loci: more closely linked loci show higher linkage disequilibrium.

probably show heterozygous advantage, and the same process may have caused the exceptional pattern of evolution in silent and amino acid changing bases within codon triplets.

The HLA system also provides examples of linkage disequilibrium. Particular combinations of genes are found in greater than random proportions. North European populations, for example, characteristically demonstrate an excess of the *A1B8* haplotype. Figure 11.2a provides a more general picture of the linkage disequilibrium values for all combinations of *B* alleles and the allele *A1*. An analogous graph could be constructed for each *A* allele. In Figure 11.2a, $D = 0.07$ for *A1B8*. If *A1* and *B8* combined in their population proportions, *A1B8* would have a frequency of about 0.023 (2.3%). In reality, this haplotype is found in about 9.3% of individuals ($0.093 - 0.023 = D = 0.07$). In all, the HLA-*A* and HLA-*B* loci are associated with about six clear cases of linkage disequilibrium; *A1B8* and *A3B7* are the most striking of these cases. The reason why these haplotypes are found in greater than random proportions is unknown, but it is generally believed to be due to selection in favour of the gene combinations. Selection is not the only possible reason for linkage disequilibrium, however, as the next section will reveal.

Existance of Linkage Disequilibrium

Recombination breaks down non-random genie associations. In some cases, like *Papilio memnon* and the HLA genes, non-random associations continue to exist. What causes the linkage disequilibrium? In *Papilio* and in at least some of the HLA associations, this condition probably results from selection. If selection favours individuals with particular combinations of alleles, then it produces linkage disequilibrium. Selection is not the only possible cause for linkage disequilibrium, and a full study of a real case must examine three other main possibilities.

The first factor is that the equilibrium has not yet been reached. A number of generations are required for recombination to do its randomizing work and, particularly for tightly linked genes (both the *Papilio* supergene and the HLA loci are tightly linked), linkage disequilibrium can persist for some time. In the human HLA system, the average linkage disequilibrium between more closely linked loci is larger.

A second process that may affect linkage disequilibrium is random drift in a finite population. Random processes have the interesting property that they cause persistent—not just transitory—linkage disequilibrium. If random sampling produces by chance an excess of a haplotype in a generation, linkage disequilibrium will have arisen. This statement is true for all four haplotypes; random sampling that produces an excess of any of them will disturb the state of linkage equilibrium. Any haplotype could be "favoured" by chance, so the disequilibrium is equally likely to have $D>0$ or $D<0$. As a population approaches linkage equilibrium, all random fluctuations in haplotype frequencies will tend to occur in a direction away from the linkage

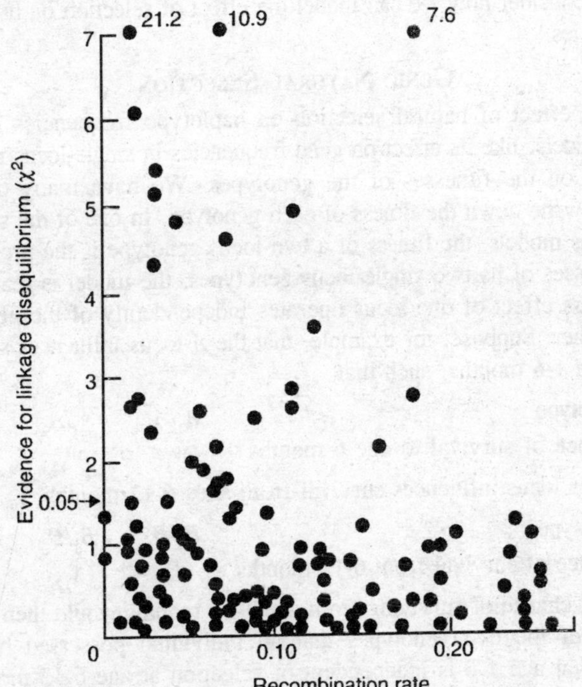

Fig. 11.3. Linkage disequilibrium (on the y-axis) among gel electrophoretic samples of pairs of genes in Drosophila.

equilibrium values. If a population is located far away from the point of linkage equilibrium, random sampling is equally likely to move it toward or away from the equilibrium. Most natural populations are probably near linkage equilibrium, and the balance between the random creation of linkage disequilibrium and its destruction by recombination, in small enough populations, is such that linkage disequilibrium will persist.

The third possibility is non-random mating. If individuals with gene A1 tend to mate with B_1 types rather than B_2 types, A_1B_1 haplotypes will have excess frequency over that seen with random mating. (The exact effect depends on whether it is homozygous A_1/A_1 individuals that mate non-randomly, or the homozygotes and the A_1/A_2 heterozygotes, and on whether they mate preferentially only with B_1/B_1 homozygotes or with B_1/B_2 heterozygotes as well. The general effect of non-random mating on linkage disequilibrium is not complicated.)

The three processes other than selection probably account for some cases of linkage disequilibrium in nature. The process that has most interested evolutionary biologists, however, is natural selection. Let us now consider how we can model the effect of selection on haplotype frequencies.

Genic Natural Selection

The effect of natural selection on haplotype frequencies in two-locus models, like its effect on gene frequencies in single-locus models, depends on the fitnesses of the genotypes. We have many possible ways to write down the fitness of each genotype. In one of the simplest two-locus models, the fitness of a two-locus genotype is the product of the fitnesses of its two single-locus genotypes; the model is realistic if the fitness effect of one locus operates independently of the genotype at the other. Suppose, for example, that the *A* locus influences survival from age 1-6 months, such that

genotype	A_1/A_1	A_1/A_2	A_2/A_2
chance of survival to age 6 months	w_{11}	w_{12}	$w2_2$

The other locus influences survival from ages 6-12 months

genotype	B_1/B_1	B_1/B_2	B_2/B_2
chance of survival from 6-12 months	x_{11}	x_{12}	x_{22}

The total chance of surviving from age 1-12 months would then be the product of the two genotypes that an individual possessed because selection at age 1-6 is independent of selection at age 6-12 months:

	A_1/A_1	A_1/A_2	A_2/A_2
B_1/B_1	$w_{11} \times x_{11}$	$w_{12} \times x_{11}$	$w_{22} \times x_{11}$
B_1/B_2	$w_{11} \times x_{12}$	$w_{12} \times x_{12}$	$w_{22} \times x_{12}$
B_2/B_2	$w_{11} \times x_{22}$	$w_{12} \times x_{22}$	$w_{22} \times x_{22}$

These fitnesses are called *multiplicative*. An individual's fitness for its two genotypes is found by multiplying the fitnesses of each of its one-locus genotypes. The genotypes are independent, in the sense that the effect of one genotype on survival is independent of the other locus. An individual with the genotype A_1/A_1 has a chance of surviving from age 1—6 months of w_{11} whcther its genotype at the other locus is B_1/B$_1$, B_1/B_2, or B_2/B_2.

Additive fitness interactions are another possibility. Imagine that the two loci encode two digestive enzymes. The enzyme produced by the *A* locus digests one kind of food, while the enzyme produced by the *B* locus digests another kind of food. The genotypes might enable a certain number of joules of energy to be obtained by feeding in a particular environment:

genotype	A_1/A_1	A_1/A_2	A_2/A_2	B_1/B_1	B_1/B_2	B_2/B_2
energy obtained	w_{11}	w_{12}	w_{22}	x_{11}	x_{12}	x_{12}

The fitness of an individual might be proportional to the total energy it obtains by feeding. Its fitness for the two loci would then be found by adding the fitnesses of its two genotypes together.

	A_1/A_1	A_1/A_2	A_2/A_2
B_1/B_1	$w_{11} + x_{11}$	$w_{12} + x_{11}$	$w_{22} + x_{11}$
B_1/B_2	$w_{11} + x_{12}$	$w_{12} + x_{12}$	$w_{22} + x_{12}$
B_2/B_2	$w_{11} + x_{22}$	$w_{12} + x_{22}$	$w_{22} + x_{22}$

The next step is to derive a recurrence relation between the frequency of a haplotype in one generation and in the next generation. However, the algebra is tiresome and we do not need to work through it here. The procedure follows the same outline as for the single-locus case, with the additional factor of recombination. The recurrence relation for haplotype frequency takes account of the frequency and fitness of all genotypes in which a haplotype is found. It also adds and subtracts the number of copies gained and lost by recombination; we multiply by $(1 - r)$ the frequency of the double heterozygotes containing the haplotype and by r the frequency of the double heterozygote that can generate it if recombination occurs. The Mendelian rules are then applied, and the frequency of the haplotype in the next generation results.

Which kinds of selection cause linkage disequilibrium? The question is important because, as we have seen, two-locus models are particularly needed when linkage disequilibrium exists. The two types of fitness (multiplicative and additive) that we have just considered both result in equilibria) haplotype frequencies with zero linkage disequilibrium. (The possibility arises only if the equilibrium has both loci polymorphic. If one gene is fixed at either locus, $D = 0$ trivially. The fitnesses as written above (w_{11}, w_{12}, and so on) were frequency-independent. A doubly heterozygous equilibrium then requires heterozygous advantage at both loci: $w_{11} < w_{12} > w_{22}$, $x_{11} < x_{12} > x_{22}$.) In conclusion, if linkage disequilibrium exists at any time that fitnesses are independent at the two loci, it will decay to zero as the generations pass.

The more interesting case is when the fitnesses of the two loci interact *epistatically*. Epistatic interaction means that the fitness effects of a genotype depend on the genotype with which it is associated at the other locus. The selection in the mimetic polymorphism of *Papilio memnon* is epistatic, for example.

We can reduce the situation in *P. memnon* to the most elemental theoretical form if we imagine that one locus controls whether the butterfly has a tail on its hindwing and a second controls colouration. Let T_+ (presence of tail) be dominant to T_- (absence). At the other locus, C_1 is dominant. C_1/C_1 and C_1/C_2 individuals have a colour pattern that mimics a model species with a tail, whereas C_2/C_2 individuals are coloured like a model species that has no tail. The relative fitness of each genotype depends on the genotype found at the other locus. For example, a T_+/T_+ genotype in the same butterfly as a C_2/C_2 will be less fit than a T_-/T_- with C_1/C_1. The fitnesses can be written as follows (the simplification relative to the earlier fitness matrixes arises because of dominance, and because one term exists for the fitness of both loci together rather than one term for each locus):

	T_+/T_+	T_+/T_-	T_-/T_-
C_1/C_1	w_{11}	w_{11}	w_{21}
C_1/C_2	w_{11}	w_{11}	w_{21}
C_2/C_2	w_{12}	w_{12}	w_{22}

In the case we discussed, w_{12} is the fitness of a butterfly with a tail and the colour pattern of a model lacking a tail. Therefore, $w_{12} < w_{11}$ and w_{22}. The term w_{21} is the fitness of a butterfly without a tail, but the colour pattern of a tailed model. Therefore, $w_{21} < w_{11}$ and w_{22}. Selection now favours the $T_+/-$ genotypes when they are with $C_1/-$, but not when with C_2/C_2; likewise, selection favours T_-/T_- when it is with C_2/C_2 but not when with $C_1/-$ (the dash implies it does not matter which gene is present, because of dominance). The fitness relations are epistatic, and a doubly polymorphic equilibrium can now occur. All four alleles will be present, and the haplotypes $T_+ C_1$ and $T_- C_2$ will have disproportionately high frequencies. T_+C_2 and $T_- C_1$ haplotypes are selected against, because they often find themselves in poorly mimetic butterflies. Linkage disequilibrium ($D > 0$ in this case) exists at the equilibrium.

In general, selection can only produce linkage disequilibrium at equilibrium when the fitnesses of the genotypes at different loci interact epistatically. Not all epistatic fitness interactions generate doubly polymorphic equilibria with linkage disequilibrium, but all such equilibria do have epistatic fitnesses.

We have been discussing the different sorts of fitness interactions—multiplicative, epistatic, and so on—as properties of formal models. Real genes in real organisms will have fitness interactions as well, and the more important question deals with the characteristics of these interactions. In cases like *Papilio*, epistasis is present and powerful; these cases may be isolated examples rather than representing a general condition, however. Evolutionary biologists are interested in whether fitness interactions between loci are generally epistatic and generate strong linkage disequilibrium, or whether they are generally independent and generate linkage equilibrium. These two extremes roughly correspond to a more "holistic" and a more "atomistic" (or "reductionist") school of thought (which is not to say that they correspond to two clearly demarcated camps of biologists).

No general answer has been found for this question, but it is possible to make some observations. Different loci will tend to interact multiplicatively when they have independent effects on an individual's survival and reproduction. This interaction could occur with loci and influence events at different times in an organism's life —though it is perfectly possible for such events to interact. Epistatic interactions are more likely for loci controlling closely interdependent parts of an organism. The extent to which we expect loci to interact epistatically will, therefore, loosely depend on how atomistic or holistic a view we have of the organism.

Notice that epistatic fitness interaction is not the same as mere physiological or embryological interaction. Fitness epistasis requires heterozygosity at two loci. Imagine a case in which the *A* locus controls, for example, muscle strength and the *B* locus controls metabolic rate. Muscles and metabolism interact in a physiological sense: when muscles are put to work, the metabolic rate increases. However, if the population is fixed for homozygotes at both loci (all individuals are A_1B_1/A_1B_1), then fitness epistasis cannot be present. The special condition of epistatic fitness requires heterozygosity at both loci, and the kind of fitness relations that we saw in the *Papilio memnon* example. Although this situation is often called fitness "interaction," the term interaction is used here in a technical, not a colloquial, sense.

Another, empirical method has been applied to answering the question of how commonly epistatic fitness interactions occur in nature. Linkage disequilibrium is produced by epistatic selection, and the degree of linkage disequilibrium in a population can be measured. If it is high, then epistatic selection may be common. The argument works in one direction but not the other. That is, because linkage disequilibrium may be caused by several factors, its existence does not demonstrate epistatic selection. If linkage disequilibrium is absent or low, however, we can infer that epistatic selection is unimportant in nature.

A few general surveys of the extent of linkage disequilibrium in natural populations have been made. An investigation by Maynard Smith et al. (1993) for bacteria found high levels of linkage disequilibrium in some species, such as *E. coli* (which lives in our, and other mammals', guts) but low levels in other species, such as *Neisseria gonorrhoeae*. The reason why many bacteria show linkage disequilibrium is that they reproduce asexually, and no recombination happens to decay the linkage disequilibrium, *N. gonorrhoeae* presumably has enough sexual exchange of genes to produce linkage equilibrium.

In eukaryotic organisms that are known to reproduce sexually, the evidence suggests that little deviation from linkage equilibrium appears in nature. The main evidence comes from gel electrophoretic surveys of protein polymorphisms, to see directly whether genes at different loci are associated.

Some evidence of linkage disequilibrium is found, but the results suggest that the level is low and most loci are in linkage equilibrium. One conclusion from this survey would be that, although epistatic interactions are important in particular cases (e.g., *Papilio*), they may not be of general importance at least in the evolution of sexually reproducing species.

Other biologists would disagree. They might be unconvinced by the evidence, perhaps calling it "limited" or "for a single species." The amount of interaction between loci that must go on during the development of a complex, organic body is so high that they would expect epistatic fitness interactions to be common.

Hitchhiking

When a gene is changing frequency at one locus over time, it can cause related changes at linked loci; conversely, events at linked loci can interfere with one another. Suppose, for instance, that directional selection is substituting one allele *A*' for another (*A*) at one locus, and a neutral polymorphism (*B*, *B*') appears at a linked locus. The frequency of *B* or *B*' (whichever was linked with *A*' when it arose as a mutant) will then be increased. If the new mutant *A*' arose on a *B*-bearing chromosome, then B will eventually be fixed together with the selected allele *A*' unless recombination splits them before *A* has been eliminated. The increase in the *B* allele frequency is due to hitchhiking.

Another possibility is for the polymorphism at the *B* locus to be a selectively "balanced" polymorphism, due to heterozygous advantage. Suppose that a selectively favoured mutation *A'* arises at a linked locus, and that it happens to arise on the same chromosome as a *B* allele. The polymorphism at the *B* locus will now interfere with the progress of *A'*. As *A'* increases in frequency by directional selection, it will also increase the frequency of *B*. Because *A'* is linked to *B*, it will be more likely to appear in a body with a *B/B* homozygote than will its allele *A*, and less likely to be in a *B/b* heterozygote. *B/b* has higher fitness than *B/B*, and the selection against *B/B* individuals will also work against the *A'* gene. Depending on the selection coefficients at the two loci, and the rate of recombination between them, the heterozygous advantage at the *B* locus can slow the rate at which *A'* is fixed. The *A'* gene will then have to wait for a recombinational "hit" between the two loci before it can progress to fixation.

Hitchhiking has a further interest in relation to the neutral theory of molecular evolution. It alters the neutral theory's predictions for the level of heterozygosity. We saw that observed heterozygosities are lower than those predicted by the neutral theory for plausible estimates of population size and mutation rate. Heterozygosities are not only too low, but also too uniform, given the variation in natural population sizes. Maynard Smith and Haigh pointed out that hitchhiking could bring the neutral theory more closely into line with the observations. As a favoured allele is fixed, the heterozygosity declines at any linked locus that has a neutral polymorphism; the more closely linked the polymorphic locus, the greater the reduction in its heterozygosity. We do not know to what extent hitchhiking causes heterozygosities to deviate below the neutral theory's prediction; however, the hitchhiking effect of a favourable gene must undoubtedly act to reduce heterozygosities at linked neutral loci.

Application of Linkage Disequilibrium

The linkage disequilibrium in *Papilio memnon's* mimetic polymorphism is advantageous. Natural selection favours individuals with the genie associations like $T_{-}\ W_2F_2E_2B_2$, whereas it works against recombinants like $T_{+}\ W_2F_2E_2B_2$. An individual benefits from having the haplotypes that are in excess frequency in the population. Whole populations of *P. memnon* survive better than they would if the five loci existed in linkage equilibrium.

In other cases, the opposite is true. As the favoured allele *A'* increased in frequency, and the frequency of an allele, *B*, at a linked polymorphic locus was also increased by hitchhiking. The polymorphism at the *B* locus was caused by heterozygous advantage, and the linkage disequilibrium built up by selection on the *A* locus (creating an excess of the *A*' *B* haplotype) was, therefore, disadvantageous. The individuals on average have lower fitness than if linkage equilibrium existed between the *A* and *B* loci, because the increase in the *A*' *B* haplotype reduces the proportion of *B*/*b* heterozygotes. Natural selection will favour recombinant individuals that do not have the *A*' *B* haplotype.

A third possibility is for linkage disequilibrium to be selectively neutral. An example of this case was provided by the hitchhiking of an allele at a neutral polymorphic locus with a selectively advantageous mutant at a linked locus. While the mutant is being fixed, linkage disequilibrium temporarily builds up between it and the alleles at nearby loci to which it is linked. It disappears when the mutant reaches a frequency of one.

The distinction between advantageous and disadvantageous linkage disequilibrium is crucial to understanding one of the major problems of evolutionary biology—why recombination exists.

Congeal Genome

Genetic recombination undoubtedly exists, but it is not obvious why it occurs. The problem can be seen by considering two kinds of populations near equilibrium—one at linkage equilibrium and the other with linkage disequilibrium due to epistatic selection (or coadaptation between genetic loci, like in *Papilio memnon*). We will assume that mating is random and the population is large. What effect does recombination have in the first case, which includes linkage equilibrium? None at all—the frequencies of A_1B_1, A_1B_2, A_2B_1, and A_2B_2 are repeated from generation to generation, regardless of whether recombination happens. If there is no recombination, the haplotype frequencies are necessarily constant. If recombination is present, it creates and destroys each haplotype at the same rate, which is what is meant by linkage equilibrium. In the absence of linkage disequilibrium, therefore, recombination has no effect.

Suppose that linkage disequilibrium occurs, and it is due to coadaptation between loci. Let us imagine that A_1B_1 and A_2B_2 are coadapted pairs of alleles, like the tail genes and colour genes in *P. memnon*, that are present in greater than random frequencies ($D > 0$). The actual amount of linkage disequilibrium will be a balance between the increase due to selection and the decrease due to recombination. In this case, recombination is disadvantageous. The lower the recombination rate, the stronger the association between the coadapted alleles that can be built up, and the fitter on average the individuals will be. Selection, therefore, acts to reduce the recombination rate between the loci.

It might seem, then, that recombination either makes no difference or is disadvantageous. If there is linkage equilibrium, the genotype frequencies are unaltered by recombination; if beneficial associations between genes have built up, recombination tends to destroy them. So why does recombination exist? Why are recombination rates not reduced to zero? The answer to this question has not found universal agreement. The vague answer is probably that recombination is favoured because of environmental change. This answer does not eliminate problems in specifying the kind of environmental change needed for recombination to be advantageous, and in showing that natural environments do change in the necessary way. Before we discuss environmental change in depth, we should first rule out one other possible explanation.

Natural selection cannot work on an organism if it shows no genetic variation. If no genetic variation influenced recombination rates, then recombination rates could not be reduced to zero. Is the recombination rate genetically variable? The most convincing test is to carry out an artificial selection experiment. If a trait varies genetically, an attempt to change it by artificial selection should be successful; if it does not vary genetically, the attempt should fail. Several experiments have selected successfully for both higher and lower recombination rates in fruitflies. Thus, absence of genetic variation is not the reason why recombination exists.

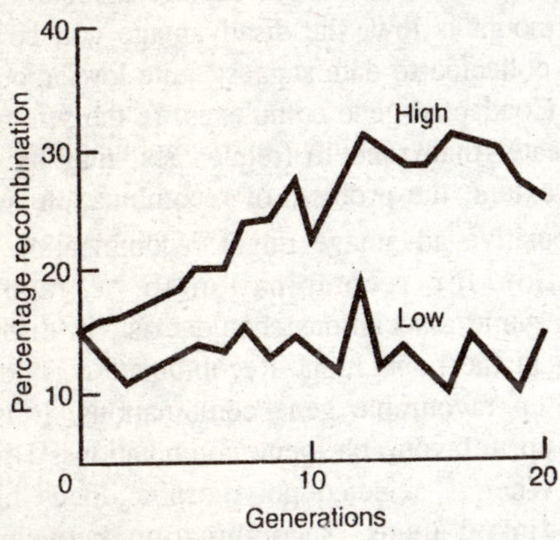

Fig. 11.4. Artificial selection can increase or decrease the rate of recombination.

Moreover, in the case of linked groups of coadapted genes, recombination can be effectively eliminated by a chromosomal inversion. We do not know whether the *P. memnon* mimicry genes arc contained within an inversion, but let us consider what would happen if they were. If a single recombination event occurs within an inverted chromosomal region, the resulting chromosomes lack certain genes and are, therefore, likely to be lethal. Only when recombination (or, more precisely, an odd number of recombination events) has not occurred in the inversion will the offspring be viable. Thus, the set of genes within the inversion are practically protected from recombination.

This point is not merely theoretical, because chromosomal inversions are known to exist. They are particularly easy to study in fruitflies, and Dobzhansky did much research on them. He emphasized the idea that chromosomal inversions protect sets of epistatically interacting genes, which is an important principle of chromosomal evolution. Notice, by the way, that the genes in an inversion will tend to behave as a "supergene"—a set of linked genes that segregate as if they were a single Mendelian locus. In a supergene, the recombinants will often die and therefore not be seen.

In summary, the recombination rate can be reduced. It is, therefore, unlikely that recombination persists in natural populations simply because natural selection cannot eliminate it for want of genetic variation.

Recombination, as we saw, is disadvantageous only when it disrupts coadapted sets of genes in linkage disequilibrium. Its disadvantage will depend on how much of this sort of linkage disequilibrium exists in nature. If the amount is low, the disadvantage will be small. As we have seen, data collected to date suggest quite low amounts of linkage disequilibrium. Coadapted gene complexes, in the sense of haplotypes occurring in greater than random frequencies, may be exceptional in nature. To this extent, the problem of recombination is reduced.

But what positive advantage might recombination provide? The general condition for recombination to be favoured is that disadvantageous genie associations should exist, in disproportionately high frequency, in the population. Recombination is selected against when it breaks up favourable gene combinations. It is selected for when it breaks up unfavourable gene combinations. Hitchhiking is a case in point. When a selected polymorphic locus interferes with selection at a linked locus, recombination between the loci is advantageous. Strobeck et al. have formally demonstrated this advantage, although the argument may, in fact, may be a special circumstance.

The process would be more common in a rapidly changing environment, because the fitnesses of different alleles would then be changing and selection attempting to adjust the frequencies of genes at many (potentially interfering) loci.

The problem of why recombination exists is closely related to the problem of why sex exists. That theory depends on the rate of deleterious mutation, and the discussion of selection-mutation balance provides a background to it. In this chapter, however, the main point is that recombination is a puzzle. It can have clear disadvantages, and it does not exist simply because natural selection cannot remove it.

ADAPTIVE TOPOGRAPHY

Wright's idea of an *adaptive topography* is particularly useful for thinking about complex genetic systems, but it is easier to begin with the simplest case of a single genetic locus. The topography is a graph of mean population fitness ($\overline{w}$) plotted against gene frequency. (Adaptive topographies can also be drawn for fitness in relation to genotype frequencies. They can even be drawn with phenotypic variables on the *x*-axis.) We have repeatedly encountered the concept of mean fitness; it is equal to the sum of the fitnesses of each genotype in the population, each multiplied by its proportion in the population. In a case in which the genotypes containing one of the alleles have higher fitness than those of the alternative, the mean fitness of the population simply increases as the frequency of the superior allele increases and reaches a maximum when the gene is fixed. When heterozygous advantage occurs, mean fitness is highest at the equilibrium gene frequency given by the standard equation; it declines on either side, where more of the unfavourable homozygotes will be dying each generation than at the equilibrium. The graph of this relation is also called a *fitness surface*.

In these two cases, natural selection carries the population to the gene frequency where mean fitness is at a maximum. With one favourable allele, the maximum mean fitness appears where the allele is fixed—and natural selection will act to fix the allele. With heterozygous advantage, the maximum mean fitness appears where the smallest number of homozygotes are dying each generation—and natural selection drives the population to an equilibrium where the amount of homozygote death is minimized. An interesting question in theoretical population genetics is whether natural selection always drives the population to the state at which the mean fitness is the maximum possible. In some cases, it does not, such as in frequency dependence. When frequency dependence maintains a polymorphism, the fitness of a genotype is highest when it is rare. When a genotype is rare, natural selection acts to increase its frequency, making it more common. It is possible to devise theoretical cases in which the mean fitness is maximized at genotype frequencies that are not at equilibrium. In these instances, natural selection acts to reduce mean fitness by moving the genotype frequencies away from those points. If natural selection does not always maximize mean fitness, a further—and still unanswered—theoretical question is whether natural selection always maximizes some other function; we will not pursue the question here. Whatever the answer to this question, there are many cases in which natural selection maximizes simple mean fitness—a useful assumption for many purposes. We can then think of natural selection as a hill-climbing process, analogous to the hills in the adaptive topography.

Now consider a second locus. Selection can take place here as well, and the fitness surface for the two loci might resemble a simple case in which one locus has heterozygous advantage and the other possesses a single favoured allele. The idea of an adaptive topography can be extended to as many loci as interact to determine an organism's fitness, but the additional loci must be imagined, rather than drawn on two-dimensional paper.

Wright believed that, because the genes at different loci interact, a real multidimensional fitness surface would often contain multiple peaks, with valleys between them. The kind of reasoning behind this theory is abstract rather than concrete. Imagine a large number of

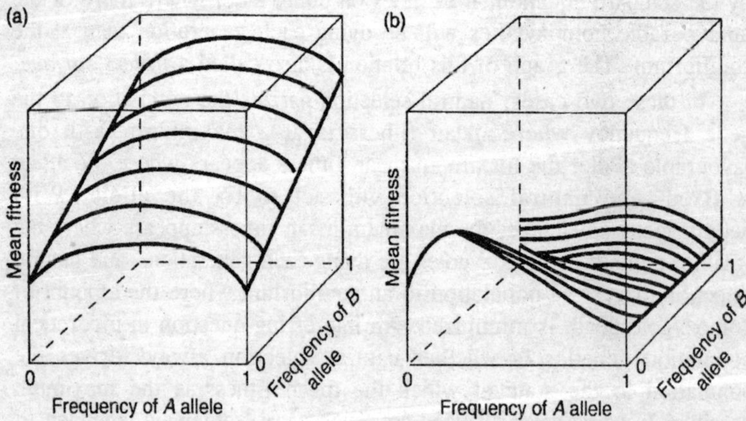

Fig. 11.5. Fitness surface for two loci. (a) heterozygous advantage at locus A and one allele has higher fitness than the allele at locus B. (b) Two-locus fitness surface with two peaks.

loci, many with more than one allele, and the alleles at the different loci interacting epistatically in their effects on fitness. Epistatic interactions, we now suppose, are common because organisms are highly integrated entities compared with the atomistic chromosomal row of Mendelian genes from which they develop. The genes will have to interact to produce an organism. As we saw above, developmental interactions among genes do not automatically generate epistatic fitness interactions among loci. The extent to which undoubted developmental interaction will produce a multiply peaked fitness surface is, therefore, open to question, but the possibility is plausible. (Wright called the genes that interact favourably to produce an adaptive peak an "interaction system.")

In the coadapted supergenes of *Papilio memnon*, the mimetic genotypes occupy fitness peaks and the recombinants occupy various fitness valleys. The actual shape of the adaptive topography in nature is, however, a more advanced question than can be tackled here. The point of this section is to define an adaptive topography, and to point out that its visual simplicity can be useful in thinking about evolution when many gene loci interact.

Shifting Balance Theory of Evolution

Wright used his idea of adaptive topographies to create his general evolutionary theory. He imagined that real topographies would have multiple peaks, separated by valleys, and that some peaks would be higher than others. When the environment changed and competing species evolved new forms, the shape of the adaptive topography for a population would change as well. The surface would also change shape when a new mutation arose. A new allele at a locus could interact with genes at other loci differently from the existing alleles, and the fitnesses of the genes at the other loci would then be altered. Thus, genetic changes would take place at other loci to adjust to the new mutant. At the same time, natural selection would be a hill-climbing process, directing the population up toward the currently nearest peak. When the surface changed, the direction to the nearest peak might change, and selection would then send the population off in the new upward direction.

Natural selection, even in so far as it is a hill-climbing process (i.e., maximizing mean fitness), is only a *locally* hill-climbing process. In theory, the local fitness peak could lie in the opposite direction from a higher, or the global, peak. Natural selection, however, will direct the population to the local peak. Now suppose that the mean

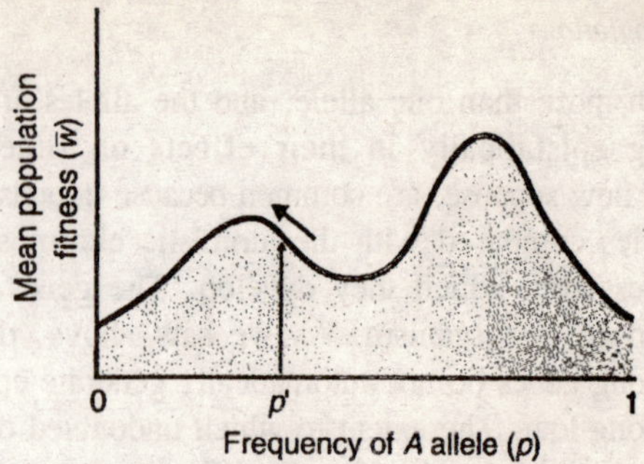

Fig. 11.6. A two-peaked fitness surface with local and global maxima.

fitness of a population measures the quality of its adaptations, such that a population with a higher mean fitness has better adaptations than a population with a lower mean fitness. Because natural selection seeks out only local peaks, natural selection may not always allow a population to evolve the best possible adaptations. A population could be stuck on a merely locally adaptive peak. However, adaptation and mean fitness cannot always be equated in this way. In the simple case in which one allele is superior to another, the organisms with the better genotype will also be better adapted. When frequency-dependence of fitness develops or when conflict emerges between group and individual adaptations, it is no longer true that the maximum of mean population fitness automatically corresponds to the best adaptation.

Wright was interested in how evolution could overcome the tendency of natural selection to become stuck at local fitness peaks. When fitness peaks correspond to optimal adaptations, the question is relevant to the evolution of adaptation; when they do not correspond, the question still has a technical interest in population genetics. Wright suggested that random drift could play a creative role. Drift will tend to make the population gene frequencies "explore" the terrain around their present position. By drift, the population could move from a local peak to explore the valleys of the fitness surface. Once it had explored to the foot of another hill, natural selection could start climbing uphill on the other side. If this process of drift and selection were repeated with different valleys and hills on the adaptive topography, a population would be more likely to reach the global peak than if it remained under the exclusive control of the locally maximizing process of natural selection.

Wright's full shifting balance theory includes ideas other than those of selection and drift within a local population. He also suggested that populations would be subdivided into many small local populations, and drift and selection would occur in each. The large number of subpopulations would multiply the chance that one subpopulation would find the global peak. If the members of a subpopulation at the highest peak were better adapted, they could produce more offspring; they would also produce more emigrants to the other subpopulations. Those other subpopulations would then be taken over by the superior immigrant genotypes. Thus, the whole species would evolve to the higher peak. As we can see, Wright's theory is an attempt at a comprehensive, realistic model of evolution. It includes everything: multiple loci, fitness interactions, selection within and between populations, drift, and migration.

As always in science, complex realistic models are useful for some purposes and simple, idealized ones for others. Realistic models may describe nature more accurately, but they are less illuminating when explaining principles or studying the consequences of particular factors. For this reason, when we consider particular factors in isolation in this book, we limit ourselves to simple and idealized models that concentrate on the factor of interest. The encumbrance of all of the other factors that may influence gene frequencies in nature is needed when we are trying to describe the real world, but not when we are trying to understand arguments.

Fisher generally disagreed with Wright over the importance of random drift in evolution. Fisher maintained that natural populations are generally too large for drift to play an important role; Wright retorted that Fisher had misunderstood the role of drift in the shifting balance theory, which was to assist, rather than counteract natural selection. Much of their argument seems to have been at cross-purposes, but many of the points are interesting.

Fisher, for instance, doubted whether natural selection actually would confine populations to local peaks. Fisher was preeminently a geometric thinker and he pointed out that, as the number of dimensions in an adaptive topography increases, local peaks in one dimension tend to become points on hills in other dimensions. In the extreme case of an infinity of dimensions, natural selection will certainly be able to hill-climb all the way to the global peak without any need for drift. Each one-dimensional or two-dimensional peak will be crossed at its top by an infinity of other dimensions, and it is highly implausible that the fitness surface will turn downhill in all of them at that point. This argument is highly interesting, although it is, of course, purely

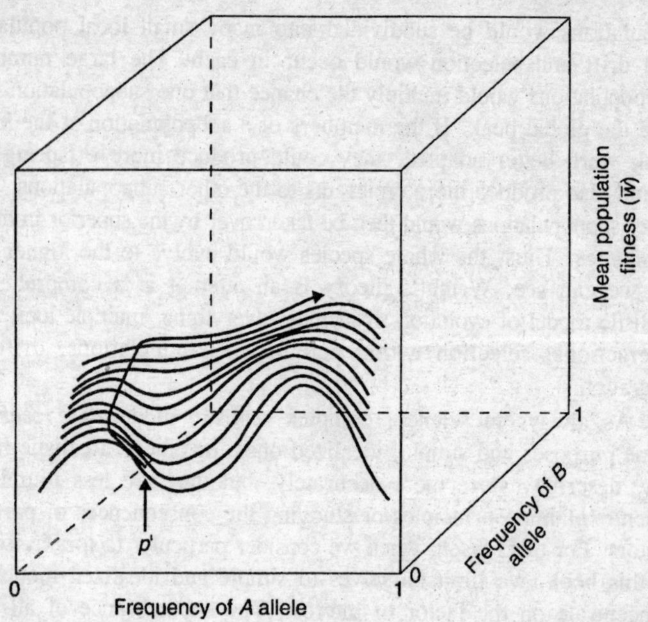

Fig. 11.7. As extra gene loci are considered (extra axes in the adaptive topography), it becomes increasingly likely that what appeared to be a local maximum in fewer dimensions will turn out to be a hillside or saddle point in more dimensions.

theoretical. It refutes Wright's *theoretical* claim that natural selection will get stuck at local peaks, but leaves open the empirical question of how important selection and drift have been in exploring the fitness surfaces of nature.

12

Gene Frequencies

At the center of Darwin's evolution theory was the concept that small heritable changes provided the *continuous variation* on which natural selection acted. However, soon after publication of *On the Origin of Species*, Francis Galton, Darwin's cousin, became convinced that a mathematical approach to heredity showed that evolution must have proceeded in sharp, discontinuous steps. By 1871, Galton had already disproved Darwin's pangenesis hypothesis to his own satisfaction by showing that transfusing blood among rabbit strains had no effect on heredity.

Similar to August Weismann's later germ plasm theory, Galton in 1875 suggested that instead of somatically acquired "gemmules," the hereditary material was passed on between generations with little or no change. It seemed apparent to Galton that parents who deviate significantly from the average for some continuous quantitative trait such as height tend to produce offspring that are closer to the average than themselves (Galton's law of regression). He surmised that continuous variation was therefore not the agency that leads to the origin of new species but that nonblending, *discontinuous variation* ("sports") provided the abrupt changes between species.

Mutationists and Selectionists

By the end of the nineteenth century, two schools of thought had established themselves in England based on their preference for continuous or for discontinuous variation. Upholding the importance of continuous variation in evolution were the mathematically oriented *biometricians*, Weldon and Pearson, and in opposition to them was Bateson and his supporters. When Mendel's 1865 paper on the genetics of peas was discovered in 1900, these two camps polarized further. Bateson and the *Mendelians* proposed that most hereditary characteristics were discontinuous and could be explained by the segregation of Mendelian factors, whereas the biometricians insisted that most characteristics were continuous and Mendelian factors were only involved in exceptional traits. To Mendelians such as De Vries and others, evolution could only be effective if selection operated on large mutations of the kind discovered in the evening primrose, *Oenothera lamarckiana*, whereas the biometricians allied themselves with Darwin's original concept, that selection acting on small differences was the primary mechanism for evolutionary change.

Apparently supporting the mutationist position were Johannsen's experimental observations that selection was ineffective in quantitatively changing the size of beans descended from homozygous *pure lines*. Furthermore, even when selection was practiced on beans descended from crosses between different pure lines, size differences among their descendants seemed to show relatively little change from the range of values initially observed in the F_2 generation of the cross. It seemed therefore that marked changes in size of beans could only come from mutations with large effect (*macromutations*), rather than from selection among the small differences observed in Johannsen's experiments. Various biologists therefore extended these views to propose that new species can arise in only one or a few mutational steps driven perhaps by mutation pressure in a particular and even nonadaptive direction, a view known as *saltation*. According to the saltationists, the slow, plodding process of Darwinian selection was no longer necessary to explain evolution.

Castle and Phillips, in contrast, demonstrated that selection could lead to entirely new coat colour patterns in hooded rats: some selected lines had "less pigment than any known type other than albino," whereas others were "so extensively pigmented that they would readily pass for the Irish type' which has white on the belly only." Although these studies identified no specific genes with quantitative effect, similar results with other organisms did point to the likelihood that selection could act on small continuous characters to produce marked changes in phenotype.

The rift between *mutationists* and *selectionists* in explaining the basic mechanisms of evolution remained until the 1920s and 1930s but the gap had already been bridged by further experimental work on *quantitative characters*. For example, Nilsson-Ehle and East showed experimentally that a number of different gene pairs (multiple factors) may affect a single quantitative character so that a wide array of possible genotypes can occur, each with a different phenotype. That is, genes that segregate in typical Mendelian patterns are responsible for many observed distributions of continuous traits. Thus, no real difficulty arose in providing a Mendelian interpretation of Darwinian selection for quantitative traits, and dependence on the introduction of mutations with very large effect no longer seemed necessary to explain most basic evolutionary changes.

The emphasis on evolution through small continuous characters does not mean that alleles with large effects on phenotype are always unadaptive. As Orr and Coyne point out, "mutations of large effect clearly play a substantial role in animal and plant breeding." Natural populations also show such effects; for example, insecticide resistance is often caused by only a few favourable mutations, and polymorphism for large and small beak size in an African seed-eating bird is the result of segregation for only two alleles at a single locus. Along with other data, especially those from plants, these findings indicate that adaptations may be caused by some genes with major phenotypic effects although these effects are not really those envisioned by macromutationists—that is, they do not create a complex organ or new species in a single stroke.

Neo-Darwinian Synthesis

Concurrent with the disputes between selectionists and mutationists, Yule, Castle, Hardy, and Weinberg were also proposing important new concepts. These workers emphasized that populations rather than individuals were an important evolutionary focus, and researchers had lo pay attention to population gene frequencies, rather than only to whether a gene was present or absent. It became clear that the collection of gametes a population contributes to the next generation can be considered as a giant gene pool from which offspring draw their various genotypic combinations at random. In the absence of selection and other factors that could change gene frequencies, these frequencies tend to be conserved, as the *Hardy-Weinberg equilibrium* demonstrated, discussed later.

During the 1920s, Fisher, Wright, and Haldane developed in considerable detail the approach toward considering evolution as a change in gene frequencies. They pointed this out in various papers that dealt with the effects of inbreeding, the evolution of dominance, and the effect of selection on gene frequencies, as well as the effects of mutation, migration, and genetic drift. In the early 1930s these studies culminated in a variety of papers and books that laid the foundations for *population genetics*, the study of gene frequencies and their changes.

Along with these mathematical models, Chetverikoff in the Soviet Union and others elsewhere observed considerable genetic variation in natural populations on which selection could act. All these studies helped establish the concept that the population has the variability necessary to explain evolutionary genetic change through space and time, whereas an individual is extremely limited in these dimensions: populations evolve, not individuals.

This emphasis on the genetics of populations helped transform evolutionary thinking into its more modern form, often called the *Neo-Darwinian* (or *modern*) *synthesis*. At the base of this synthesis is the concept that mutations occur randomly and furnish the fuel for evolution by introducing genetic variability. We can then define evolution as an ongoing process in which random mutation introduces genes whose frequencies change through time, with natural selection usually considered as the most important, although not only, cause for such change.

The accumulation of gene frequency differences, by whatever means, eventually leads to more pronounced (racial) differences among populations in geographically different localities. When gene exchange between racial groups can no longer occur because of reproductive barriers, separate species become established. Essentially, this approach gave a genetic slant to the biological species concept by conceiving of a species as a population of individuals bearing distinctive genes and gene frequencies, separated from other species by biological mechanisms that prevent gene exchange. Furthermore, mechanisms such as mutation and selection that lead to the origin of races and species are generally no different from mechanisms that lead to the origin of higher taxa such as genera, families, and orders, although the formation of higher taxa usually takes place over longer periods of time.

In other words, the Neo-Darwinian synthesis helped explain how mutation led to variation and how selection led to adaptation. The synthesis proposed that what begins quantitatively in populations as gene frequencies and changes in gene frequencies, becomes through time and its environmental vicissitudes, qualitative changes that transform some groups into races, and races into species, which, in turn, may even lead to higher categories of diversity.

By providing the general ideological framework in which to understand these events, the Neo-Darwinian synthesis helped motivate many evolutionary biologists to ask and answer more detailed questions as to how and why particular evolutionary events and adaptations actually occurred. Such questions began at basic levels: What are the gene frequencies in populations? How do gene frequencies change? Why do they change? What genetic differences separate races? species? At what rates do these differences arise? What historical phenomena can account for these differences?

Because of this fundamental materialistic approach, which Dobzhansky and Huxley popularized in books, the Neo-Darwinian synthesis made itself widely felt in biology by effectively eliminating Lamarckian concepts and other mystical or semimystical theories such as saltation and orthogenesis described earlier. The influence of population genetics, the essential component of this synthesis, also extends to many other fields such as demography, ecology, epidemiology, plant and animal breeding, and other areas in which gene variation and distribution affect the relationships and life patterns of organisms. Although population genetics has emphasized mathematical models that cannot reflect reality in all its myriad details, it has provided logically precise concepts that help us understand many common populational features, such as the frequently observed conservation of gene and genotype frequencies and the general effects of forces that change these frequencies.

Populations and Gene Frequencies

Geneticists usually define a population as a group of sexually interbreeding or potentially interbreeding individuals. Since Mendelian laws apply to the transmission of genes among these individuals, Wright has called such a group a *Mendelian population*. The size of the population may vary, but it is usually considered to be a local group (also called *deme*), each member of which has an equal chance of mating with any other member of the opposite sex. Most of the theory and experiments have so far emphasized populations of diploid organisms, and the discussions that follow deal mostly with such cases. However, whether diploid or haploid, populations have two important attributes: *gene frequencies* (also called *allele frequencies*) and a *gene pool*.

Gene frequencies are simply the proportion of the different alleles of a gene in a population. To get these proportions, we count the total number of organisms with various genotypes in the population and estimate the relative frequencies of the alleles involved. Of course, except for gametes and occasional mutation, the genetic complements of all cells in a multicellular organism are the same. We may therefore adopt the convention that a haploid organism has only one gene at any one locus, a diploid has two, a triploid three, and so on.

For example, we can presume that the difference between humans who can and cannot taste the chemical phenylthiocarbamide resides in a single gene difference between two alleles, *T* and *t*. Since the allele for lasting, *T*, is dominant over *t*, two genotypes (homozygous *TT* and heterozygous *Tt* represent tasters and the nontasters are *tt*. A population of 200 individuals composed of 90 *TT*, 60 *Tt*, and 50 *tt* will therefore have a total of 400 alleles at this locus. 240 of these are *T* (a frequency of .60), and 160 are *t* (a frequency of .40). We can also calculate the same gene frequencies from the frequencies of the three genotypes, according to the formula: frequency of a gene = frequency of homozygotes for that gene + 1/2 frequency of heterozygotes, who each contain one such gene out of two.

The gene pool is the sum total of genes in the reproductive gametes of a population. It can be considered as a gametic pool from which samples are drawn at random to form the zygotes of the next generation. Thus the genetic relationship between an entire generation and the subsequent generation is very similar to the genetic relationship between a parent and its offspring. Since the frequencies of genes in the new generation will hinge, to some degree at least, on their frequencies in the old, we might say that gene frequencies rather than genes are inherited in populations. In what form can we express and analyze these gene frequency relationships between generations?

One of the first attempts at using the concept of gene frequencies occurred in the dispute mentioned earlier between the biometricians and Mendelians. Some argued that dominant alleles, no matter what their initial frequency, would be expected to reach a stable equilibrium frequency of three dominant individuals to one recessive, since this was the Mendelian segregation pattern for these genes. Also offered as evidence that Mendelian dominants and recessives were not segregating properly in populations was the observation that many dominant alleles such as that for human brachydactyly (short fingers) were very rare when they should have been frequent.

Although it was widely accepted at first, in 1908 both Hardy in England and Weinberg in Germany disproved the argument against normal Mendelian segregation in populations. They demonstrated that gene frequencies do not depend on dominance or recessiveness but remain essentially unchanged from one generation to the next under certain conditions.

Conservation of Gene Frequencies

The principle Hardy and Weinberg discovered may be simply illustrated using the tasting example previously mentioned. Let us, for instance, place on an island a group of children in the ratio just given: .45 *TT*/.30 *Tt*/.25 *tt*, where gene frequencies are therefore .60 *T* and .40 *t*. Let us also assume that the number of individuals in this newly formed population is large and that tasting or nontasting has no effect on survival (viability), fertility, or attraction between the sexes.

As these children mature, they will choose their mates at random from those of the opposite sex regardless of their tasting abilities. We can then predict matings between any two genotypes solely on the basis of the genotypic frequencies in the population. Nine different types of matings can occur, of which three matings are reciprocals of others (e.g., *TT* × *tt* = *tt* × *TT*). In all, therefore, these six different mating combinations will produce offspring in the ratios shown.

Note that although *random mating* has altered the frequencies of genotypes, the gene frequencies among the offspring have not changed. For *T* the offspring gene frequency is equal to .36 + 1/2 (.48) = .60, and the frequency of *t* is .16 + 1/2 (.48) = .40, exactly the same as before. Under these conditions, no matter what the initial frequencies of the three genotypes, the gene frequencies of the next generation will be the same as those of the parental generation. For example, if the founding population of this island contained .40 *TT*, .40 *Tt*, and .20 *tt*, the gene frequency for *T* would be .40 + 1/2 (.40) and .20 + 1/2 (.40) for *t*, the same as before. However, despite the different initial genotype frequencies, offspring are again produced in the ratio .36 *TT*/ .48 *TT*/.16 *tt*, or a gene frequency of .60 *T*/.40 *t*.

Two important conclusions follows:

1. Under conditions of random mating (*panmixia*) in a large population where all genotypes are equally viable, gene frequencies of a particular generation depend on the gene frequencies of the previous generation and not on the genotype frequencies.
2. The frequencies of different genotypes produced through random mating depend only on the gene frequencies.

Both these points mean that by confining our attention to genes rather than to genotypes, we can predict both gene and genotype frequencies in future generations, providing outside forces are not acting to change their frequency and random mating occurs between all genotypes. To continue our previous illustration, we may predict that under these conditions the initial gene frequencies in taster-nontaster populations will not change in the next or succeeding generations. Also, after the first generation the genotype frequencies will also remain stable, that is, at *equilibrium*.

The theory describing this genotypic equilibrium, based on stable gene frequencies and random mating, is known as the *Hardy-Weinberg principle* (or *law*) and has served as the founding theorem of population genetics. Perhaps its main contribution to evolutionary thought lies in demonstrating that genetic differences in a randomly breeding population tend to remain constant unless acted on by external forces: a point contrary to the pre-Mendelian concept that heredity involves a blending of traits that become more dilute with each generation of interbreeding.

The general relationship between gene frequencies and genotype frequencies can be described in algebraic terms by means of the Hardy-Weinberg principle as follows: if p is the frequency of a certain gene in a panmictic population (e.g., T) and q the frequency of its allele (e.g., t), so that $p + q = 1$ (i.e., there are no other alleles), the equilibrium frequencies of the genotypes are given by the terms $p^2(TT)$, $2pq(Tt)$, and $q^2(tt)$. If the gene frequencies of T and t are $p = .6$ and $q = .4$, respectively, the equilibrium frequencies will then be

$$(.6)^2(TT) + 2(.6)(.4)(Tt) + (.4)^2(tt) = 36\ TT + .48\ Tt + .16\ tt$$

You can visualize this relationship by drawing a checkerboard in which the genotype frequencies stem from random union between alleles that are in the frequencies of p and q. The same results also derive from the *binomial expansion* $(p + q)^2 = p^2 + 2pq + q^2$. Therefore with any given p and q and random mating between genotypes, one generation of a population in which generations do not overlap is enough to establish equilibrium for the frequencies of genes and genotypes. Once established, the equilibrium will persist until the gene frequencies are changed

The genotypic frequencies at Hardy-Weinberg equilibrium for a two-allele locus, where the frequency of each allele ranges from zero to 1. Note that the frequency of heterozygotes never exceeds .50, but is significantly higher than the frequency of homozygotes for a rare allele (e.g., when a is .1, aa is .01 but Aa is .18).

The presence of two alleles at a locus is only one example for which we may want information. There are often more than two alleles at a locus, and we must then consider each allelic frequency as an

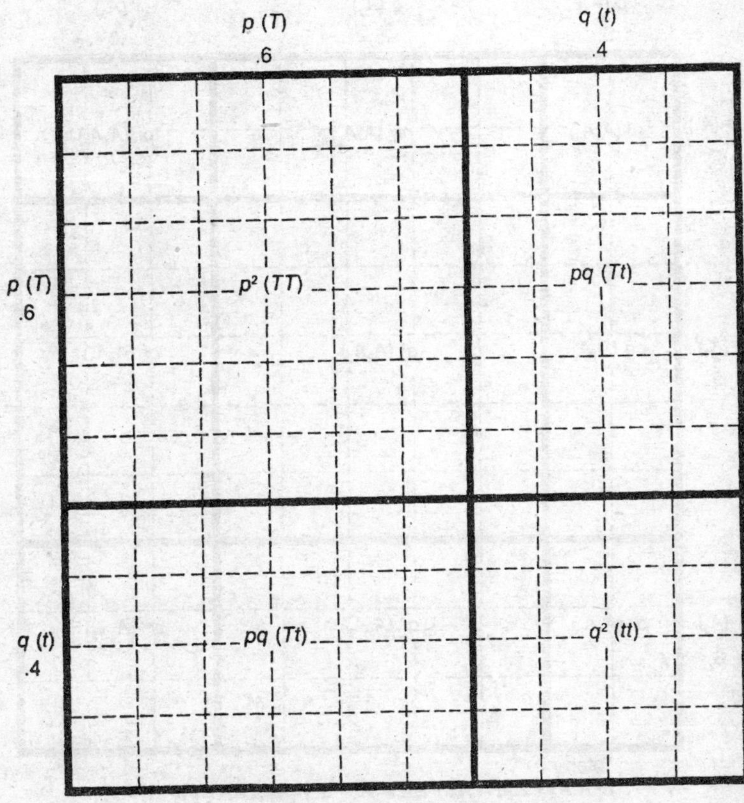

Fig. 12.1. Genotypic frequencies generated under conditions of random mating for two allels, T and t, at a locus when their respective frequencies are $p = .6$ and $q = .4$. Equilibrium genotypic frequencies are therefore .36 TT, .48 Tt, and .16 tt.

element in a multinomial expansion. For example, if there are only three possible alleles at a locus, A_1, A_2, and A_3, with respective frequencies p, q, and r, so that $p + q + r = 1$, the *trinomial expansion* $(p + q + r)^2$ determines the genotypic equilibrium frequencies. The six genotypic values are then

$$p^2 A_1A_1 + 2pq A_1A_2 + 2pr A_1A_3 + q^2 A_2A_2 + 2qr A_2A_3 + r^2 A_3A_3$$

Since each haploid gamete contains only a single allele for any one gene locus, zygotic combinations will depend only on the frequency of each allele, and as when there are only two alleles, equilibrium is established in a single generation of random mating.

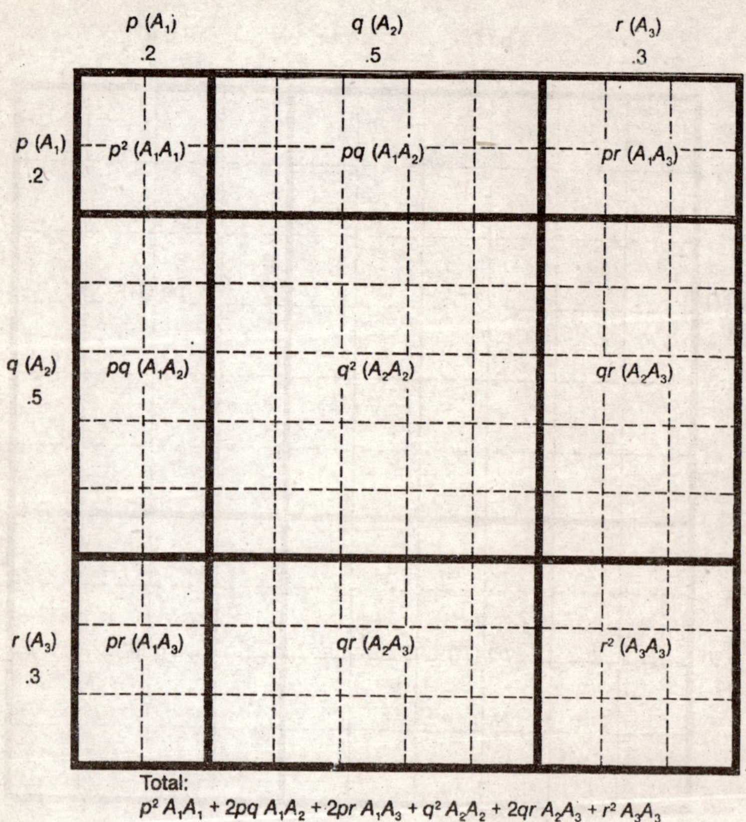

Fig. 12.2. Genotypic frequencies generated under conditions of random mating when there are three alleles, A_1, A_2, and A_3, present at a locus.

Attainment of Equilibrium at Two or More Loci

Establishment of equilibrium in one generation holds true as long as we consider each single gene locus separately without being concerned about what is happening at other gene loci. If, however, we consider the products of two independently assorting gene pair differences simultaneously—for instance, *Aa* and *Bb*— the number of possible genotypes increases to 3^2 (i.e., *AABB*, *AABb*, *AaBB*, *AaBb*, etc.). As expected, more terms are now involved in the multinomial expansion, so that if we call *p*, *q*, *r*, and *s* the gene frequencies of *A*, *a*, *B*, and *b*, respectively, the equilibrium ratios of their genotypes are expressed as $(pr + ps + qr + qs)^2$, or p^2r^2 *AABB*, $2p^2rs$ *AABb*, $2p^2s^2$ *AAbb*, $2pqr^2$ *AaBB*, . . . , q^2s^2 *aabb*.

This equilibrium formula depends on the terms pr, ps, qr, qs, which are the equilibrium frequencies of the gametes *AB*, *Ab*, *aB*, and *ab*, respectively. Once the gametic frequencies reach these equilibrium values, the equilibrium genotypic frequencies will also have been reached. The problem of attainment of equilibrium therefore resolves itself to the time that it takes for the gametic frequencies to reach these values. If we begin only with heterozygotes (*AaBb* × *AaBb*) in which the frequencies of all genes are the same (i.e., $p = q = r = s = .5$), all four types of gametes (*AB*, *Ab*, *aB*, *ab*) are immediately produced at equilibrium frequencies (.25), and genotypic equilibrium is reached within one generation.

However, an initial population of heterozygotes is the only condition in which equilibrium is reached so rapidly. To take an extreme case, if we begin with the genotypes *AABB* and *aabb*, only two types of gametes are produced (*AB* and *ab*) and equilibrium for all genotypes cannot be reached in the next generation since many genotypes are missing (i.e., *AAbb*, *aaBB*, etc.). In general, we may ask two questions:

- What are the expected equilibrium frequencies of gametes?
- How rapidly are these frequencies achieved?

To deal with these questions, we can divide gametes into those in *repulsion* (*Ab* and *aB*) and those in *coupling* (*AB* and *ab*). Since the frequencies of genes in gametes in repulsion are equal to the frequencies of genes in gametes in coupling, we would expect the products of the frequencies of both types of gametes to be equal at equilibrium: $(Ab) \times (aB) = (AB) \times (ab)$.

For example, if the frequencies of *A* and *B* are .6 each, and the frequencies of *a* and *b* are .4 each, then at equilibrium $(.24)(.24) = (.36)(.16)$, or both products equal .0576. If the coupling and repulsion products in the initial population differ, this difference therefore represents the change in gametic frequencies that must occur for equilibrium values to be reached. If we call this difference *disequilibrium*, or d, and it is positive so that $(Ab)(aB) - (AB)(ab) = + d$, then at equilibrium this fraction will have been added to each of the coupling gametes and subtracted from each of the repulsion gametes, It d is negative, the reverse operation will occur. In both cases, disequilibrium will have diminished to zero.

Until the final gametic ratios are reached, half the difference from equilibrium reduces each generation, so that within four to five generations more than 90 percent of this difference from equilibrium frequency has been attained by all gametes, or less than 10 percent of disequilibrium value remains. For three gene pairs, the speed of approach to equilibrium is even further diminished, and it becomes slower still as more gene pairs are involved.

As you might expect, linkage between two loci complicates the reaching of equilibrium since the chances that all the different types of dihybrid gametes will be found depend on crossover frequencies between the two loci. The closer the linkage, the longer it will take for the frequency of coupling gametes to equal the frequency of repulsion gametes. In other words, such *linkage disequilibrium* depends on

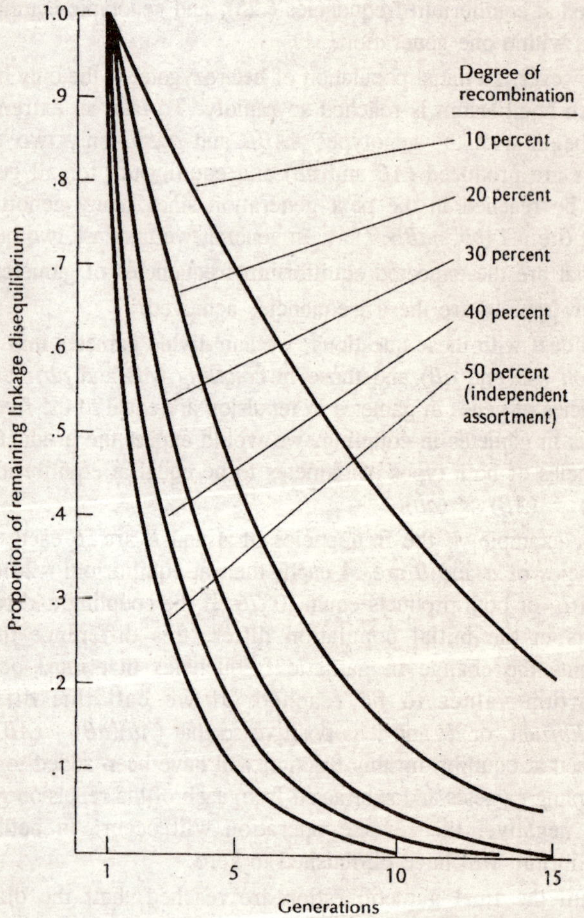

Fig. 12.3. The proportion of linkage disequilibrium that remains in various generations (starting from an initial value of 1.00) when different degree of recombination occur between two loci.

recombination frequency, and lower recombination frequencies between linked loci delay the attainment of equilibrium. This does not mean that the eventual equilibrium values for linked genes will differ from those attained in the absence of linkage; *d* depends on gametic frequencies and not on linkage. Thus, once equilibrium is attained we have no way of distinguishing linked or unlinked genes except through tests for departures from independent assortment (recognizing, however, that even genes on different chromosomes may show linkage disequilibrium if the chromosomes do not assort independently).

Despite these theoretical considerations, not all gametes of linked loci in natural populations reach equilibrium frequencies, and researchers ascribe this phenomenon to various causes. For example, some linkage disequilibrium appears for genes between which recombination is extremely rare. Researchers have also discovered a number of cases in which linkage disequilibrium seems to be maintained by advantages conferred on certain linked allelic combinations. Thus, the third chromosome gene arrangements common in *Drosophila pseudoobscura* probably represent linked groups of genes that are advantageous under particular environmental conditions. Because these genes are included within inversions that restrict recombination, their linkage can be preserved for relatively long periods of time, thereby forming *coadapted gene complexes*.

Sex Linkage

For sex-linked genes, the number of possible genotypes is increased because of the difference in number of sex chromosomes between the homogametic and heterogametic sexes. If females are chromosomally XX and males XY, five genotypes can occur for a sex-linked pair of alleles *A* and *a*: three in females (*AA*, *Aa*, *aa*) and two in males (*A* and *a*). If we assign the frequencies *p* and *q* to *A* and *a*, respectively, the equilibrium genotypic values in females are the same as for an autosomal gene, p^2 *AA*, $2pq$ *Aa*, and q^2 *aa*, but are expressed directly in hemizygous males as *pA* and *qa* genotypes. Thus, at equilibrium the sex-linked gene frequencies are the same in both sexes, although the genotypes differ.

Assuming all genotypes are equally viable for the sex-linked gene, a difference in gene frequencies between males and females indicates that the population is not at equilibrium. For example, in a population with the proportions .20 *A*/.80 *a* in males and .20 *AA*/.60 *Aa*/.20 *aa* in females, the frequency of *A* is .2 in males and .5 in females. We can then calculate equilibrium frequencies of all five genotypes by considering that since there is only one X chromosome in males and two in females, the average frequency of a sex-linked gene in a breeding population with equal numbers of males and females is the sum of one-third of its frequency in males plus two-thirds of its frequency in females, or $p = 1/3\ (p_{males}) + 2/3\ (p_{females}) = (p_{males} + 2p_{females})/3$. In the present example, this translates into an *A* (*p*) frequency of [.2 + 2(.5)]/3 = 1.2/3 = .4, and an *a*(*q*) frequency of .6. The equilibrium genotypic values expected are therefore .16 *AA*/.48 *Aa*/.36 *aa* in females, and .4 *A*/.6 a in males.

However, in contrast to single autosomal loci with two alleles, equilibrium values for these genotypes will not be reached in a single generation. Because males inherit their X chromosomes only from their mothers, the frequency of a sex-linked gene among them is the same as its maternal frequency, whereas the frequency of the gene among daughters is an average of paternal and maternal frequencies, because they each inherit one paternal and one maternal X chromosome. Therefore, if the females in a founding population had a frequency of *A* equal to .5, but the males had an *A* frequency of only .2, the daughters would have an *A* frequency of (.2 + .5)/2 = .35, while their brothers would have the .5 frequency of their mothers.

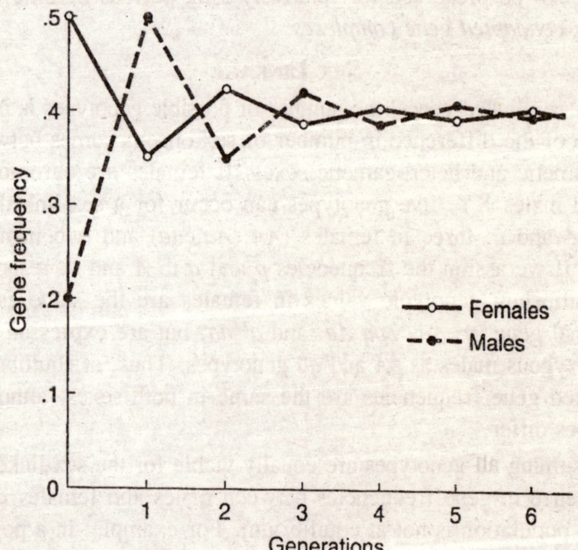

Fig. 12.4. Frequencies of the sex-linked gene A in males and females in successive generations under conditions of random mating when the initial frequency of A is .2 in males and .5 in females.

Thus, in the first generation of random mating the daughters will not reach the *A* equilibrium value of .4, and the sons will exceed it. In the second generation, the difference from equilibrium values will diminish, but this time the sons will not achieve equilibrium (*A* = .35) and the daughters will exceed it [*A* = (.35 +.5)/2 = .425]. Each succeeding generation will show a similar reversal but nevertheless achieve a successively closer approximation to the final equilibrium values.

EQUILIBRIA IN NATURAL POPULATIONS

Where we can score all segregants of a gene at a single locus in a natural population, we can reliably estimate gene frequencies, and we can then easily compare the observed genotype frequencies to their expected equilibrium values. To take a simple example, codominance at the *MN* blood group locus (using only two alleles, *M* and *N*) enabled Boyd to classify 104 American Ute Indians into genotype frequencies of .59 *MN*, .54 *MN*, and .07 *NN*. Since the gene frequencies are .59 + .17 = .76 for *M*, and .07 + .17 = .24 for *N*, the expected genotype frequencies are $(.76)^2$ = .58 for *MM*, 2(.76)(.24) = .36 for *MN*, and $(.24)^2$ = .06 for *NN*. The close correlation between observed and expected genotypic values thus indicates that this population has reached Hardy-Weinberg equilibrium.

When researchers know the genotypes of mated couples in the population, they can also test the assumption of random mating. Under random mating the frequencies of the different mating combinations should depend only on the frequencies of their genotypes. An actual set of data that Matsunaga and Itoh collected provided the blood types of 741 couples (or 1,482 individuals) in a Japanese town that showed genotypic frequencies of .274 *MM*, .502 *MN*, and .224 *NN*. (Since the gene frequencies are .525 *M* and .475 *N*, the expected equilibrium genotypic frequencies are .276 *MM*, .499 *MN*, and .225 *NN*, again indicating equilibrium for this locus.)

Table demonstrates random mating among these individuals, with the number of observed matings of different combinations given in the last column. To the left of this column, the expected mating combination frequencies are calculated on the basis of the gene frequencies, p for *M* and q for *N*. For example, since the frequency of the genotype *MM* is p^2 at equilibrium, the random mating combination *MM* × *MM* should be p^4. When we calculate the frequencies of all expected mating combinations this way, comparing observed and expected agrees very well with the assumption of random mating.

A more common case in natural populations is when the effect of one allele at a locus is completely dominant over another so that we cannot phenotypically distinguish the heterozygous genotype (e.g., *Aa*) from the homozygous dominant (e.g., *AA*). Under such circumstances we cannot obtain gene frequencies directly, as in codominance, because we do not know two of the genotypic frequencies (*AA*, *Aa*), and instead we must rely on the recessive homozygote (*aa*), whose phenotypic frequency coincides with its genotypic frequency. That is, if we assume that such a population (p^2 *AA*, $2pq$ *Aa*, q^2 *aa*) has reached Hardy-Weinberg equilibrium, the recessive homozygotes are present in a frequency q^2 equal to the square of the recessive gene frequency, q. If, let us say, q^2 is .49, then q is $\sqrt{49} = .70$, and the frequency of the dominant allele p is $1 - q$, or .30. The homozygous dominants, therefore, have the frequency $p^2 = (.30)^2 = .09$, and the heterozygotes have the frequency $2pq = 2(.30)(.70) = .42$.

One consequence of this analysis is that when recessive phenotypes are rare, it is surprisingly common to find that the *carrier heterozygotes*, phenotypically disguised as dominants, are present in relatively high frequency. Albinism, for example, affects only about 1 in 20,000 humans in some populations, or $q^2 = 1/20{,}000 = .00005$. The gene frequency, q, of the albino gene is therefore .007, and the frequency, p, of the nonalbino allele is .993. The frequency of heterozygous albino carriers is therefore $2(.993)(.007) = .014$, or approximately 1 in 70 individuals. Thus, there are $.014/.00005 = 280$ times as many heterozygotes for this trait as there are homozygotes. Similarly high proportions of carriers of other recessive traits point to the difficulty of eliminating rare harmful recessive alleles, since they are carried mostly in the unexpressed heterozygous condition.

When more than two alleles are present at a locus, the Hardy-Weinberg equilibrium is based on a multinomial expansion. For that example of three alleles (A_1, A_2, A_3), we expect six genotypes, and we can calculate the gene frequency of each allele (p, q, r) from the following equations:

$$p = \frac{2(A_1A_1) + (A_1A_2) + (A_1A_3)}{2N}$$

$$q = \frac{2(A_2A_2) + (A_1A_2) + (A_2A_3)}{2N}$$

$$r = \frac{2(A_3A_3) + (A_1A_3) + (A_2A_3)}{2N}$$

where A_1A_1, A_1A_2, A_1A_3, and so on refer to the numbers of genotypes in each category, and N refer to the total number of individuals scored.

A system of this type appears in human populations bearing different forms of the red blood cell enzyme acid phosphatase, which researchers can score into six different phenotypes, *AA*, *BB*, *CC*, *AB*, *BC*, or *AC*, as determined by all possible combinations of the alleles *A*, *B*, and *C* at a single locus. Investigations of a Brazilian population indicate that the observed phenotypic frequencies of the acid phosphatase combinations conform closely to those the Hardy-Weinberg equilibrium predicts. Although exceptions arise, conformities to the Hardy-Weinberg equilibrium seem quite common for both autosomal and sex-linked genes and appear in a variety of sexually outbreeding organisms. In general, as noted earlier, these studies emphasize that populational gene and genotype frequencies do not change without cause.

Inbreeding

One set of conditions that interferes with the Hardy-Weinberg equilibrium is *nonrandom mating*. An important example occurs when related individuals of similar genotype mate preferentially with each other in a phenomenon called *inbreeding*. (An extreme form of inbreeding is when two gametes of a single individual unite to form a fertile zygote, *self-fertilization*.) Although the effect of inbreeding will not change the overall gene frequency, it will lead to an excess of homozygous genotypes. In the case of a rare recessive allele, inbreeding will thus cause it to appear in greater homozygous frequency than under random mating, thus providing selection with an increased opportunity to act on rare recessives.

We usually quantify inbreeding by an *inbreeding coefficient*, *F*, which measures the probability that the two alleles of a gene in a diploid zygote are identical-descended from a single ancestral allele. Having identical alleles, of course, means homozygosity, and *F* can range from one (complete homozygosity) to zero (complete heterozygosity).

By this definition, if inbreeding occurs at a particular locus with only two alleles, *A* and *a*, there will be some identical homozygotes (F). Of this inbred proportion, some will be *AA* and some *aa*, the frequencies of each depending on their respective population gene frequencies p and q. Thus inbreeding will produce pF *AA* and qF *aa* genotypes. In addition to these, however, the remaining individuals in this population $(1 - F)$ will bear genotypes whose frequencies are determined according to the Hardy-Weinberg equilibrium of p^2 *AA*,

$2pq\ Aa$, and $q^2\ aa$. The three genotypes will therefore have the following frequencies:

$$AA = p^2(1 - F) + pF = p^2 - p^2F + pF = p^2 + pF(1 - p)$$
$$= p^2 + pqF$$
$$Aa = 2pq(1 - F) = 2pq - 2pqF$$
$$aa = q^2(1 - F) + qF = q^2 - q^2F + qF = q^2 + qF(1 - q)$$
$$= q^2 + pqF$$

It is now easy to see that the increase in the frequency of each homozygote type by a factor of pqF flows from an equivalent fall in the heterozygote frequency ($- 2pqF$). Note also that this reduction in heterozygotes affects the gene frequencies p and q equally, so that only the genotypic frequencies change. When inbreeding is absent, $F = 0$, and the preceding equations reduce to the Hardy-Weinberg frequencies $p^2\ AA$, $2pq\ Aa$, and $q^2\ aa$. When inbreeding is complete, $F = 1$, $2pq - 2pqF = 0$, and the only remaining genotypes are pAA and qaa.

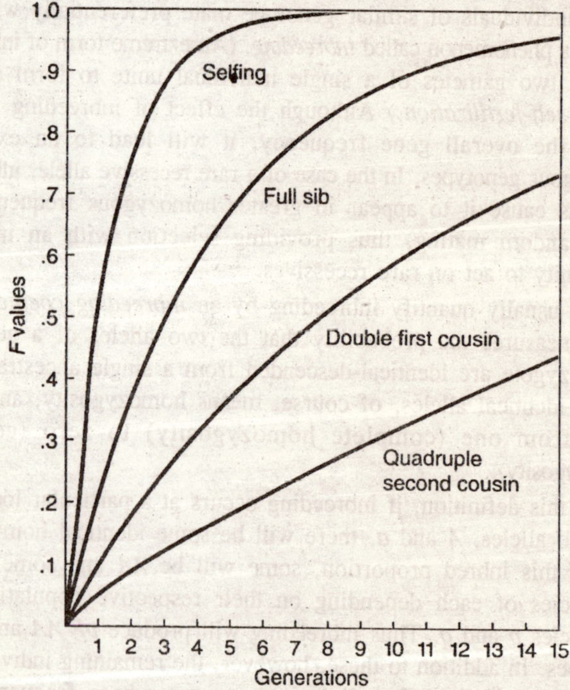

Fig. 12.5. Inbreeding coefficients at generations 1 to 15 for four different systems of inbreeding.

The mating system in which inbreeding is greatest is under self-fertilization (e.g., hermaphrodites), where F is equal to .5 in the first generation and approaches 1 within four or five generations. Any other mating scheme slows the rate of inbreeding. Similarly, population size can also affect inbreeding, since the smaller the size, the greater the opportunity for related individuals to mate.

Although some degree of inbreeding occurs in most outbreeding populations, significant amounts of inbreeding can cause *inbreeding depression* in which rare deleterious recessives may now appear with increased homozygous frequency. Thus, if a recessive disease with genotype *aa* occurs with frequency q^2 in a random outbred population, its frequency will increase by pqF in an inbred population as just derived. The ratio of inbred to outbred frequency for the homozygous recessive will therefore be

$$\frac{q^2 + pqF}{q^2} = \frac{q(q + pF)}{q^2} = \frac{q + pF}{q}$$

Obviously, if q is large and F is small, the inbreeding increment pF will be relatively small, and the increased frequency of homozygous recessives will hardly be noticeable. However, if q is very small

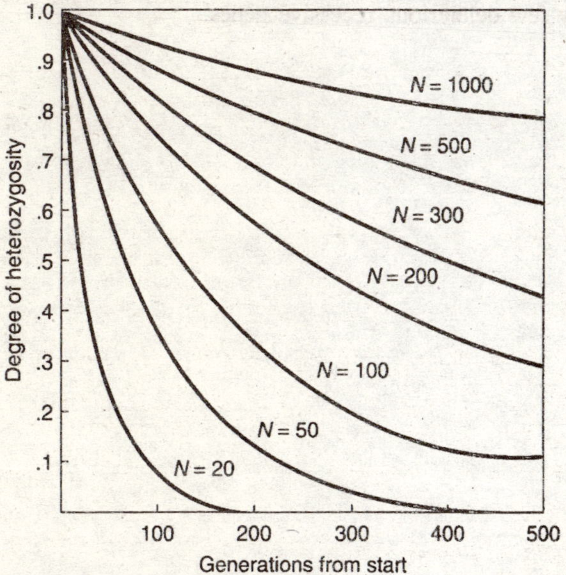

Fig. 12.6. Degrees of heterozygosity remaining in populations of different sizes after given generations of random union between gametes.

(rare) and p is large, then pF provides a notable increase in recessives even when F is fairly small. For example, if q is .5, first-cousin mating (F = .0625) will produce an inbred-to-outbred ratio of homozygotes of

$$\frac{q+pF}{q}=\frac{.5+(.5)(.0625)}{.5}=\frac{.53125}{.5}=1.06$$

However, if q is .005, this ratio increases to .067/.005 = 13.4. When q = .0005, the increase of homozygotes because of first-cousin mating is .0630/.0005, or 126 times that of randomly bred populations.

Should homozygous recessives have a quantitative effect on one or more traits, inbreeding would cause the measured values of these traits to tend in the direction of recessive values. Thus, in various outbred populations, such as corn, inbreeding depression can reduce height, yield, and other characters. Inbreeding depression, however, is not a universal phenomenon in all species, certainly not in many species that are normally self-fertilized and have eliminated most or all of their harmful recessives. On the whole, ample evidence shows that most normally cross-fertilizing species deteriorate on consistent inbreeding, although some strains may escape because they carry relatively few deleterious recessive genes.

13

POPULATION GENETICS

Examination of the data of the fossil record has led some workers to espouse a theory ('punctuated equilibria') in which evolutionary change is hypothesized to be concentrated in relatively short periods of geological time. These metastable periods are then supposed to be followed by long periods without change. Such theories are based on limited data, both with regard to the lengths of time embraced and the biology of the characters involved. Punctuated equilibria cannot be directly applied to evolutionary theories that are based on the biology and genetics of contemporary populations. The accomplishments of studies on the genetics of natural populations in this century are reviewed. Empirically and theoretically, the approach to evolution through population biology (Neo-Darwinism) is highly robust. All evolutionary processes including varying rates and periods of stasis can be understood by recourse to studies of contemporary organisms. There is no reason to believe that any basic principle can be observed in the fossil record that it not also reflected in population biology.

On Punctuated Equilibrium

> The patterns . . . speak for themselves. They are there for all to see. We have not simply invented them to fit a set of preconceived notions. (Eldredge & Tattersall 1982)

> I remain skeptical . . . but satisfied that the authors could make radioactive decay appear punctuated—all you have to do is to look closely enough. (White 1983)

The theory of punctuated equilibria was proposed more than a decade ago. The idea turned out to be highly provocative, resulting in a flood of seminars, discussions and symposia. Articles either strongly pro or strongly con were numerous. This debate, in my opinion, has unfortunately been of poor quality and clarification has not been served by the considerable participation of reporters from various scientific journals, as well as from the non-technical press and proponents of creationism.

Provocation of one's peers, of course, is an important means of communication in the advancement of science, especially when the significance of new and little-understood data is apparently perceived by only a few persons. The history of science is replete with stories of the manner in which advocates of the 'conventional wisdom' have clung steadfastly to classical ideas that seemed at the time to be utterly unassailable. On the contrary, there is nothing more exciting for a scientist than the uncovering of data that appear to cause a restructuring of knowledge along lines that simultaneously clarify and simplify. Only a few workers, however, are destined by play this kind of role. The retrospective glamour of a Wegener arguing for continental drift or a McClintock announcing mobile genetic elements is a pervasive motivating influence for both peers and students who view the pioneer days from the comfort of current acceptance. Many eager scientists pressure their own thoughts to take theoretical directions that often outstrip the scope of their data. To some, like the great geneticist and evolutionist, Richard Goldschmidt, the overturn of what to him appeared as the orthodox complacency of the establishment became almost obsessive. It fuelled his drive as a scientist. Late in his career, he began one of his challenging articles with a quotation from a letter written by Orville Wright, six months before the first successful flight at Kitty Hawk:

> '. .. but if we all worked on the assumption that what is accepted as true is really true, there would be little hope of advance.'

Whether they turn out to be right or wrong, such people are a necessary part of the social fabric of science; they require us to re-examine facts and entertain hypotheses that from some perspectives seem quite absurd. But today's 'absurdities' sometimes become tomorrow's laws.

A scientist who will not force himself to put his peers' theories at least to the mind's test condemns himself to be merely a pedestrian plodder. An annoying aspect is the necessity to discard most of these notions. Only the great among us are generous enough with their thought processes to remark, as Haldane once did after hearing one of Crodowaldo Pavan's ideas: '... interesting - interesting - even if untrue ...'

So, what role is the theory of punctuated equilibria to play? Must we dismiss Eldredge and Gould or must we restructure our basic thinking around their ideas?

Palaeontological Theory

In approaching these questions an important circumstance must be considered: the idea of punctuated equilibrium was proposed by palaeontologists from a consideration of the data of palaeontology. Were the debate over the idea to remain entirely within that data base, I believe the discussion of it would have been clearer and less acerbic. But the proposers chose to consider the idea as an alternative to what they labelled 'phyletic gradualism'. Although at first the gradualism idea was ascribed only to certain palaeontologists, it was later broadened and, thus, became the basis for a criticism of the whole structure of Neo-Darwinism, Thus the latter area, basically one of contemporary population biology, was also equated with undifferentiated gradualism. 'Punctuation' was proposed as the new and ex- citing alternative.

The trouble with this way of framing up the argument is that Neo-Darwinism is virtually equivalent, in moderns terms, to the sophisticated quantitative science of population genetics. The data of population genetics are wholly foreign to those of palaeontology. They must be considered separately and should have been considered separately by Eldredge and Gould.

In punctuated equilibrium, as I have pointed out earlier, evolutionary change is supposed to be concentrated in a 'brief metastatic period, the punctuation phase. Using the shortest punctuation that I could find in the examples cited, the geological data suggest that it must have extended for about 55 000 years. The standard error on such values is not clear. Now, this may be a very short time from the palaeontological point of view, but from the perspective of modern population genetic biology which considers generations as the building blocks of change, it is substantial. The generation time in a relatively slow-breeding organism like man, for example, is generally taken as 30 years. This would provide 1833 generations for the operation of Neo-Darwinian forces such as mutation, recombination, selection, immigration and population size effects. *Drosophila melanogaster* is much faster; we may use the figure of 32 generations per year. In this case, the palaeontological punctuation under discussion would have extended over 1,760,000 generations. A great deal of genetic change can be incorporated into a population in ten or a hundred generations; and a thousand or a million is a sumptuous amount.

Dramatic changes in the phenotypic and genotypic characters in populations can be experimentally accomplished in as few as five generations of natural selection. A minor genetic alteration to an experimental population of *D. melanogaster* caused a tripling of population size and a drastic change in phenotype. Under challenge from the insecticide DDT, resistance of an outbred population of *D. melanogaster* was increased by natural selection more than 2000 times in 3 years. Many such instances could be cited. Some Neo-Darwinists refer to such episodes as times of 'quantum' change; they are part and parcel of the data of population genetics and should not be considered especially unusual or exceptional. Episodes of rapid change have also been associated with genetic revolutions, founder effects and hybridizations. They acquire importance in the present discussion as they emphasize the existence of dramatic shifts in the *rate* of change, the subject of the symposium.

Much longer periods of geological time have been assigned to the equilibrium phase of the punctuated equilibrium idea. During these periods, it is alleged, the morphological attributes of the relevant fossils as seen in the fossil record do not undergo change. This is an aspect of the theory to which population genetics cannot contribute, since the vast periods of geological time involved cannot be tested or modelled experimentally from the point of view of contemporary population biology. As will be discussed below, however, theoretical population genetics may be able to make a contribution to the equilibrium concept. On the other hand, palaeontological data are far from robust on this point since the preservation of hard parts or casts do not provide much information on physiology, reproductive cycles or behaviour.

As the advocacy of punctuated equilibria has intensified, various questionable corollaries have been increasingly stressed. One suggests that some kind of macromutational change affecting a major developmental pathway is the key event that occurs during the punctuation phase. The original proponent of this idea, of course, was the genetic evolutionist, Goldschmidt (1940). Employment of this idea, however, especially as the key to a 'saltatory' change instrumental in the formation of species, goes back to Hugo de Vries and T. H. Huxley, if not further. Gould and Eldredge (1977) have extracted several delicious quotes from Huxley's writings, indicating that the geneticist William Bateson agreed with him on saltation but that Darwin had expressed 'disgust' at the idea.

Although downgraded by Simpson (1944 and later) the saltation idea has frequently been used by others, particularly by those to whom 'population thinking' is somewhat foreign. As a field naturalist, Darwin's approach to evolutionary change was definitely microevolutionary, that is, he invoked the accumulation of small increments of change in populations as the *modus operandi* of evolution. He did not, however, insist on gradualism and considered rapid rates of evolutionary change to be important; it was the abrupt change in phenotypes (saltations) that he felt were not important in natural populations.

Saltation, especially in species formation, appears to have few advocates among modern Neo-Darwinian population geneticists, either those who take an empirical or a theoretical approach. The feeling is widespread that homoeotic mutants, which all recognize as real, so disturb developmental pathways as to strongly decrease fitness in their carriers. This is not to say that a gene of moderately large effect may not increase in the population after the accumulation of modifier genes. The latter process, in which genetic recombination probably plays an important role, can be accomplished only over a number of generations.

The degree to which the punctuated equilibrium enthusiasts reject Neo-Darwinism is shown by the manner in which the origin of species is diagrammed. Thus, Stanley (1979) proposes that species formation be diagrammed by 'rectangular' phyletic branches, a procedure that appears to exclude the possibility of transitional stages.

Population Genetics and Evolution

Mention has been made earlier in this discussion of the tendency on the part of some palaeontologists to describe Neo-Darwinism in simplistic terms as a concept that reflects some basic preoccupation with phyletic gradualism. I believe the best way to counter such oversimplification is to state the nature of the Neo-Darwinian approach and to recount a few of the major 20th century discoveries of population genetics.

In contrast to those of palaeontology, the data of population genetics come from genetical analysis of individual organisms drawn directly from living populations. Unlike the palaeontologist, the geneticist has methods that can deal quantitatively with descent with change from generation to generation in both phenotype and genotype. The phenotypic attributes available for study include biochemical, developmental, behavioural, physiological and soft-part characters not available to the palaeontologist.

Biometrical or quantitative genetics involves methods borrowed by the evolutionist from the plant and animal breeder. Applied breeding is an elegant and sophisticated science that is able to partition environmental and genetic variance with great precision. This opens the way to an understanding of population alterations over a series of generations. Further, the science, dealing as it does with quantitative characters, finds itself clarifying the inheritance of those very characters that the systematist and the palaeontologist themselves stress as important in evolution. There are good and important reasons why these characters do not show simple single-factor Mendelian inheritance.

Although useful in theoretical constructs, evolutionary schemes based on simple allelic substitution (e.g. the industrial melanism case) should not serve as general examples of what is happening in most natural populations. Indeed, even the case of industrial melanism is far from being a single-locus effect: clear genome-wide modifier genes and associated heterotic effects caution us against the strict application of single-locus evolution as a model.

Increasingly, Neo-Darwinian evolutionists distinguish between the phyletic origin of adaptations and the origin of species. Both theoretically and in actual experiment, the origin of adaptations follows the Neo-Darwinian path of mutation, recombination and selection as it occurs within single populations of interbreeding organisms. Population genetics, however, is less able to deal with the origin of species, in which a sundering of a pre-existing gene pool must somehow be instituted. For this reason, theories as to the origin of species are less well grounded in data than is the intrapopulational origin of adaptations.

Over the past 50 years, a substantial literature on the genetics of natural populations of various species of *Drosophila* has accumulated. Although the subject has been reviewed repeatedly, the 'non-drosophilist' may have some difficulty seeing the woods for the trees. I now list, with brief comments, what appear to be the main results of these extensive studies of this genus as well as population studies of some other genera of fast breeding, cross-fertilizing diploid organisms.

The genetic structure of natural populations

1. Natural populations contain abundant cryptic genetic variability sheltered in the heterozygous state. This variability has many components: there are oligogene visibles, chromosomal aberrations, protein-coding loci and specific regulatory and modifying genes. In the past few years, as DNA sequencing has proceeded, extensive heterozygosity has been observed at this, the ultimate source of genetic variation.

2. Most of this variability consists of genes of individually small phenotypic effect. The evidence for a substantial role for what Mather has called polygenes continues to be extended.
3. Genes of larger effect, if experimentally isolated, can be seen to have been strongly influenced in their expression by modifying or regulatory genes in the natural genetic environment.
4. Mutability in natural populations is not a species-wide constant but varies with the local population or stock, or even individual organism studied. In a wide variety of plants and animals much mutational activity can be ascribed to the effects of mobile genetic elements. Their effects are often suppressed in inbred populations but expressed in certain out-crossed populations. A major effort is now under way in genetics research to separate 'spontaneous' mutability from mutability mediated by mobile genetic elements.
5. Many naturally occurring variants are deleterious in the homozygous state yet tend to remain in the population rather than being eliminated by normalizing selection. This widespread observation has led to the balance theory of the genetic structure of populations. Much of the variability is not fixed but exists in complex heterozygous balances capable of continual revision by meiotic recombination and renewed selection each generation. Included in this category are major chromosomal aberrations such as inversions and translocations. Although some species appear to lack them, others accumulate them as segregating genetic variants that are apparently also held in the population by balancing selection. Observations of isofemale lines (laboratory stocks started in the laboratory from a single wild female that was inseminated in nature) show that much of the inversion and soluble protein variation and indeed the DNA sequence variation as well, may be maintained over 50 generations or more even in the uniform laboratory environment. These balances appear to be virtually obligatory in some species; inbreeding and selection frequently lead to lowered fitness. As early as 1955, Dobzhansky wrote: 'The adaptive norm is an array of genotypes, heterozygous for more or less numerous gene alleles, gene complexes and chromosomal structures.'

This characterization is even more true today than it was then, before modern electrophoretic and sequencing methods had been invented.

Hartl (1980) has stressed another aspect of our modern view of population structure. He calls attention to the interaction of the components of this variability:

> . . .the genetic underpinning of fitness is extremely complex, involving alleles at many loci that influence fitness in two ways—one through the main affect of each locus (which typically may be an effect on some quantitative trait such as body size, growth rate, or developmental time) and the other, through pleiotropic effects on other traits related to fitness.

Over the years, Wright has emphasized gene interaction. Recently he wrote (Wright 1982):

> . . . each gene . . . (has) multiple pleiotropic effects because of interaction of its product with those of others . . . the effects of these interactions are not in general additive. These are the inevitable consequences of the complex network of biochemical and developmental reactions that intervene between primary gene action and the ultimate effects subject to selection.

The implications of this situation for evolution are clear. Periods of intense selection, recombination, mutability and drift can greatly speed up evolutionary change, especially if they operate in small or attenuated subpopulations. Under these circumstances rapid ('quantum') evolutionary change can occur. These changes can result in the formation of new adaptive modes or species or both.

The periods of quantum change, extending over perhaps 50 to 100 generations, may be followed by a return to a more stable state characterized by a balanced genetic variability.

However, not all evolutionary change appears to be concentrated in these bursts of evolutionary activity. As Wright emphasizes, every character important to the life of the organism appears to require maintenance by the all-pervading action of the principal Darwinian force, that is, natural selection. Thus, if selection is removed, the pertinent character will tend to decay back to the level at which selection again operates. The active 'decay' in the genetic material is caused by the inexorable process of random mutational change in the DNA. Without natural selection as a force leading to conservation, the character will start to become unstable.

All of these concepts have grown out of modern Neo-Darwinism and are available for confirmation and further direct study in living contemporary populations. They may be applied to the intrapopulational origin of adaptive traits as well as to the mode of origin of new isolated populations and species.

The facts that have led some palaeontologists to espouse punctuated equilibria are unfortunately not robust enough to warrant meaningful comparisons to the principles that have been established by the Neo-Darwinian evolutionist and population geneticist. Accordingly, only rough, imprecise analogies of questionable significance can be drawn between 'punctuation' and 'quantum' change and between 'equilibrium' and 'balanced genetic polymorphism'. Despite some claims to the contrary, the major edifice of Neo-Darwinism is unaffected by these arguments and is stronger than ever in 1985.

14

POLYGENIC MECHANISM

The Mendelian genetics of beak size in finches is unknown, but the character shows evolutionary changes as the food supply changes through time. We begin by looking at beak size in Darwin's finches as an example of the kind of character studied by quantitative genetics. We then move on to the theoretical apparatus used to analyze characters that are controlled by large numbers of unidentified genes. The influences on these characters are divided into environmental and genetic factors, and the genetic influences are divided into those that are inherited and influence the form of the offspring and those that are not. A number called heritability expresses the extent to which parental attributes are inherited by offspring. With the theoretical apparatus in place, we can then apply it to a number of evolutionary questions, including directional selection, in both artificial and natural examples, and stabilizing selection. We investigate whether a balance between selection and mutation can explain the amount of genetic variation seen in nature, and consider an experiment in which the effect of selection is minimized to reveal the power of deleterious mutation.

Fourteen species of Darwin's finches live in the Galapagos archipelago, and many of them differ most obviously in the sizes and shapes of their beaks. A finch's beak shape, in turn, influences how efficiently it can feed on different types of food. Grant has been studying these finches since 1973, and his best evidence to demonstrate that beak size influences feeding efficiency comes from a comparison of two species, the large-beaked *Geospiza magnirostris* and the smaller *G. fortis*, feeding on the same kind of hard fruit.

The large-beaked *G. magnirostris* can crack the fruit (called the *mericarp*) of caltrop (*Tribulus cistoides*) transversely, taking on average only 2 seconds and exerting an average force of 26 kgf. It can then easily eat all 4-6 seeds of the smashed fruit, taking only 7 seconds. The smaller *G. fortis* is not strong enough to crack *Tribulus* mericarps and instead twists open the lower surface, applying a force of only 6 kgf and taking 7 seconds on average to reach the seeds inside. Only one or two of the seeds can be obtained in this way, and it takes an average of 15 seconds to extract them. Thus, *Geospiza magnirostris* usually has an advantage with these large, hard types of food.

Smaller finches are probably more efficient with smaller types of food, although this capability is more difficult to show. Both large and small finches on the Galapagos eat small seeds, although there is an indirect reason to believe that smaller finches do so more efficiently. From the evidence we have examined so far, we should predict that natural selection would favour larger finches when large fruits and seeds are abundant. The prediction should apply both within and between species. A *G. magnirostris* looks like an enlarged *G. fortis*, and a larger individual *G. fortis* can probably deal with a large food item more efficiently than can a smaller conspecific, much as an average specimen of *G. magnirostris* is more efficient than an average *G. fortis*. When large seeds are common, we might expect the average size of a population of *G. fortis* to increase between generations, and vice versa when large seeds are rare—if beak size is inherited.

If beak size is inherited . . . but is it? Beak size is inherited if parental finches with larger than average beaks produce offspring with larger than average beaks. Grant measured the sizes of parental and offspring finches in several families and plotted the latter sizes against the former. Large-beaked parental finches do, indeed, produce large-beaked offspring: thus, beak size is inherited. It therefore makes sense to test the prediction that changes in beak size should follow changes in the size distribution of food items. The test was carried out by Grant for the species *Geospiza fortis*, on one of the Galapagos Islands, Daphne Major. Since the study began, this species has undergone two major, but contrasting, evolutionary events.

In the Galapagos, the normal pattern of seasons calls for a hot, wet season from about January to May to be followed by a cooler, dryer season through the rest of the year. In early 1977, for some reason, the rain did not fall. Instead of the normal progression of a wet season followed by a dry season, then another wet season and dry season, the dry season that began in mid-1976 continued until early 1978—one whole wet season was missed. The finch population of Daphne

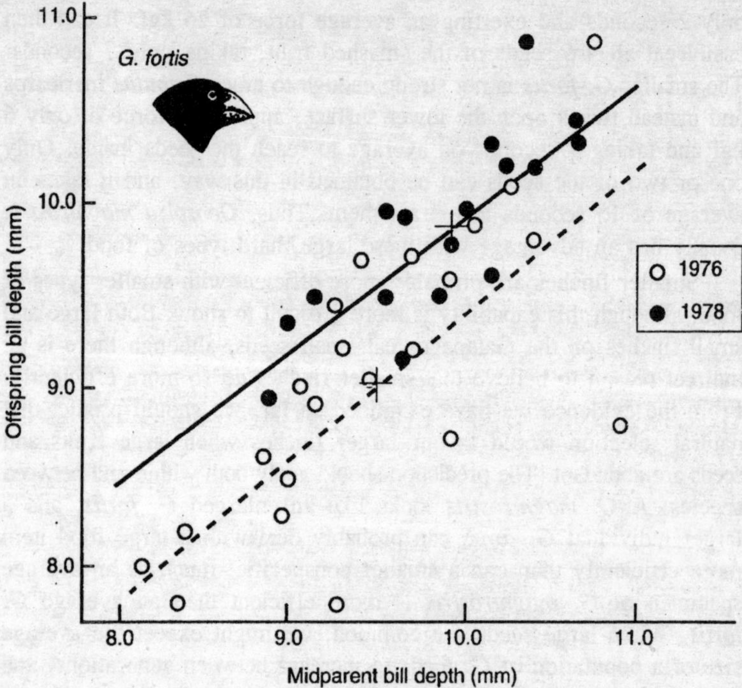

Fig. 14.1. Parents with larger than average beaks produce offspring with larger than average beaks in Geospiza fortis on Daphne Major.

Major collapsed from about 1200 to about 180 individuals, with females being particularly hard hit; at the end of 1977, the sex ratio was about five males per female. As the sex difference shows, not all finches suffered equally. Smaller birds died at a higher rate. The reason, again, relates to the food supply. At the beginning of the drought, the various types of seeds were present in their normal proportions. *G. fortis* of all sizes take small seeds, and, as the drought persisted, these smaller seeds were relatively reduced in numbers. With time, average available seed size became larger. Now the larger finches were favoured, because they could eat the larger, harder seeds more efficiently; and the average finch size increased as the smaller birds died off. (Females died at a higher rate than males because females are on average smaller.) Size, as we have seen, is inherited. The differential mortality in the drought, therefore, caused an increase in the average size of finches born in the next generation. *G. fortis* born in 1978 were about 4% larger on average than those born before the drought.

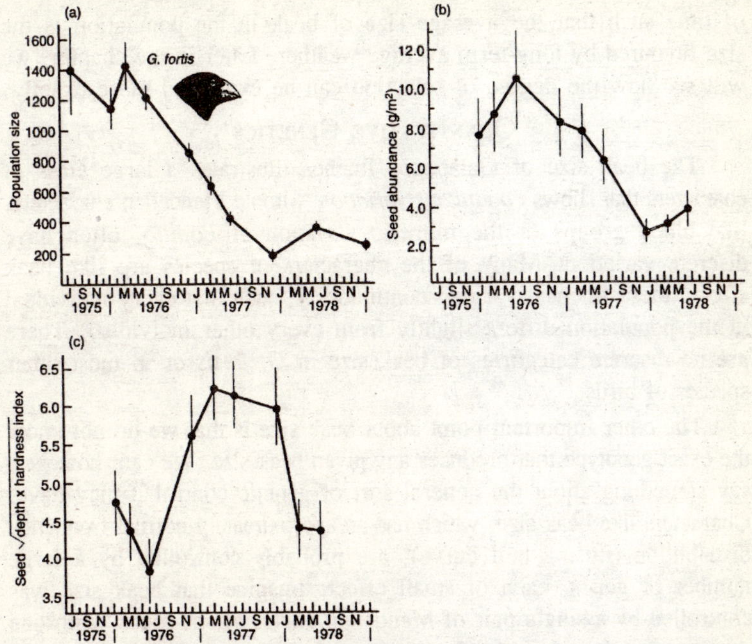

Fig. 14.2. During a drought in 1976-1977, (a) the population of Geospiza fortis decreased on the island of Daphne Major in the Galapagos archipelago, due to (b) the decline of the food supply. (c) The average size of the seeds available as food increased during the drought.

Four years later, in November 1982, the weather reversed. The rainfall of 1983 was exceptionally heavy and the dry volcanic landscape became covered with foliage in the periodic disturbance called El Nino. Seed production was enormous. The theory developed for 1976-1978 could now be tested. The conditions had reversed: the direction of evolution should go into reverse as well. In the year after the 1983 El Nino event, an excess of small seeds appeared. If the smaller finches could, in fact) exploit them more efficiently, the smaller finches should survive relatively better. Grant again measured the sizes of *G. fortis* on Daphne Major in 1984-1985 and found that the smaller birds were indeed favoured; finches born in 1985 had beaks about 2.5% smaller than those born before the El Nino downpours. His theory, that seed sizes control beak size in these finches, was confirmed.

The fluctuations in the direction of selection on beak shape—with smaller beaks favoured in some years and larger ones in others—probably results in a kind of "stabilizing" selection over a long period of time such that the average size of beak in the population is the size favoured by long-term average weather. Later in the chapter, we will see how the degree of selection can be expressed more exactly.

QUANTITATIVE GENETICS

The beak size of Galapagos finches illustrates a large class of characters that shows *continuous variation*. Simple Mendelian characters, like blood groups or the mimetic variation of *Papilio*, often have discrete variation. Many of the characters of species are like beak size in these finches, varying continuously, such that every individual in the population differs slightly from every other individual. There are no discrete categories of beak size in *G. fortis* or in most other species of birds.

The other important point about beak size is that we do not know the exact genotype that produces any given beak size. We can, however, say something about the general sort of genetic control it may have. Characters like beak size, which has an approximately normal frequency distribution (i.e., a bell curve), are probably controlled by a large number of genes, each of small effect. Imagine that beak size was controlled by a single pair of Mendelian alleles at one locus, with one dominant to the other, so that *AA* and *Aa* is long and *aa* is short. In this case, the population would contain two categories of individuals. Now suppose that beak size was controlled by two loci with two

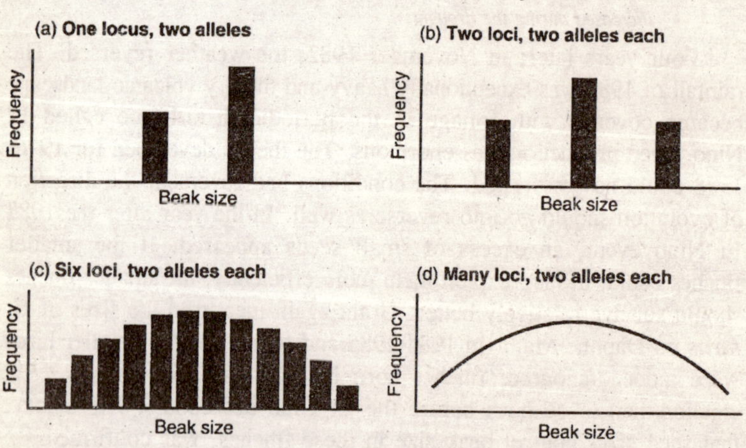

Fig. 14.3. (a) The phenotypic character—beak size, for example—is controlled by one locus with two alleles (A and a). A is dominant to a. (b) The character is controlled by two loci with two alleles each; A and B are dominant to a and b. (c) Control by six loci, with two alleles each. (d) Control by many loci, with two alleles each.

alleles each. Beak size might now have a background value (e.g., 1 cm) plus the contribution of the two loci; an *A* adds 0.1 cm and a *B* subtracts 0.1 cm. *A* and *B* are dominant to *a* and *b*. Then *aaBb* and *aaBB* individuals have 0.9 cm beaks, *aabb*, *AaBB*, *AaBb*, *AABB*, and *AABb* have 1 cm beaks, and *AAbb* and *Aabb* have 1.1 cm beaks. The frequency distribution if all alleles have frequency one-half and the two loci are in linkage equilibrium. The distribution now has three categories and has become more spread out. It becomes even more spread out if beak size is influenced by six loci and it begins to look normal for 12 loci.

Thus, when a large enough number of genes influence a character, it will have a continuous, normal frequency distribution. The normal distribution can result if a large number of alleles are present at each of a small number of loci influencing the characters, or if fewer alleles appear at a larger number of loci. In this chapter, we will mainly discuss the theory of quantitative genetics as if there were many loci, each with a small number of alleles. Although this genetic system may well underlie many continuously varying characters, the theory applies equally well with a few (even one) loci containing many alleles at each.

Mendel had noticed in his original paper in 1865 that multifactorial inheritance (i.e., the character is influenced by many genes) can generate a continuous frequency distribution. This theory was not well confirmed until later work was carried out, particularly by East, Nilsson-Ehle, and others, circa 1910. Quantitative genetics is concerned with characters influenced by many genes, called *polygenic characters*. For a quantitative geneticist, five to 20 genes is a small number of genes to be influencing a character; whereas many quantitative characters may be influenced by more than a hundred, or even several hundred, genes. For characters influenced by a large number of loci, it ceases to be useful to follow the transmission of individual genes or haplotypes (even if they have been identified) from one generation to the next. The pattern of inheritance, at the genetic level, is too complex.

There is an additional complication. So far we have considered only the effect of genes. The value of a character, like beak size, will usually be influenced by the environment in which the individual grows up. Beak size is probably related to general body size, and all characters related to bodily stature will be influenced by the amount of food an organism happens to find during its life. If we take a set of organisms with identical genotypes and allow some to grow up with abundant

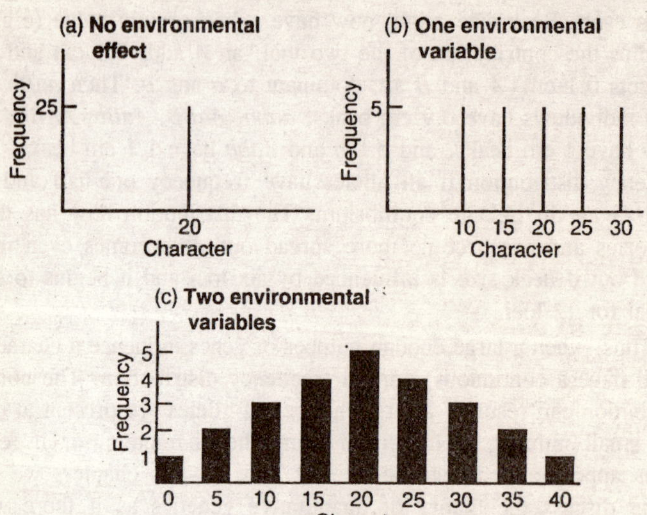

Fig. 14.4. Environmental effects can produce continuous variation. (a) Twenty-five individuals in the absence of environmental variation all have the same phenotype, with a value for a character of 20. (b) Influence of one environmental variable. (c) Influence of a second environmental variable.

food and others with limited food, the former will typically become larger. In nature, each character will be influenced by many environmental variables, some tending to increase it, others to decrease it. Thus, if we consider a class of genotypes with the same value of a character before the influence of the environment and add the effect of the environment, some of the individuals of each genotype will be made larger and others smaller in various degrees. This variation will produce a further "spreading out" of the frequency distribution. Any pattern of discrete variation in the genotype frequency distribution is likely to be obscured by environmental effects and the discrete categories converted into a smooth curve.

The small effects of many genes and environmental variables are two separate influences that tend to convert the discrete phenotypic distribution of characters controlled by single genes into continuous distributions. A continuous distribution for a character could, in principle, result from either process; quantitative genetics is mainly concerned with characters influenced by both factors. Quantitative genetics employs higher-level genetic concepts that are genetically less exact than those of one- or two-locus population genetics, but which are more useful for understanding evolution in polygenic characters.

Instead of following changes in the frequency of genes or haplotypes, we now follow changes in the frequency distribution of a phenotypic character. Quantitative genetics is important because so many characters have continuous variation and multi-locus control.

Genetic and Environmental Variations

Quantitative genetics contains an unavoidable minimum of formal concepts that we need to understand before we can put it to use. Those formalities are the topic of this section and the next.

To understand how a quantitative character like beak size will evolve, we must "dissect" its variation. We tease apart the different factors that cause some birds to have larger beaks than others. Suppose, for example, that all the variation in beak size was caused by environmental factors—that is, all birds have the same genotype and they differ in their beak sizes only because of the different environments in which they grew up. Beak size could not then change during evolution (except for non-genetic evolution due to environmental change). For the character to evolve, it must be at least partly genetically controlled. We need to know how much beak size varies for environmental reasons, and how much for genetic reasons. However, even if different finches vary in their beak size for genetic reasons, it does not necessarily mean it can evolve by natural selection. As we will see, genetic influence must be classified even further, into components that allow evolutionary change and those that do not.

In quantitative genetics, the value of a character in an individual is always expressed as a deviation from the population mean. Thus, beak size will have a certain mean value in a population, and we can talk about environmental and genetic influences on an individual as deviations from that mean. The procedure is easy to understand if the population mean is viewed as a "background" value, and the influences leading to a particular individual phenotype are then expressed as increases or decreases from that value. Suppose there is one locus with two alleles influencing beak depth. *AA* and *Aa* individuals' beaks are 1 cm from top to bottom, *aa* individuals 0.5 cm. In this system, the environment has no effect. If the population average was 0.875 cm (as it would be for gene frequency of $a = {}^{1}/_{2}$), then the beak phenotype of *AA* and *Aa* individuals would be written as +0.125 cm and that of *aa* as –0.375 cm. In general, we symbolize the phenotype by P. In this case, $P = +0.125$ for *AA* and *Aa* individuals and $P = -0.375$ for *aa*.

Clearly, the value of P for a genotype depends on the gene frequencies. When the frequency of *A* is ${}^{1}/_{2}$, the genotypic effects are those just given. When the frequency of A is $^1/_4$, however, the population mean would be 0.71875 cm. For aa, P is now −0.21875; for AA and Aa, P = +0.28125.

In this example the phenotype is controlled only by a genotype, whose effect can be symbolized by G. G, like P, is expressed as a deviation from the population mean. In this case, for an individual with a particular genotype (because the environment has no effect)

$$\text{mean of population} + P = \text{mean of population} + G$$

The background population mean cancels from the equation, and can be ignored. We are then left (in this case in which the environment has no effect) with $P = G$.

The value of a real character will usually be influenced by the individual's environment as well as its genotype. If the character under study is related in some way to size, for example, it will probably be influenced by how much food the individual found during its development, and how many diseases it has suffered. These environmental effects are measured in the same way as for the genotype, as a deviation from the population mean. If an individual grew up in an environment that caused it to develop a larger beak than average, its environmental effect will be positive. In contrast, the environmental effect would be negative if it grew up in an environment giving it a smaller than average beak. The phenotype can then be expressed as the sum of environmental (E) and genotypic influences:

$$P = G + E$$

This simple relationship is the fundamental model of quantitative genetics. For any phenotypic character, the individual's value for that character (expressed as a deviation from the population mean) results from the effect of its genes and environment.

We must look further into the genotypic effect. We need to consider how to subdivide the genotypic effect, and why the subdivision is necessary. The main point can be seen in the one-locus example we have already used. The A gene is dominant, and both AA and Aa birds have 1 cm beaks (P = +0.125). (Because we are investigating the genotypic effect, we will ignore environmental effects, so that $P = G$.) Suppose we take an AA individual and mate it to another bird drawn at random from the population. The gene frequency = $^1/_2$ and the random bird is AA with chance $^1/_4$, Aa with chance $^1/_2$, and aa with chance $^1/_4$. Regardless of the mate's genotype, all offspring will have beak phenotype P = +0.125, because A is dominant. Now suppose we take an Aa individual and mate it to a random member of the population. The average beak of their offspring is $(^1/_4 \times +0.125) + (^1/_2 \times 0) + (^1/_4 \times -0.125) = 0$. That is, the average offspring beak size is the same as the population average. Thus, for two genotypes with the same beak size ($P = +0.125$), one produces offspring with beaks like their parents, while the other produces offspring with beaks closer to the population average.

As we can see, some genotypic effects are inherited by the offspring and some are not. We can divide the genotypic effect into a component that is passed on and a component that is not. The component that is passed on is called the *additive effect* (A) and the component that is not is called (in this case) the *dominance effect* (D). The full genotypic effect in an individual is the sum of the two:

$$G = A + D$$

The additive effect has the greatest impact. The parent deviates from the population mean by a certain amount, and its additive genotypic effect is the part of that deviation potentially passed on to its offspring. However, when an individual reproduces, only half of its genes are inherited by its offspring. The offspring inherit only half the additive effect of each parent. Thus, the additive effect A for an individual is equal to twice the amount by which its offspring deviate from the population mean, if mating is random. For the *AA* parent, therefore, the additive effect is +0.25. (The full quantitative genetics of the *AA* individuals are: $G = +0.125$, $A = +0.25$, and $D = -0.125$.) The offspring of *Aa* birds deviate by zero, so their additive effect is twice zero (or simply zero). Thus, the amount by which *Aa* heterozygotes deviate from the population mean is entirely due to dominance, and is not inherited by their offspring. (For *Aa* individuals, $G = +0.125$, $A = 0$, $D = +0.125$.)

The division of the genotypic effect into additive and dominance components tells us what proportion of the parent's deviation from the mean is inherited, and reveals how the non-inheritance of the *Aa* individuals' genotypic effect is due to dominance. In practice, quantitative geneticists do not know the genotypes underlying the characters they study; they know only the phenotype. They might, for instance, focus on the class of birds with 1 cm ($P = +0.125$) beaks. In the example given above, the additive component of their phenotypic value depends on the frequencies of the *AA* and *Aa* genotypes. If all birds with 1 cm beaks are *Aa* heterozygotes, then none of the offspring will inherit their parents' deviation; if they are all *AA*, then half of the offspring will inherit the deviation.

Why is the additive effect of a phenotype so important? Once the additive effect for a character has been estimated, that estimate plays much the same role in quantitative genetics as the exact knowledge of Mendelian genetics in a one- or two-locus case. We use this estimate to predict the frequency distribution of a character in the offspring, given a knowledge of the parents. In a one-locus genetic model, we know the genotypes corresponding to each phenotype, and can predict the phenotypes of offspring from the genotypes of their parents. In the case of selection, the gene frequency in the next generation is easy to predict if we know selection allows only *AA* individuals to breed. In two-locus genetics, the procedure is the same: If the next generation is formed from a certain mixture of *Ab*/*AB* and *AB*/*AB* individuals, we can calculate its haplotype frequencies if we know the exact mixture of parental genotypes.

In quantitative genetics, we do not know the genotypes. We have only measurements of phenotypes, like beak size. If we can estimate the additive genetic component of the phenotype, then we can predict the offspring in a manner analogous to the procedure when the real genetics are known. When we know the genetics, Mendel's laws of inheritance tell us how the parental genes are passed on to the offspring; when we do not know the genetics, the additive effect tells us which component of the parental phenotype is passed on. Thus, estimating the additive effect is the key to understanding the evolution of quantitative characters. The estimates are practically made by breeding experiments. In the case of finches with I cm beaks in a population of average beak size 0.875 cm, the additive effect can be measured by mating finches with 1 cm beaks to random members of the population. The additive effect is then two times the offspring's deviation from the population mean.

The genetic partitioning that we have undertaken so far applies reasonably well for one locus. In that case, dominance is the main reason why the genotypic effect of a parent is not exactly inherited in its offspring. When many loci influence a character, *epistatic interactions* between alleles at different loci can produce additional effects. Epistatic interactions, like dominance effects, are not passed on to offspring—they are not additive. They depend on particular combinations of genes and the effect disappears when the combinations are broken up (by genetic recombination). An example of an epistatic interaction would be for individuals with the haplotype A_1B_2 to show a higher deviation from the population mean than the combined average deviations of A_1 and B_2; the extra deviation is epistatic. Other non-additive effects can arise because of gene-environment interaction (when the same gene produces different phenotypes in different environments) and gene-environment correlation (when particular genes are found more often than random in particular environments).

A full analysis can take all of these effects into account. In a full analysis, such as the simple case given here, the aim is to isolate the additive effect of a phenotype. The additive effect is the part of the parental phenotype that is inherited by its offspring.

Variance of Characters

We return now to the frequency distribution of a character. We will continue to express effects as deviations from the population mean. If we consider an individual some distance from the mean, some of its deviation will be environmental, some genetic. In the genetic component, some parts will be additive, some dominant, some epistatic. These terms have been defined so that they add up to give the exact deviation of the individual from the mean. Any individual has its particular phenotypic value (*P*) because of its particular combination of environmental experiences and the dominance, additive, and interaction effects in its genotype. The different combinations of *E*, *D*, and *A* in different individuals explain why the character shows a continuous frequency distribution in the population.

The variation seen for the character in any one population could exist because of variation in any one of, or any combination of, these effects. Thus, the individual differences could all be due to different environmental effects, with every individual having the same value for *G*. Alternatively, it could be 25% due to the environment, 20% due to additive effects, 30% due to dominance effects, and 25% due to interaction effects. The proportion of variation related to the different

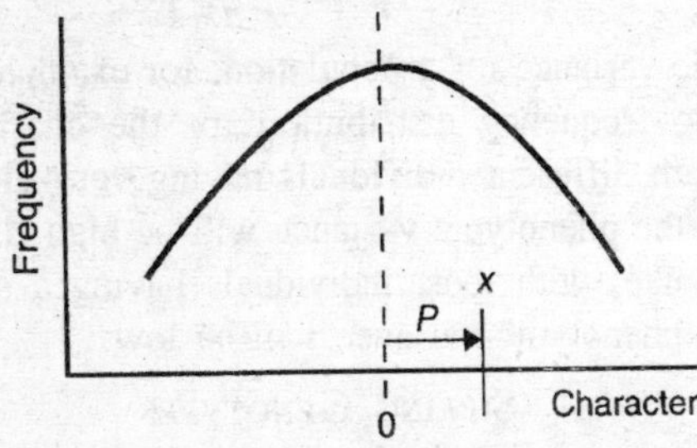

Fig. 14.5. The x-axis of the continuous distribution for a character can be scaled to have a mean of 0.

effects matters when we consider how a population will respond to selection. If the variation exists because different individuals have different values of E, selection will not generate a response. On the other hand, if the variation is mainly additive genetic variation, the response will be large. The proportion of the variation resulting from different values of A in different individuals tells us whether the population can respond to selection.

The variability in the population related to any particular factor, such as the environment, is measured by the statistic called the *variance*. Variance is the sum of squared deviations from the mean divided by the sample size minus one. For all values x of a character like beak size in a population, the variance of x is

$$V_x = \frac{1}{n-1}\Sigma(x_i - \bar{x})^2$$

We have seen how the total phenotype, genetic effect, environmental effect, and so on, can be measured for an individual. The measurements (P for phenotypic effect, and so on) are expressed as deviations from the mean. We can easily calculate what their variances are for a population.

phenotypic variance $= V_P = \frac{1}{n-1}\Sigma P^2$

environmental variance $= V_E = \frac{1}{n-1}\Sigma E^2$

genetic variance $= V_G = \frac{1}{n-1}\Sigma G^2$

dominance variance $= V_D = \frac{1}{n-1}\Sigma D^2$

additive variance $= V_A = \frac{1}{n-1}\Sigma A^2$

The phenotypic variance for a population, for example, expresses how spread out the frequency distribution for the character is. If it is spread out, with different individuals having very different values of the character, the phenotypic variance will be high. If the distribution is a narrow spike, with most individuals having a similar value for the character, phenotypic variance will be low.

Sibling Genotypes

Grant measured the depth of the beak in many parent-offspring pairs in *Geospiza fortis*, and plotted the results on a graph. Each point indicates one offspring and the average of its parents. Beak depth in *G. fortis*, like many characters in most species, shows a correlation between parent and offspring. This correlation exists between relatives as well, for just as parents and offspring are similar to one another, so are siblings and more distant relatives similar to some extent.

Similarity among relatives may be either environmental or genetic, and the two effects may have varying importance in different kinds of species. In humans, in which parents and offspring live together in social groups, much of the similarity results from the family's common environment (i.e., gene—environment correlation). At the other extreme, in a species like a bivalve mollusc in which eggs are released into the sea at an early stage, relatives will not necessarily grow up in similar environments, and shared environments will not exert such a strong influence on similarity among relatives. Darwin's finches probably lie somewhere between these two extremes. It is easy to understand similarity among relatives that is caused by similar environments. In so far as relatives grow up in correlated environments, and environmental variation exists in a character, relatives will be more similar than non-relatives.

The similarity between relatives due to their shared genes is evolutionarily more important. It is possible to deduce this type of correlation among any two classes of relatives from the variance terms we have already defined. We will consider only one case—the correlation between parents and offspring—to see how it occurs. To simplify matters, we will assume that the environments of parent and offspring are not correlated. This assumption allows us to ignore environmental effects (because any environmental effect in the parent will not appear in the offspring—if the parent is larger than average because it benefited from a good food supply, that does not mean its offspring will share those benefits). Any correlation between parent and offspring will then be due to their genetic effects.

The genetic value of the character in the parent, we have seen, consists of several components of which only the additive component is inherited by the offspring. When mating is random, half of that additive component of the individual parent is diluted. At a locus, a parent has an additive deviation from the population mean in both its genes. When an offspring is formed, one of the parental genes goes into the offspring together with another gene drawn at random from the population (we are assuming random mating). The average value of the character in the offspring is, as we saw earlier, half of the additive value of the parent ($^1/_2 A$); the average genetic value in the parent is $A + D$. The correlation between parent and offspring is the covariance between the two:

$$\text{cov}_{OP} = \Sigma \frac{1}{2} A(A+D)$$

where all offspring-parent pairs are summed. The covariance can also be expressed as:

$$\text{cov}_{OP} = \frac{1}{2} \Sigma A^2 + \frac{1}{2} \Sigma AD$$

$^1/_2 \Sigma AD = 0$ because A and D have been defined to be uncorrelated. That is, if an individual has a large value of A, we know nothing about whether its value of D will be large or small. Thus, the equation can be written as

$$\text{cov}_{OP} = \frac{1}{2} \Sigma A^2 + \frac{1}{2} V_A$$

To describe this relation in words, the covariance of an offspring and one of its parents is equal to half the additive genetic variance of the character in the population.

The expression for the covariance between one parent and its offspring is true for each parent. It requires only a small step (though we will not go into it here) to show that the same expression ($^1/_2 V_A$) also gives the covariance between offspring and the mid-parental value. Other expressions can be deduced, by similar arguments) for the covariance between other classes of relatives. The formulae are useful for estimating the additive variances of real characters. However, the estimates become most interesting, for the evolutionary biologist, when expressed in terms of the statistic called *heritability*.

Phenotypic Variance

The similarity between relatives in general, and between parents and offspring in particular, is governed by the additive genetic variance of the character. If a character has no additive genetic variance in a population, it will not be passed on from parent to offspring. For instance, many of the properties of an individual phenotype are accidentally acquired characters, such as cuts, scrapes, and wounds; if we measure these in parent and offspring they will show no correlation. Thus, $V_A = 0$ for such characters. Moreover, some characters, such as the number of legs per individual in a natural population of, for example, zebra, show practically no variation of any sort; for those characters, V_A trivially is zero. Additive variance is, therefore, often discussed as a fraction of total phenotypic variance. This fraction is called the *heritability* (h^2) of a character.

$$h^2 = \frac{V_A}{V_P}$$

Heritability is a number between zero and one. If heritability is one, all variance of the character is genetic and additive. Given that $V_P = V_E + V_A + V_D$, all terms appearing on the right other than *VA* must then be zero. In so far as the factors other than- additive variance account for the variance of a character, heritability is less than one.

Heritability has an intuitive meaning. Consider a parent that differs from the population by a certain amount. If its offspring also deviate by the same amount, heritability is one; if the offspring have the same mean as the population, heritability is zero; if the offspring deviate from the mean in the same direction as their parents but to a lesser extent, heritability is between zero and one. Heritability, therefore, is the quantitative extent to which offspring resemble their parents, relative to the population mean.

How can we estimate the heritability of a real character? One method is to cross two pure lines. This approach is mainly of interest in applied genetics, where the problem might be to breed a new variety of crops; it holds little interest for the evolutionary biologist. The two other main methods involve measuring the correlation between relatives and the response to artificial selection, respectively.

Figure 14.1 is an example that uses the correlation between relatives to determine heritability. The slope of the graph, which shows the beak size in offspring finches in relation to the average beak size of the two parents, is equal to the heritability of beak size in that population. The reason is as follows. The slope of the line is the regression of offspring beak size on mid-parental beak size. The regression of any variable *y* on another variable $x = \mathrm{cov}_{xy}/\mathrm{var}_x$. The covariance of offspring and mid-parental value = $^1/_2 V_A$ and the variance of the mid-parental beak size is equal to $^1/_2 V_p$. (The variance is half of the total population variance because *two* parents have been drawn from the population and their values averaged; if Figure 14.1 had the value for one parent on the *x*-axis, its variance would be V_p). The regression slope simply equals V_A/V_p, which is the character's heritability. For beak depth in *Geospiza fortis* on Daphne Major, the regression and therefore heritability is 0.79.

Artificial Selection

How can quantitative genetics be applied to understand evolution? We will consider two of the many ways possible here: directional selection and stabilizing selection. A third kind of selection—disruptive selection—will not be discussed here. This section focuses on *directional selection*, which has been studied extensively through artificial selection experiments. Artificial selection is important in applied genetics, as it provides the means of improving agricultural stock and crops.

To increase the value of a character by artificial selection, we can use any of a variety of selection regimes. One simple form is truncation selection, in which the selector picks out all individuals whose value of the character under selection is greater than a threshold value, and uses them to breed the next generation. What will be the value of the character in the offspring generation? Let us define *S* as the mean deviation of the selected parents from the mean for the parental population; *S* is also called the selection differential. The response to selection (*R*) is the difference between the offspring population mean and the parental population mean. In this case, calculating the response to selection is found by regressing the character value in the offspring on that in the parents, where the parents are the individuals that were selected to breed. We plot the offspring's deviation against the parental deviation from the population average to produce a graph. The slope of the graph for parents and offspring is symbolized

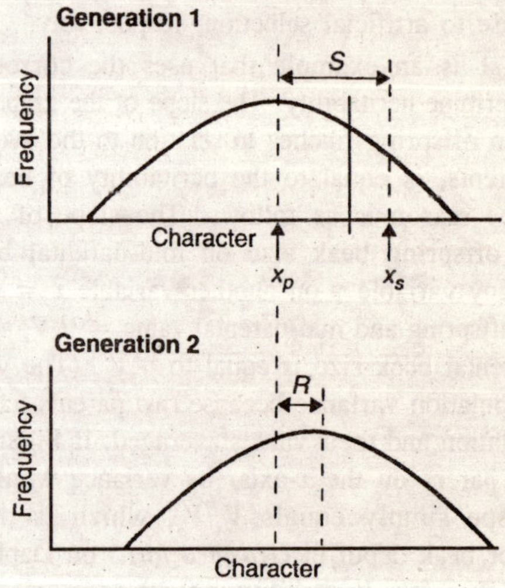

Fig. 14.6. Truncation selection.

by b_{OP} and, as we saw in the previous section, for any character $b_{OP} = h^2$; that is, the parent-offspring regressional slope equals the heritability. Therefore

$$R = b_{OP}S$$

or

$$R = h^2S$$

This result is important because it shows that the response to selection is equal to the amount by which the parents of the offspring generation deviate from the mean for their population multiplied by the character's heritability. (The response to selection or the parent-offspring regression can be used to estimate the heritability of a character; for a selected population, they are two ways of looking at the same set of measurements.)

A real example of directional selection may not take the form of truncation selection. In truncation selection, all individuals above a certain value for the character breed and all individuals below do not breed. All selected individuals contribute equally to the next generation. It is possible that there could be no sharp cut off, and that individuals with higher values of the character contribute increasing numbers of offspring to the next generation. However, the same formula for evolutionary response works for all forms of directional selection. The difference between the mean character value in the whole population and in those individuals that actually contribute to the next generation (if necessary, weighted by the number of offspring they contribute) is the "selection differential." This value can be plugged into the formula to find the expected value of the character in the next generation.

A population can respond to artificial selection only for as long as the directional selection lasts. Consider, for example, the longest-running controlled artificial selection experiment. Since 1896, corn has been selected, at the State Agricultural Laboratory in Illinois, for (among other things) either high or low oil content. Even after 76 generations the response to selection for high oil content was not exhausted. However, it will eventually stop. As the corn is selected for increased oil content, the genotypes encoding for high oil content will increase in frequency and be substituted for genotypes for lower oil content. This process can proceed only until all the individuals in the population have the same genotype for oil content. At the loci controlling oil content, no additive genetic variance will then be left; heritability will have been reduced to zero and the response to artificial selection will terminate. In the Illinois corn experiment, the process has not yet run its full course. The heritability of oil content in both

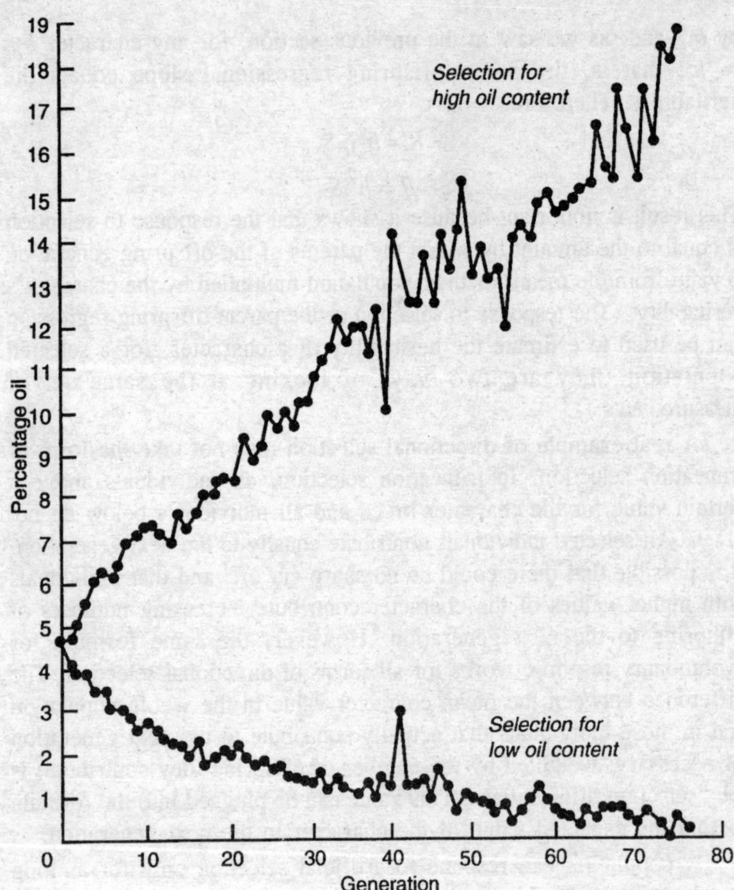

Fig. 14.7. Response of corn (Zea mays) artificially selected for high or low oil content.

the high and low selected lines has decreased through the experiment, but it remains above zero—which is why the population continues to respond to selection.

In other artificial selection experiments, the complete process has been recorded. Figure 14.8 shows the response of a population of fruitflies to consistent directional selection for increased numbers of scutellar chaetae (i.e., bristles on a dorsal region of the thorax). Initially the population responded. Then, as the additive genetic variation was used up (or, as its heritability declined), the rate of change slowed to a stop in generations 4-14. It also appeared that, if selection was continued after the response stopped, the population suddenly began to

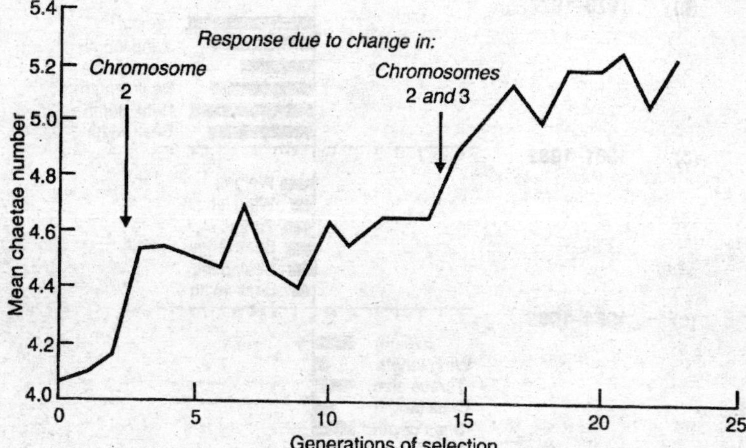

Fig. 14.8. Artificial selection for increased numbers of scutellar chaetae in Drosophila melanogaster.

respond again after an interval (in generations 14-17). The renewed bout of change is attributed to a rare recombinant or mutation that reinjected new genetic variation into the population.

The relation between response to selection (*R*), heritability, and selection differential (*S*) enables us to calculate any one of the three variables if the other two have been measured. For example, fishing has selected for small size in salmon, because larger fish are selectively taken in the nets. The selection differential S can be estimated from three measurements: the average size of salmon caught in the nets, the average size of salmon in the population at the mouth of the river (before it is fished), and the proportion of the population that is removed by fishing. All three have been measured and lead to the estimates that the salmon who survive to spawn are roughly 0.4 lb smaller than the population average. The response (*R*) for this population—the average size of the salmon decreased by about 0.1 lb between each two-year generation. We can, therefore, estimate the heritability, $h^2 = 0.1/0.4 = 0.25$.

In the finches, Grant measured the response to selection (*R*) and heritability for several characters related to body size, and used these values to estimate selection differentials. We saw that in *Geospiza fortis* that heritability of beak size is about 80%. After the bout of selection for large size in 1976-1977, the finches were approximately 4% larger. We can estimate the selection differential as $S = 0.04/0.8 = +5\%$. As the direction of selection reversed from favouring larger

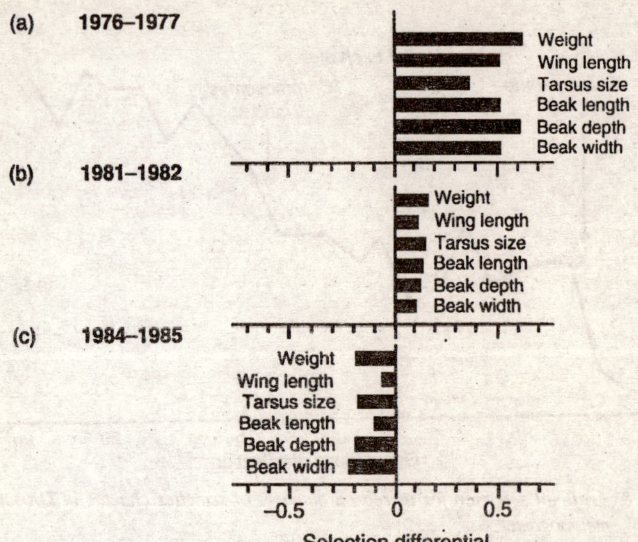

Fig. 14.9. (a) In the drought of 1976-1977, Geospiza fortis individuals with larger beaks survived better and the average size of the finch population increased. (b) In the normal years of 1981-1982, having a larger beak provided a slight advantage, but much smaller than during the drought. (c) After the 1983 El Nino, in 1984-1985, finches with smaller beaks survived better and the average size of the finch population decreased.

beaks in 1976-1978 and smaller beaks in 1983-1985, the selection differentials reversed from positive values in 1976-1977 to negative values in 1984-1985.

In summary, the response to directional selection is controlled by the heritability of a character. The relation between the selection differential, response, and heritability can be used to estimate one of the variables if the other two are known. Sustained selection will exhaust the heritable variation, and the response will then stop.

Relation between Genotype and Phenotype

Scharloo selected a population of fruitflies for increased relative length of the fourth wing vein. The figure shows the frequency distribution of vein lengths in the population for 10 generations. A length of 60-80 has been reached by about generation 5. At this stage the frequency distribution (amid the scatter often encountered in real experiments) starts to show a consistent bimodality. This condition is clearest in generations 5-7, with only the high peak being maintained in generations 8-10. The experiment suggests that more complicated

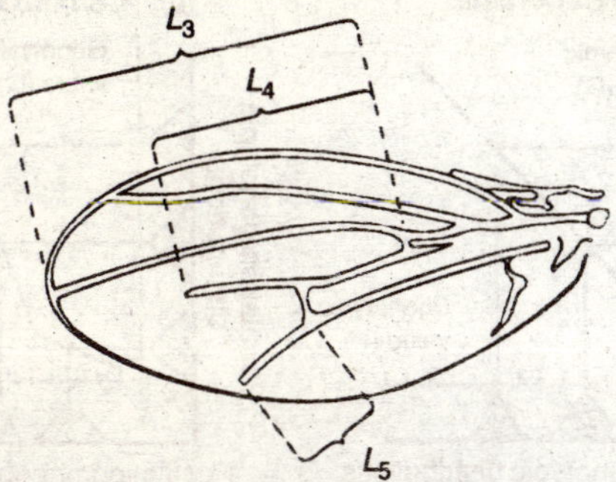

Fig. 14.10. The main veins in the wing of Drosophila.

events can occur in artificial selection experiments than we have seen so far. What is going on?

The key to understanding the shape of the response is the relation between genotype and phenotype. A simple response, such as that for oil content or bristle number, results when an approximately linear relation exists between genotype and phenotype. Genotype here is expressed as a metrical variable. The easiest way to think of this relation is to imagine that the character is controlled by many loci; at each, some alleles (+) cause the phenotypic character to increase, and others (-) cause it to decrease. The more + genes an individual possesses, the higher its genotypic value. When we select for an increase in the character, we pick the individuals with more + genes, enabling the value of the character to increase smoothly between generations in the manner of the Illinois corn oil experiment.

The approximately linear form is not the only possible relation between genotype and phenotype. The bimodal response is thought to result from a threshold relation between genotype and phenotype. The graph has been rotated 180° relative to the form; the *x*-axis (genotype) is drawn down the page on the left. The genotype is believed to control the amount of some vein-inducing substance. Vein length appears across the top of the graph. The relation between substance and vein length is hypothesized to contain a jump at vein length 60-80 (where the artificial selection response becomes bimodal).

Imagine the course of selection for longer wing veins with this threshold relation between genotype and phenotype. The population starts at the top left of the graph with relatively short veins. Initially, some variation appears in the population for genotype, with an associated normal distribution of vein lengths. Artificial selection produces flies with a higher concentration of the vein-inducing substance and with

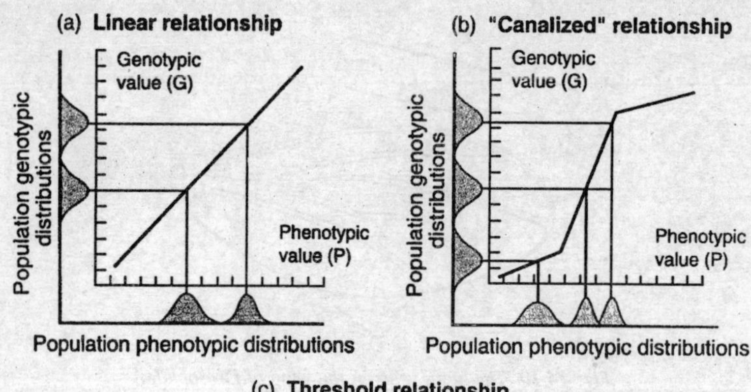

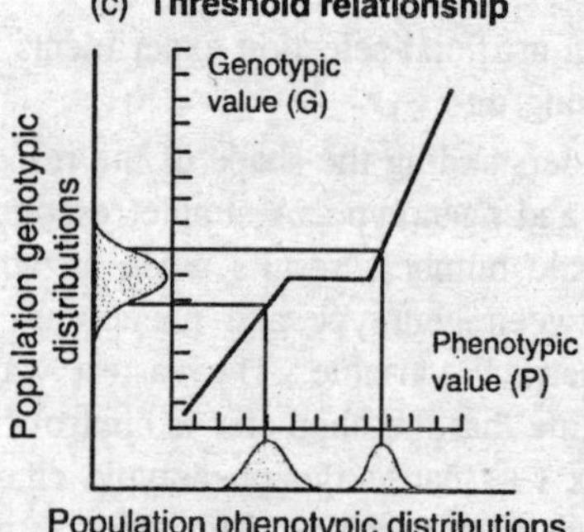

Fig. 14.11. Relation between genotype and phenotype

correspondingly longer wing veins. This process continues until the population reaches a vein length of about 60-80. At that point, the normal (unimodal) variation for the amount of vein-inducing substance generates a bimodal distribution of vein lengths. The bimodal distribution will later disappear, as the population passes beyond the jump in the genotype-phenotype mapping function. Hence, we have the observed response to selection. The relation of genotype and phenotype for vein length is only a hypothesis, but it does show how, in theory, a bimodal response to selection could arise.

The main points to take away from this discussion are that when the genotype-phenotype relation has the linear form, a simple response to artificial selection occurs. The population changes until the genetic variation is exhausted. However, we have no reason to think that this process reflects the typical genotype-phenotype relation. When the relation is more complex, the response to artificial selection can be interestingly different, as the bimodal response to selection on wing vein length in fruitflies illustrates.

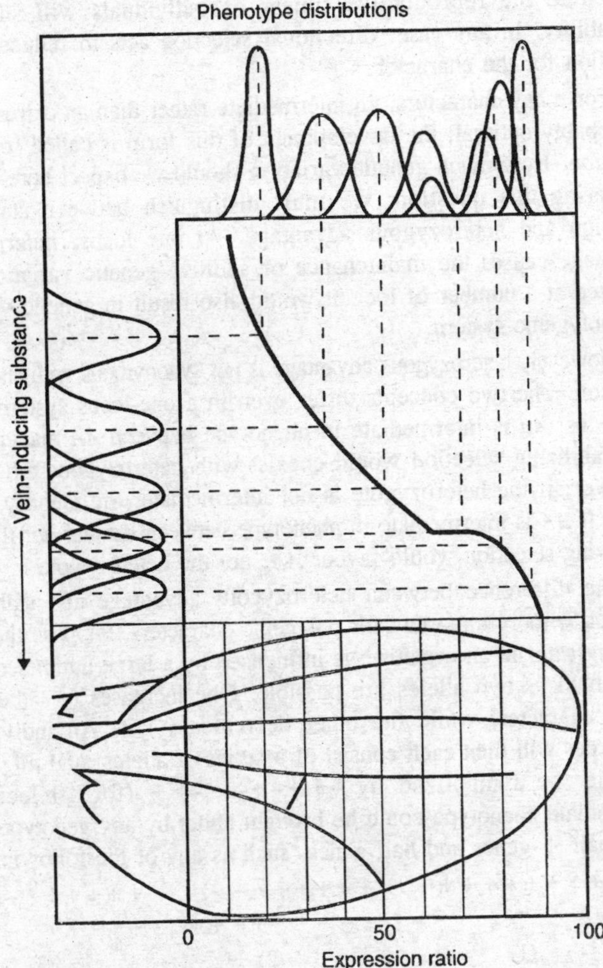

Fig. 14.12. Model of relation between genotype and phenotype for Drosophila fourth wing vein.

Genetic Variability

Natural populations have been undergoing the natural analogy of artificial selection for many generations. They can be thought of as being at the end of, rather than the beginning or anywhere else in, the selection experiment. For characters on which directional selection has been working, the heritability will have been reduced to a low level. The more strongly selection works for a character, the more its heritability should decline, and we might expect that characters closely related to the reproductive fitness of individuals will show low heritability. In any case, directional selection acts to reduce genetic variation for the character.

For many characters, an intermediate rather than an extreme form is probably optimal; the development of this form is called *stabilizing selection*. How much genetic variation should we expect here? Before answering this question, we must distinguish between stabilizing selection and heterozygous advantage. At one locus, heterozygous advantage causes the maintenance of additive genetic variation; if it operated at a number of loci, it would also result in genetic variation in a polygenic system.

However, heterozygous advantage is not synonymous with stabilizing selection. The two concepts differ even in a one-locus system. If, at one locus, *Aa* is intermediate in phenotype between *AA* and *aa*, then any stabilizing selection would coexist with heterozygous advantage. However, if the heterozygote is not intermediate, the identity breaks down. If *AA* is intermediate in phenotype between *Aa* and *aa*, then any stabilizing selection would favour *AA*, not the heterozygote.

The difference between heterozygous advantage and stabilizing selection is more important for polygenic characters. Suppose the value of a character in an organism is influenced by a large number of loci. At each locus two alleles are possible. One increases (+) the value of the character, while the other decreases (–) it. An individual's haplotypes will then each consist of a series of alleles, and might, for example, be symbolized by –++–+–––++ (for 10 loci). An intermediate phenotype could be brought about by any genotype made up of half + genes and half-genes, such as any of the following:

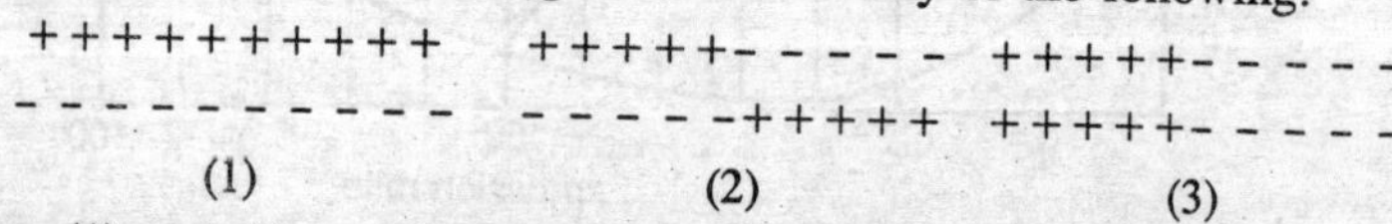

Cases (1) and (2) would produce heterozygous advantage. Selection could also favour the intermediate genotype (3), however, and in a population containing only this haplotype there is no heterozygous advantage.

How much genetic variability do we expect for a character subject to stabilizing selection? In a population containing the four haplotypes, the three genotypes (1), (2), and (3) will all have the same fitness, and we might expect a population to retain considerable genetic variation as these genotypes interbreed and produce a variety of offspring types. However, over a long period of time, genotypes like (3) that breed true will show an advantage, because all of this genotype's offspring have the optimal phenotype, whereas some of the offspring of genotypes (1) and (2) do not. In a population consisting of these three genotypes, selection slightly favours genotype (3). If the environment were constant for a long time, always favouring the same phenotype, selection should eventually produce a uniform population with a genotype like (3).

We can take the argument a stage further. Genotype (3) is not the only true-breeding homozygote that can produce the intermediate optimal form. All of the following also have this characteristic:

+-+-+-+-+-	++--++--+-	+++---++--
+-+-+-+-+-	++--++--+-	+++---++--
(4)	(5)	(6)

Suppose we have a population made up of genotypes (3)-(6), and selection favours the intermediate phenotype. What will happen now? Evolution will again tend toward a population with only one genotype, and that genotype should be a multiple homozygote. The reason behind this result is that any one of the homozygotes that has a slightly higher frequency than the others will have an advantage. Suppose, for example, that genotype (4) had a higher frequency than (3), (5), and (6). All genotypes will now be most likely to mate with a genotype (4). When genotype (4) mates with genotype (4), their offspring will all have the favoured phenotype, identical to their parents. On the other hand, when genotypes (3), (5), and (6) mate with genotype (4), the offspring contain potentially disadvantageous genotypes. The offspring of a mating between a genotype (3) and a genotype (4), for example, will be +-+-+-+-+- / ++--++--+- and has the favoured intermediate phenotype. However, its offspring will contain disadvantageous recombinants. The end result is for selection to produce a uniform population, in which minority genotypes are selected against because they do not fit with the majority form. Selection eventually reduces the genetic variability to zero, even with stabilizing selection.

The argument leads to the following conclusion. If we start with a set of genotypes that influence a character subject to stabilizing selection, and if the environment remains constant for a long enough period, then selection ought eventually to produce a population containing only one, multiply homozygous, genotype.

Stabilizing Selection

The conclusion of the previous section is contradicted by observable facts. Heritabilities can be measured for real characters, and many show significant genetic variation. It suggests that typical values for heritability occur in the range 0.1-0.5. These estimates are for lab populations, and most reliable heritability estimates are unfortunately artificial. Some measurements have been taken from nature as well, such as Grant's measurements, for the Galapagos finch, and these natural observations agree with the conclusion—real characters often have heritabilities higher than zero.

If stabilizing selection tends to eliminate heritable variation, why does it exist? The question is currently a matter of theoretical controversy, and we will look at two possible explanations: mutation-selection balance, and selective processes (like heterozygous advantage) that maintain variation.

Selection-Mutation Balance

Some genetic variation exists in a population as mutation creates new deleterious genes and selection eliminates them. For any one locus, the amount of variation maintained by this selection-mutation balance is low because mutation rates are low. For a polygenic character, however, mutation rates should be approximately multiplied by the number of loci influencing the character. A character controlled by 500 loci will have 500× the mutation rate of a one-locus character. The amount of variability that can be maintained is proportionally increased.

How much genetic variation will exist? This question has been approached from two directions. One approach, which was revived and developed by Lande in the 1970s, considers stabilizing selection on a continuous character (such as body size) controlled by many loci. Mutations at any of the loci can influence the character; because a genotype may be above or below the optimal value for the trait, a small random mutation has a 50% chance of being an improvement. The other approach, which was revived and developed by Kondrashov and Turelli in 1992, does not consider stabilizing selection on a phenotypic character, but supposes mutations occur at many loci and the great majority (many more than 50%) are deleterious. The result is a balance between selection and deleterious mutation at many loci.

We do not have space to go far into either theoretical system, or their relative merits. We will not discuss the other theoretical system at all, but we can look at the experimental evidence measuring mutation rates of deleterious mutation at many loci. The results of this examination have general evolutionary importance.

Deleterious Mutations

The first large-scale experiments to illustrate the destructive power of mutation when selection is minimized were performed by the Japanese geneticist Terumi Mukai. (They were indeed large-scale—Mukai's first experiment in 1964 made use of about 1.7 million flies.) The aim of the experiment was to reduce the effect of selection to a minimum for a number of generations, and measure the decline in fitness. Fitness declines because mutations accumulate—mutations that would normally be eliminated by selection. Mutations can be of varying degrees of destructiveness, from lethal to neutral. The rate of lethal mutations had been measured before Mukai's work, as had the rate of visible mutations (if the effect of a mutation is visible, it is likely to be a large change). These measurements are of only limited interest because lethal and other large mutations probably constitute only a minority of all deleterious mutations. Geneticists suspected the generation of a large class of "slightly deleterious" mutations, which cannot be individually detected at the phenotypic level, and with fitness effects appearing in between the lethal and the neutral mutations. Mukai's experiment revealed the importance of those mutations empirically.

Mukai designed his experiment to transfer a chromosome through a number of generations without recombination or selection, to reveal the way mutations accumulate on it. The experiment used fruitflies (*Drosophila melanogaster*). Fruitfiles have four pairs of chromosomes; they are distinguishable and referred to as chromosomes 1-4. The second chromosome, which was the subject of Mukai's study, makes up about 30% of the total genome. Mukai carried out genetic crosses to create an individual male fly that was homozygous for its entire second chromosome (symbolized +/+): the particular second chromosome came from a fly caught in nature and was not peculiar in any way—the fitness of the homozygotes was in the normal range for homozygotes of natural chromosomes. After creating 104 lines from the original chromosome, Mukai bred the flies through a number of generations, preventing recombination and minimizing selection.

Recombination needed to be eliminated so that the changes in the chromosome would be only due to mutation: if the chromosome could also recombine, some changes would be due to exchange with other versions of the 2nd chromosome. Mukai prevented recombination in the females by means of genetic tricks; no recombination occurs in male fruitflies.

Ideally, Mukai have prevented all selection as well, but that goal is impossible. Instead he minimized it, by two procedures. First, he picked only one male, at random, to father the next generation in each line. Second, the flies were reared in optimal conditions for fly growth, so that as many as possible eggs should have been able to grow up as adults. Selection would not have been reduced to zero, because some of the flies probably had such poor genotypes that they failed to develop; this kind of selection in not excluded in the experiment. Nevertheless, a high proportion of suboptimal flies should have grown up and had a fair chance of being picked to father the next generation. Thus, very little selection operated against slightly deleterious mutations.

As the generations passed, we could have expected mutation to cause the average quality of the flies to decline. Mukai measured the viability (as one component of fitness) of the flies every 10 generations. He sampled individuals from each line and used them to generate flies that were homozygous (+/+) for the chromosome that had accumulated the mutations, or heterozygous for it and a standard lab chromosome (+/standard). The relative survival of the two types of fly were then measured. Any lower survival in the homozygotes relative to the heterozygotes should be due to the accumulated mutations,

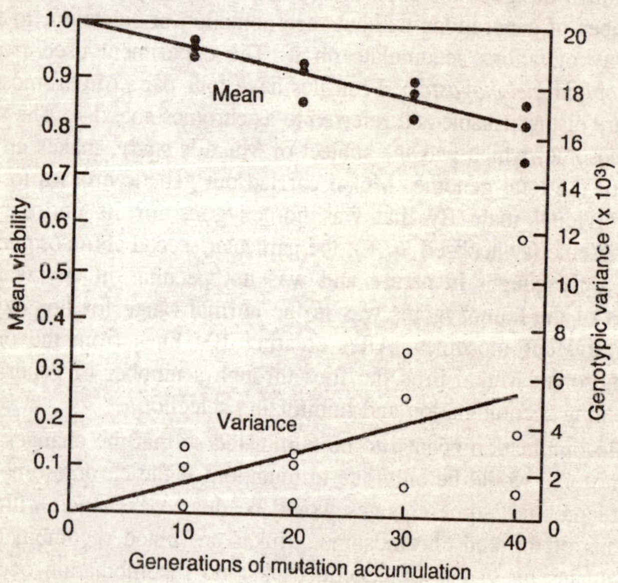

Fig. 14.13. The mutational meltdown in fruitflies protected from selection.

which appear in double dose in the homozygote but only single dose in the heterozygote (allowance has to be made for dominance). In fact, the relative viability of +/+ flies steadily declined.

The result of Mukai's experiment is important for two reasons. First, it reveals the destructive power of mutation. After 40 generations, enough mutations have accumulated in the + chromosome to reduce by 15% the viability of a homozygous +/+ fly relative to a heterozygote. When selection on a population is stopped (or at any rate minimized), the quality of the individuals in it deteriorates over the generations. This fact can be seen even more dramatically, if in less scientific form, by anyone who looks at the flies after a number of generations of the Mukai experience. Mukai himself did not describe his experimental flies, but after 50 generations or so in an similar experiment, the flies are noticeably less vigorous than normal flies.

Second, the quantitative rate of decline in relative fitness can be used to estimate the actual rate of deleterious mutation. We will not work through the full estimation procedure here, except to say that it makes use of the variance in fitness. If the decline in fitness were caused by many mutations of small effect, the variance among lines would be small, whereas the variance would be large if it resulted from a smaller number of mutations of large effect. The actual variance can be used to derive the magnitude, and thus the implied rate, of the mutations responsible for the observed decline in fitness. Several estimates can be made, depending on the treatment of the numbers, but a low estimate is that the second chromosome in Mukai's experiment had a total mutation rate of 0.15 per generation (corresponding to 0.5 per whole genome per generation). That rate applies to all slightly deleterious mutations; thus, each mutation will have a particular effect on fitness, but we can estimate the average effect on fitness of all mutations. The total reduction in fitness, calculated by multiplying the mutation rate and the average fitness effect per mutation, was -0.003 per generation. The number is approximately correct, because after 10 generations the relative fitness of a +/+ fly is about 3-4% lower than at the start of the experiment.

We can conclude that mutation introduces a large amount of genetic variation each generation, and for polygenic characters it is not implausible that selection-mutation balance could maintain quite large amounts of genetic variation at equilibrium. We can also draw one other incidental moral from Mukai's experiment concerning the role of natural selection in the living world. We have met natural selection as a force of change, in peppered moths, pestiferous insects, and

Darwin's finches; we have encountered it as a force for constancy in human birth weight; and we have seen how it leads to the evolution of adaptation. Natural selection is all of these things, but now we have another way of looking at the process. We have just seen what happens to a population in which the effect of selection is minimized. We can imagine by extension how much worse things would be if no selection operated at all. Natural selection in a sense acts to support living things against a ferocious headwind of deleterious mutations that, in the absence of selection, would soon destroy them.

Conclusion

One- and two-locus population genetics is used for characters controlled by one or two loci and whose genetics is known. Quantitative genetics provides the techniques to understand evolution in characters that are influenced by a large number of genes, and for which the exact genotype (or genotypes) producing any given phenotype are unknown. It is possible that the majority of characters have this kind of genetics, in which case quantitative genetics would be appropriate for understanding the majority of evolution. At any rate, it is a highly important set of techniques.

In this chapter, we have seen how quantitative genetics divides up the variation in a character to recognize the component that controls how offspring resemble their parents; the component is called the additive genetic effect. The additive genetic effect plays the same role in quantitative genetics as a knowledge of Mendelian genetics in one- and two-locus population genetics.

The response to selection can be analyzed by means of the heritability of a character, which is the fraction of its variation due to additive genetic effects. However, even with simple directional selection, the exact response depends on the underlying genetic control. For example, the possible threshold relation between the genotype and phenotype for the wing veins of the fruitfly generates an interesting bimodal response to selection. Here, the heritability of the character would show strange changes as the character evolved. Directional selection unambiguously should continue to alter a character until its heritability is reduced to zero. With stabilizing selection, it might be thought that many genotypes could be maintained if they all produce the same intermediate phenotype. However, even here it can be argued that all but one of the genotypes should eventually be eliminated by selection. The argument appears to be contradicted by the facts, and it remains unsettled how mutation and the exact form of stabilizing selection stop populations from becoming genetically uniform.

15

EVOLUTIONARY RATES

A genetic system is defined by its basic mode of inheritance (bisexual versus unisexual, diploid versus haploid) and, when referring to a specific phenotype, by its genetic architecture (number and relative importance of loci, the pattern of dominance, epistasis and pleiotropy, linkage relationships, etc.). Both influence the rate of evolution, but this chapter will focus only on basic mode of inheritance.

The interaction of cladogenesis and mode of inheritance is illustrated by examining the impact of founder events in *Drosophila*. Basic theory predicts that X chromosomes (a haplo-diploid genetic system) should be more sensitive to founder effects than autosomes (a bisexual diploid system). An examination of the genetic basis of evolved differences in both natural and experimental founder populations reveals a strong preferential involvement of X-linked loci over autosomal loci. Further evidence for the increased sensitivity of X chromosomes to founder events is provided by the fact that X chromosomal inversions are preferentially fixed over autosomal inversions in Hawaiian *Drosophila*, a group for which founder events are important, but there is no preferential fixation in continental *Drosophila*. Basic theory also predicts that mitochondrial DNA (mtDNA) should be more sensitive to founder events than nuclear DNA. This prediction is tested by contrasting two lineages of Hawaiian *Drosophila*, one for which founder events have apparently been important in speciation and the second for which founder events were not important. The pattern of nuclear versus mtDNA evolution is discordant between these two lineages, and the difference is shown to be due to a greatly accelerated rate of mtDNA evolution in the lineage subjected to inter-island founder events.

The importance of anagenesis in determining rates of evolution is illustrated by contrasting rates of evolution in mtDNA and nuclear DNA in the hominoid primates. The human lineage has had a large increase in generation time and a simultaneous expansion of population size over at least the last two million years. Theory predicts that these trends should cause an evolutionary slowdown in all genetic systems. This prediction is supported by a statistical analysis that indicates that mtDNA, nuclear DNA and karyotypes all have slower rates of evolution in humans than in African apes.

In summary, the rate of evolution of genetic systems is sensitive to both cladogenesis and anagenesis, but the degree of sensitivity may differ among the genetic systems. The heterogeneity in evolutionary rates does not mean that molecular data cannot be used to estimate divergence times; rather, it is shown that more accurate estimates are possible when rate heterogeneity is taken into account. Moreover, patterns of differences and uniformity among different genetic systems can therefore be used to make inferences concerning the events that have influenced the evolution of the group under study.

A genetic system is defined by its basic mode of inheritance (bisexual versus unisexual, diploid versus haploid) and, when referring to a specific phenotype, by its genetic architecture (number and relative importance of loci, the pattern of dominance, epistasis and pleiotropy, linkage relationships, etc.). Templeton (1982a,b) has shown that the type of genetic architecture strongly interacts with the mode of speciation and the type of anagenetic evolution in determining rates of evolution. Because genetic architecture has been discussed elsewhere, this chapter will focus only on basic mode of inheritance and its relationship to evolutionary rates.

Previous work on the impact of basic mode of inheritance on evolutionary rates has traditionally dealt with the problem of the evolution of sex. Thus, there is an extensive literature comparing rates of evolution in sexually reproducing, diploid species versus asexual haploids or parthenogenetic diploids. However, intraspecific comparisons are also possible because most organisms have several different genetic systems rather than just one. For example, in *Drosophila* and mammals, four distinct genetic systems are readily identified. First, there is the autosomal system that is inherited as a bisexual diploid. Secondly, the X chromosome is bisexually inherited, but is diploid in females and haploid in males. Thirdly, the Y chromosome is a paternally inherited haploid and finally, the mitochondrial DNA (mtDNA) is inherited as a maternal haploid. Unlike comparisons across species, the genetic systems imbedded within the same species obviously share many common evolutionary constraints; yet, their differing modes of inheritance allow these constraints to display a variety of potential impacts. The purpose of this chapter is to illustrate the impact of both cladogenetic and anagenetic evolutionary constraints on the rates of evolution of diverse genetic systems, showing that this impact across genetic systems imbedded within the same species can be homogeneous in some cases but heterogeneous in others.

CLADOGENESIS AND EVOLUTIONARY RATES

Founder Events and Genetic Systems

The impact of cladogenesis on the evolutionary rates of diverse genetic systems will be illustrated by speciation events induced by founder effects. Carson and Templeton (1984) have recently reviewed the evidence and theory for founder-induced speciation, and they have concluded that speciation events are likely only when the founder event is followed by a rapid and large increase in population size.

Neï et al. (1975) examined the genetic impact of such founder-flush events. They showed that founder events of size 10 have very little impact on overall levels of genetic variation and fixation when followed by a rapid increase in population size. Even the most extreme founder event possible, a population of two individuals, carried over a minimum of 65% of its genetic variation. The role of the founding number was investigated in more detail by Senner (1980). Restricting attention only to the case when the founding event was followed by a large increase in population size, Senner found an inflection point at about four individuals. Above four, most genetic variation is preserved and little fixation occurs. Below four, the amount of fixation is greatly influenced by the actual number of founding individuals, with decreasing numbers of founders resulting in increased fixation.

It is important to realize that both the Nei et al. (1975) and Senner (1980) models refer to bisexual, diploid genetic systems. Hence, the inflection point of about four individuals actually translates into an inflection point of about eight haploid genomes. A founder event consisting of N individuals will consist of diverse numbers of haploid genomes for the various genetic systems imbedded within those individuals. For example, much of the speciation in the Hawaiian *Drosophila* is most likely due to founder events involving a single gravid female. Suppose the female had been inseminated by M males, so that the founder size in terms of number of individuals is $N = M$

+ 1. Then, the number of autosomal genomes present in the founder population is $2M + 2$, the number of X chromosomes is $M+2$, the number of Y chromosomes is M, and the number of mitochondrial genomes is 1. Obviously, the various genetic systems can show extremely different sensitivities to the same founding event. For example, if $M = 3$, the autosomal system should be relatively insensitive to the founding event, but the X, Y and mtDNA will show progressively increasing sensitive in terms of loss of genetic variation and increased fixation rates.

X Chromosomal versus Autosomal Genetic Systems

Experimental founder events

For several years, I have been investigating the capacity for parthenogenetic reproduction in a natural, and normally bisexually reproducing, population of *Drosophila mercatorum*. As argued elsewhere, these screenings for parthenogenetic capacity constitute an experimental investigation of founder events that result in the most extreme inbreeding possible. The parthenogenetic lines were established from the virgin female offspring of single wild-caught females. Hence, the sources of genetic variation available for selection in the establishment of any given parthenogenetic strain is the genetic variation brought into the laboratory by a single gravid female. Moreover, most wild-caught females are multiply inseminated, with the probability of two offspring of the same female having different fathers being 0.32. Hence, these parthenogenetic strains are established from isofemale founding lines for which the founder effect will be much greater for the X chromosome than for any autosome.

Once a parthenogenetic strain is established, a sexual analogue can be bred. Hence, one can focus directly on the genetic impact of these experimental and extreme founder events rather than parthenogenesis *per se*. The most extensive work to date has been done on one of the first parthenogenetic lines established, K28-0-Im. Several isolating barriers evolved in this line, so it is an ideal experimental model of speciation. First, this line is characterized by extremely strong premating isolation from certain other strains of *D. mercatorum*. Chromosomal contrasts have been performed that demonstrate that the sexual isolation depends upon a strong epistatic interaction between two chromosomes: the X chromosome and the acrocentric II autosome. No other chromosomes are involved.

Given that mating occurs, other isolating barriers are unmasked. Many of the F_2 and backcross males lack motile sperm and are sterile.

The genetic basis of sterility is similar to that of premating isolation; sterile males are produced by simultaneous hemizygosity of the K28-0-Im X chromosome and homozygosity for the K28-0-Im acrocentric II autosome. Although the same chromosome pair is involved in both premating isolation and male sterility, these traits are separable genetically.

Finally, there is an F_2 breakdown in viability associated with the phenotypic syndrome known as *abnormal abdomen* (*aa*). Extensive genctic and molecular studies have been performed upon the *aa* syndrome. Genetically, this syndrome is associated with two necessary genetic components which are found on the X chromosome less than 1 map unit apart. Molecularly, the syndrome also depends upon two necessary components. First, a majority of the 28S ribosomal genes must have a 5 kb insertion into their coding region. The ribosomal genes are found in a tandem cluster of about 200 copies on the X chromosome. Second, during the formation of polytene tissue (such as fat bodies), there must be no preferential replication of non-inserted over inserted 28S genes. Many X chromosomes have the ability to replicate preferentially the non-inserted sequences both intra- and interchromosomally. If such a preferential replication occurs, *aa* is suppressed. In addition to these major and necessary X-linked elements, the expression of *aa* can be modified by genes located on the autosomes. Although the major genetic elements for *aa* are on the X chromosome, this syndrome is also separable through genetic recombination from the sexual isolation and male sterility X-linked elements.

Notice that all three isolating barriers have one or more major genes on the X chromosome, but most autosomes are not involved or have only minor modifiers. Given that the X chromosome constitutes only about 20% of the *mercatorum* genome, this preferential involvement of X-linked genetic systems is indicative of the increased sensitivity of the X chromosome to founder events over that of the autosomes. An alternative explanation is that genes involved with isolating mechanisms are clustered upon the X chromosome. This later explanation seems unlikely, since a broader review of the genetic basis of isolating mechanisms in the genus *Drosophila* shows extensive autosomal involvement.

Natural speciation events

There is evidence for this same pattern of preferential involvement of X-linked genetic systems in natural speciation events associated with founder effects. Founder-induced speciation is probably the dominant mode of speciation among the Hawaiian *Drosophila*. Two of these Hawaiian species are *D. silvestris* and *D. heteroneura*. These species are morphologically distinct and are reproductively isolated by premating barriers that can be broken down in the laboratory. Since no postmating barriers are present, it is possible to perform genetic analyses of the traits differentiating these two species. One of the more dramatic differences is in head shape. *Drosophila heteroneura* has a hammer head, whereas *D. silvestris* has a round head. Head shape is controlled by a major gene or genes on the X chromosome epistatically interacting with several minor autosomal modifiers. The species also differ in carina and groove (face) colour, with *D. heteroneura* being yellow and *D. silvestris* black. Both aspects of face colour are controlled by an X-linked locus interacting with one autosomal gene. Two different aspects of wing pattern are each controlled by interactions between an X-linked and autosomal locus. Differences in the pigmentation of the mesopleurae are due to an autosomal locus. Hence, five of the six characters analyzed involve major genes on the X chromosome. Thus, both experimental and natural founder populations display a preferential involvement of X-linked loci for their evolved differences.

Further evidence for the preferential sensitivity of X-linked genetic systems to founder events can be obtained by looking at the pattern of fixation of chromosomal inversions among the Hawaiian picture-winged *Drosophila*. Of 127 inversions fixed in 103 species, 59 (46%) are on the X chromosome despite the fact that there are four autosomes of about equal size. One might explain this by assuming that the X chromosome has a greater rate of inversion mutation, but this seems unlikely. If fixation rate is merely a consequence of mutation rate, the proportion of fixed to polymorphic inversions should be the same across all chromosomes. The contingency chi-square test of this hypothesis is 23.19 with four degrees of freedom, a result that is significant at the 0.00001 level of significance. Excluding the X chromosome from the analysis, the resulting contingency chi-square is 0.66 with three degrees of freedom. Hence, the autosomal inversions are being fixed at a rate that is proportional to their polymorphism rates, but the X chromosomal inversions are being fixed at a rate that is far in excess of their polymorphic rates. More specifically, the data indicate that X chromosomal inversions are being fixed at a rate 4.9 times that of autosomes. As a further check on the inference that these differences can be attributed to founder events rather than some peculiar property of the X chromosome, also presents the comparable data set on inversion evolution in the *repleta* group of *Drosophila* as summarized in Wasserman (1982). This is a speciose continental group for which there is no evidence for extensive founder-induced speciation. In great contrast to the Hawaiian *Drosophila*, only 10% of the fixed inversions are on the X chromosome, and the contingency chi-square between fixed versus polymorphic inversions as a function of chromosomal location is 3.62 with four degrees of freedom. Hence, in the *repleta* group, the rate of inversion fixation is proportional to rate of inversion polymorphism, and there is no discrepancy between X versus autosomal inversions. The contrast between Hawaiian versus *repleta* group *Drosophila* provides strong support for the theoretical prediction that X-linked genetic systems are far more sensitive to founder events than autosomal genetic systems.

Mitochondria DNA versus Nuclear Genetic Systems

The theory outlined above predicts that mtDNA should be more sensitive to single-female founder events than nuclear systems. To test this prediction, DeSalle (1984) performed an extensive study of mtDNA restriction site evolution in two lineages of Hawaiian *Drosophila*, the alpha and beta lineages of the *planitibia* subgroup. To understand why these two lineages where chosen, it is first necessary to review some pertinent facts concerning the geological history of the Hawaiian Islands. Most of the major islands of the Hawaiian chain are separated by deep ocean channels. The fluctuating sea levels that occurred during the Pleistocene could never have connected these major islands. Hence, the traditional arguments for single female founder events are applicable to instances of speciation involving transfers between these major islands. Most speciation in the beta lineage falls into this category. However, Maui, Molokai and Lanai (known collectively as Maui Nui) are islands that were connected to one another at various times in the past as a function of fluctuating sea levels. Hence, the *Drosophila* on these islands, which includes the alpha lineage, were much more likely to be separated by subdivision of large populations rather than by founder events.

Behavioural studies provide further evidence that founder events were not important in speciation processes occurring on Maui Nui. Kaneshiro (1976) discovered a pattern of mating asymmetry between ancestor and descendant species that exists among many Hawaiian *Drosophila*. Basically, derived females will mate with ancestral males, but ancestral females will not readily mate with derived males. This pattern has also been observed in non-Hawaiian *Drosophila* for which biogeographical evidence indicates founder-induced speciation. In addition, this mating asymmetry evolves in experimental founder events involving both Hawaiian and non-Hawaiian *Drosophila*. This pattern is not observed in contrasts between continental *Drosophila* species for which the biogeographical evidence indicates founder events to be unlikely and is not present in *Drosophila* speciation experiments that do not involve founder effects. Consequently, this pattern of mating asymmetry is one that appears to be specifically associated with founder-induced speciation in *Drosophila* for the theoretical reasons why this specificity is expected). Most Hawaiian *Drosophila*, including the beta lineage, display this asymmetry. The only exceptions discovered so far involve species from the Maui Nui complex, thereby implying that speciation on Maui Nui does not involve founder effects.

To see if mtDNA is more sensitive to founder events than nuclear genetic systems, comparisons were made between DeSalle's genetic distances on mtDNA using Ewens' (1983) 'p' statistic with Roger's genetic distances calculated by Johnson et al. (1975) on isozyme data. The Ewens' and Roger's distances are not directly comparable, so the absolute values of the distances are not important, but rather the relative pattern within and between lineages.

The pattern of distances within and between lineages can be summarized by taking the appropriate averages. For the isozyme data, the average distances within lineages are 0.10 and 0.27 for alpha and beta respectively, and the average intralineage distance is 0.62. Hence, in terms of the nuclear genetic distances, the interlineage distance is considerably larger than intralineage distances. Moreover, as Carson (1976) has shown by overlaying these genetic distances on the known geological time scale of these islands, the distances in flies from both lineages can be explained by the same rate of evolution. In contrast, the average intralineage p-distances for the mtDNA are 0.028 and 0.053 for alpha and beta respectively, with the interlineage distance being 0.059. This pattern differs dramatically from that observed for the isozymes. Note that the interlineage distance is only slightly larger than the within beta distances. This pattern implies that the beta lineage has a much faster rate of mtDNA evolution than the alpha.

The hypothesis that the beta lineage is evolving more rapidly than the alpha lineage for mtDNA can be tested directly using the statistical procedure outlined by Templeton (1983b, 1986b). This procedure uses the character-state data rather than distances. A great advantage of character-state data over distance data is that it is often possible to reconstruct the ancestral intermediates between the taxa being compared. This property has long been used in constructing intermediate states in the inversion phylogenies of *Drosophila*, and the logic is identical in reconstructing intermediate states of restriction sites. This is an important property because it allows one to estimate the number of mutations accumulated along any branch segment with an ancestral-state estimation procedure that does not assume the molecular clock *a priori*.

A second desirable property of this test is that it is non-parametric. Hence, no assumptions are made about the underlying sampling model induced by the evolutionary process. This is important because we simply do not know at this time what the appropriate sampling model ought to be; yet, the conclusions based on parametric tests of the molecular clock are very sensitive to the sampling model assumed. However, recent work indicates that the molecular clock does not hold.

The details of this test are given in Templeton (1983b, 1986b), so they will not be repeated here. Nevertheless, some of the limitations of this approach should be mentioned. First, the test requires a knowledge of the topology of the phylogenetic tree. Fortunately, Templeton (1983b, 1986b) gives statistical criteria for judging how well a topology has been estimated. If more than one topology cannot be statistically distinguished, it is best to test the molecular clock under each topology to check for the robustness of the conclusions with respect to the ambiguities regarding the true topology. DeSalle (1984) used these statistics on the Hawaiian species being considered, and concluded that there were two statistically indistinguishable phylogenies.

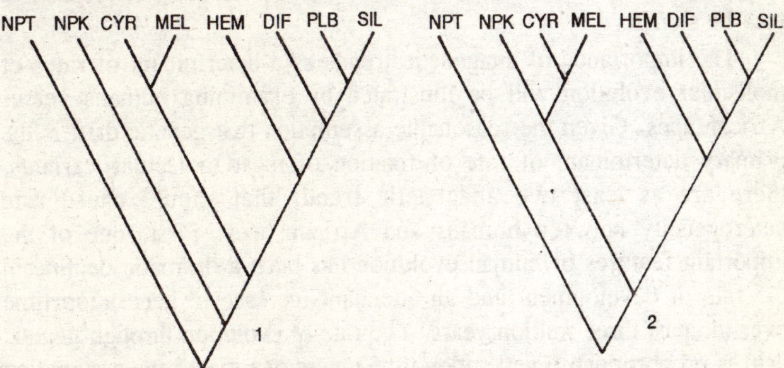

Fig. 15.1. Alternative phylogenies for the Hawaiian Drosophila species under consideration.

Second, given a topology, the ancestral states themselves may not have been estimated in an unambiguous fashion. The reason for this is that some alternative ways of allocating mutational events or even the number of events are basically equally probable. To prevent this estimation ambiguity from eroding the power of the test, it is important to only test data sets for which convergences of any sort are rare. Fortunately, DeSalle's data set falls into this category.

The pattern is very clear cut; the beta lineage is evolving at a greatly accelerated rate compared to the alpha lineage, Hence, founder events do seem to accelerate the rate of mtDNA evolution over that of nuclear genetic systems. When these results are coupled with the earlier results concerning X-linked genetic systems, it is clear that the mode of cladogenesis interacts with the nature of the genetic system to determine rate of evolution.

ANAGENESIS AND GENETIC SYSTEMS

The debate concerning punctuated equilibrium has focused much attention on the mode of speciation as a primary determinant of rate of morphological evolution. In contrast, most models of molecular evolution have ignored the details of speciation mechanisms and explained rates of molecular evolution solely in terms of continuously operating anagenetic mechanisms, primarily genetic drift. As shown by the contrast of the alpha and beta lineages given above, this disregard for speciation mechanism by molecular evolutionists is clearly not justified. Templeton et al. (1981) give additional examples of the importance of speciation in determining rates of molecular evolution, and Gillespie (1984) has recently performed a statistical analysis that indicates an episodic pattern of molecular evolution. However, anagenesis is still important in determining evolutionary rates, as will now be shown.

The importance of anagenetic trends as a determinant of rates of molecular evolution will be illustrated by examining humans versus African apes. Given the reasonable assumption that genetic drift is the primary determinant of rate of fixation of most molecular variants, there are at least two anagenetic trends that should cause rate heterogeneity between humans and African apes. First, one of the important features of human evolution has been a dramatic decline in the rate of development and an attendant increase in generation time over the past three million years. The rate of evolution through genetic drift is proportional to generation time unless one makes the assumption that mutation rates are proportional to absolute time. This assumption is not consistent with the empirical evidence among eukaryotes. For example, estimated mutation rates in *Drosophila*, mice, corn and humans are generally within an order of magnitude of one another even though the generation lengths of these organisms differ by three orders of magnitude. Although mutation rates are hard to estimate accurately, it is difficult to reconcile such grossly discrepant results with the idea that mutation rates are constant in absolute time in higher eukaryotes. Indeed, Kimura (1979), long one of the leading advocates of mutation rate constancy in absolute time, has abandoned that position and now models his version of the neutral theory in terms of mutation rates that are constant on a per generation basis. Hence, the straightforward prediction of the anagenetic trend to longer generation times in humans is that the rate of molecular evolution in humans should be slower than that of African apes. Moreover, this slowdown should affect all genetic systems.

Another trend in the human lineage, but not in the African apes, has been a consistently expanding population size starting at the very least with *Homo erectus* which had a fossil distribution covering the entire Old World and not just Africa. Nei (1976) has shown that expanding population size decelerates the rate of molecular evolution under absolute neutrality, and Kimura (1979) has shown that the deceleration is even more severe when one includes mutations with very small selection coefficients. Once again, this slowdown should affect all genetic systems in which molecular evolution proceeds by random fixation of neutral or nearly neutral mutations.

These predictions can be tested by comparing the rates of molecular evolution in humans versus African apes for a number of genetic systems. First, Templeton (1983b, 1986b) has already tested the rate of mtDNA evolution in these primates and has shown that there is a statistically significant slowdown in the rate of mtDNA evolution in humans relative to African apes. This slowdown also appears to exist for nuclear genetic systems as well. For example, Marks (1982) concluded that humans have a slower rate of karyotypic evolution than the African apes. More recently, Sibley and Ahlquist (1984) performed extensive studies on nuclear DNA-DNA hybridization between these species. Testing the molecular-clock hypothesis is far more difficult and less clean-cut with distance data than with character-state data and, unfortunately, the DNA-DNA hybridization data is inherently a distance measure. Nevertheless, inferences can be made concerning the clock, but they must be made carefully rather than from a casual inspection of the data.

For example, Sibley and Ahlquist (1984) argued that their data is consistent with the clock because it satisfied the triangular inequality. They then quoted Farris (1981) to the effect that this metricity implies 'clocklike' behaviour. However, what Farris (1981) really said was that metricity is a necessary, but not sufficient, condition for clocklike behaviour. Commenting on Sibley and Ahlquist's quotation of Farris, Farris and Kluge corrected this error and pointed out that 'a distance may be metric, yet show strong variation in rates of evolution'.

A better way of testing the molecular-clock hypothesis is to contrast the distances between two closely related species with a third, more distantly related, outgroup species. This is known as the relative rate test. The relative rate test, however, is biased in favour of the clock unless one can correct for the covariances of the distances or estimate the ancestral distance in a manner that is not biased towards the clock. Unfortunately, the covariances for DNA-DNA hybridization distances cannot be calculated unless one invokes a parametric model of unknown validity. The second course of action, estimating the ancestral distance states, is flawed by the fact that distance estimation algorithms are biased in favour of, or explicity assume, the clock. For example, Nei et al. (1985) recently gave formulae for estimating the position and variance of branch points for various types of distance data. However, their equations are only valid if the molecular clock is exactly true, so their estimation procedure cannot be used to test the clock.

Although it is difficult to test the clock hypothesis with the Sibley and Ahlquist data set, it can easily be shown that these data contain internal inconsistencies which cannot be reconciled with the hypothesis of rate constancy. For example, the distance between humans and chimps is significantly smaller (at the 0.1% level) than that between humans and gorillas, but in contrast the genetic distance between gorillas and chimps is significantly smaller (at the 1% level) than that between humans and gorillas. These conclusions are based on *t*-tests that were uncorrected for the covariance between distances, and hence these tests are strongly biased in favour of the clock. Despite this bias, these significant *t*-test results cannot be reconciled with the clock. If the clock were true, the first significant test would imply that humans and chimpanzees form a clade, but the second test implies that gorillas and chimpanzees form a clade. Since there can be only one phylogenetic topology, the molecular clock obviously cannot hold for all these species. In a survey of the available molecular data, Templeton (1985) showed that there is one and only one branching order that is statistically compatible with all the molecular data; namely, the tree that clusters the African apes together as a clade. As further discussed by Templeton (1985), the pattern of internal inconsistencies found in the data of Sibley and Ahlquist (1984) can be readily explained by a slowdown in the human lineage under this topology because, although humans are the most distantly related in a topological sense, the human slowdown in rate of molecular evolution could yield very similar expected distances among all three species.

Thus, both the nuclear and mtDNA genetic systems appear to have been slowed down in humans relative to the African apes. This pattern supports the predictions discussed above based on the well-documented anagenetic trends in the human lineage of increasing population size and generation times.

Discussion

The examples given in this chapter illustrate that rates of molecular evolution are not constant, but vary as a function of both cladogenesis and anagenesis. This variation in rates of molecular evolution does not mean that molecular data cannot be used for making inferences about times of divergence. The molecular data can still yield much information concerning divergence times, and indeed, the identification of specific heterogeneities can allow times of divergence to be estimated in an even more precise manner. These estimated numbers of mutations will be used as a distance measure. This distance can be calibrated in absolute time by noting that the divergence between *D. planitibia* and *D. silvestris* involved a founder event on the Hualalai Volcano on the Island of Hawaii, which is about 400,000 years old. Assuming this founder event occurred shortly after the formation of Hualalai, the *silvestris/planitibia* divergence is set equal to 400,000 years. Another potential calibration point is the divergence between the alpha and beta lineages. The chromosomal and biogeographical data indicate that these lineages diverged on West Maui, which is 1.3 million years old. However, for now, only the *planitibia/silvestris* calibration will be used.

Using this calibration with no correction for the rate heterogeneity indicated one obtains the divergence times. The divergence times of *differens* and *hemipeza* are consistent with the available biogeographical data; a result that is not too surprising given the fact that the clock holds within the lineages. However, the divergence between the alpha and beta sublineages is estimated as 835,000 years, a figure that is much too small. Moreover, the estimated times of divergence within the alpha lineage are low (74,000 years for *crytoloma* and *melanocephala*, and 185 000 years for *neoperkensi*).

Because highly significant rate heterogeneity was detected between the alpha and beta lineages, but not within them, the divergence times were next estimated by using the original calibration on the *silvestris/planitibia* node only within the beta sublineage. The rate of mtDNA evolution within the alpha lineage was regarded as being 3.23 times slower than that within the beta. This figure was obtained by dividing the average number of mutations on the branches leading to beta species from the alpha-beta node by the average number of mutations on the branches leading to alpha species from this node. Using the calibration corrected for rate heterogeneity produces the estimated divergence times. These estimated times make much more sense in the light of the known biogeographical data. For example, the alpha-beta split is now estimated to have occurred at 1.27 million years; a figure that is in excellent agreement with the 1.3 million year age of West Maui. Recall also that the behavioural and biogeographical evidence mentioned above implies that the alpha-lineage flies did not speciate through interisland founder events, but rather through the more traditional erection of geographical barriers between Maui Nui subpopulations. Since the *neoperkinsi* strain currently lives on Molakai (1.5 to 1.8 million years) and the *cyrtoloma* and *melanocephala* strains on East Maui, the speciation events separating these groups could only have occurred after the formation of East Maui if this interpretation of their speciation pattern is correct. Note that the estimated divergence times in part B are consistent with this interpretation, since it indicates that the first split in these alpha species occurred shortly after the formation of East Maui.

The results reported in this chapter not only reveal that cladogenesis and anagenesis can affect the rate of molecular evolution, but in addition the results show that sometimes the impact on evolutionary rates is uniform on all genetic systems within a species (as shown by the human example), but in other cases the impact is quite heterogeneous (the *Drosophila* examples). Hence, different genetic systems imbedded within a single species are differentially sensitive to some of the historical contraints that shaped or even created that species. By simultaneously studying several different genetic systems, patterns emerge that can be used to make inferences about the history of the species being examined that could not be made just by studying each genetic system in isolation. For example, Charlesworth et al. (1982)

argued that founder events were unlikely in Hawaiian *Drosophila* because they have maintained high levels of isozyme heterozygosity. However, Charlesworth et al. (1982) noted that this observation does not exclude founder events followed by a population flush because a depletion of isozyme heterozygosity is only expected if the population 'bottleneck' is 'prolonged over several generations'. Consequently, the isozyme data, considered in isolation, are consistent with two hypotheses: one, no founder or bottleneck effects occurred; or two, founder events occurred followed by large increases in population size. This ambiguity in interpretation is clearly resolved when the mtDNA data is overlaid on the isozyme data, for now the no-founder hypothesis can be rejected on the basis of the dramatic rate of increase of mtDNA. Consequently, a simultaneous examination of evolutionary rates among various genetic systems offers a potentially powerful tool for making inferences about the cladogenetic and anagenetic events that shaped the evolution of the species we observe today.

INDEX